Current Topics in Complement II

For other titles published in this series, go to
www.springer.com/series/5584

John D. Lambris
Editor

Current Topics in Complement II

Springer

Editor
John D. Lambris
Department of Pathology
Laboratory of Medicine
Johnson Pavilion 410
University of Pennsylvania
Philadelphia, PA 19104
USA
lambris@upenn.edu

ISBN: 978-0-387-78951-4 e-ISBN: 978-0-387-78952-1
DOI:10:1007/978-0-387-78952-1

Library of Congress Control Number: 2008925169

Printed on acid-free paper

9 8 7 6 5 4 3 2 1

springer.com

Preface

Nearly 110 years after the term 'Complement' was coined by Paul Ehrlich, this fascinating cascade within the innate immune system still offers many surprises. Reports about new connections to diseases, links to other physiological pathways, novel complement-binding molecules, and microbial evasion proteins are emerging almost on a weekly base. With last years' FDA approval of the monoclonal antibody Eculizumab as the first truly complement-specific drug (by targeting C5), this important field regained even more momentum. Several promising clinical candidates covering a wide area of potential treatment applications are in the pipelines of both industrial and academic groups. This indicates an increasing interest in complement as a therapeutic target. In view of these exciting discoveries, scientists from around the world convened at the 4th Aegean Conferences Workshop on Complement Associated Diseases, Animal Models, and Therapeutics (June 10-17, 2007) in Porto Heli, Greece, to discuss recent advances in this rapidly-evolving field. This volume represents a collection of topics on the "novel" functions of complement, pathophysiology, protein structures, and complement therapeutics discussed during the conference.

My sincere thanks to the contributing authors for the time and effort they have devoted to writing what I consider exceptionally informative chapters in a book that will have a significant impact on the complement field. I would also like to express my thanks to Rodanthi Lambris for her assistance in collating the chapters and preparing the documents for publication and I gratefully acknowledge the generous help provided by Dimitrios Lambris in managing the organization of this meeting. Finally, I also thank Andrea Macaluso and Melanie Wilichinsky of Springer Publishers for their supervision in this book's production.

John D. Lambris, Ph.D.
Dr. Ralph and Sallie Weaver Professor of Research Medicine, University of Pennsylvania

Contents

Contributors

Erik Åhlin
Unit of Clinical Immunology
Uppsala University
Uppsala
Sweden

Maqsudul Alam
Advanced Studies in Genomics, Proteomics, and Bioinformatics
University of Hawaii at Manoa
Honolulu, HI 96822
USA

Mohammed A.M. Ali
Hubei Province Key Laboratory of Allergy and Immune-related Diseases
The State Key Laboratory of Virology
Wuhan University School of Medicine
Wuhan, 430071.00
P.R. China

Daniel J. Allendorf
Tumor Immunobiology Program of the James Graham Brown Cancer Center
Department of Medicine
University of Louisville School of Medicine
Louisville, KY 40202
USA

Umme Amara
Department of Traumatology, Hand-, Plastic-, and Reconstructive Surgery
University Hospital of Ulm
89075 Ulm
Germany

Paul N. Barlow
Schools of Chemistry and Biological Sciences
Joseph Black Chemistry Building
University of Edinburgh
Edinburgh EH9 2PB
UK

Mallory Bowers
Melior Discovery Corporation
Exton, PA 19341
USA

Ole-Lars Brekke
Department of Laboratory Medicine
Nordland Hospital Bodø
Bodø
Norway

Uwe Bruckner
Division of Experimental Surgery
University Hospital of Ulm
89075 Ulm
Germany

Premkumar Christadoss
Department of Microbiology and Immunology
University of Texas Medical Branch
Galveston, TX 77555
USA

Dorte Christiansen
Department of Laboratory Medicine
Nordland Hospital Bodø
Bodø
Norway

John R. Ciallella
Melior Discovery Corporation
Exton, PA 19341
USA

Katherine Cianflone
Centre de Recherche Hopital Laval
McGill University
Québec, QC G1V 4G5
Canada

E. Sander Connolly
Department of Neurological Surgery
Columbia University
New York, NY 10032
USA

Kerina J. Costantini
School of Biomedical Sciences
University of Queensland
St Lucia, QLD 4072
Australia

Wei Cui
Centre de Recherche Hopital Laval
McGill University
Québec, QC G1V 4G5
Canada

Kenji M. Cunnion
Department of Pediatrics
Eastern Virginia Medical School and Children's Specialty Group
Norfolk, VA 23507
USA

Donald R. Demuth
Department of Periodontics/Oral Health and Systemic Disease
University of Louisville School of Dentistry
Louisville, KY 40202
USA

Rossen Donev
Complement Biology Group
Department of Medical Biochemistry and Immunology
Cardiff University School of Medicine
Heath Park, Cardiff CF14 4XN
UK

Andrew F. Ducruet
Department of Neurological Surgery
Columbia University
New York, NY 10032
USA

Terje Espevik
Department of Cancer Research and Molecular Medicine
Norwegian University of Science and Technology
N-7489 Trondheim
Norway

Michael A. Flierl
Department of Pathology
University of Michigan Medical School
Ann Arbor, MI 48109
USA

David C. Fritzinger
Cancer Research Center of Hawaii
University of Hawaii at Manoa
Honolulu, HI 96813
USA

Matthew C. Garrett
Department of Neurological Surgery
Columbia University
New York, NY 10032
USA

Florian Gebhard
Department of Traumatology, Hand-, Plastic-, and Reconstructive Surgery
University Hospital of Ulm
89075 Ulm
Germany

Brian V. Geisbrecht
School of Biological Sciences
University of Missouri at Kansas City
Kansas City, MO 64110
USA

Berhane Ghebrehiwet
Departments of Medicine and Pathology
Stony Brook University
Stony Brook, NY 11794
USA

Wlliam B. Gorsuch
Brigham and Women's Hospital
Center for Experimental Therapeutics and Reperfusion Injury
Harvard University
Boston, MA 02115
USA

Benjamin J. Guikema
Brigham and Women's Hospital
Center for Experimental Therapeutics and Reperfusion Injury
Harvard University
Boston, MA 02115
USA

Gregory S. Hageman
Department of Ophthalmology and Visual Sciences
University of Iowa
Iowa City, IA 52242
USA

George Hajishengallis
Department of Periodontics/Oral Health and Systemic Disease
University of Louisville School of Dentistry
Louisville, KY 40202
USA

Richard Hansen
Department of Medicine
Tumor Immunobiology Program of the James Graham Brown Cancer Center
University of Louisville School of Medicine
Louisville, KY 40202
USA

Benjamin Hassid
Department of Neurological Surgery
Columbia University
New York, NY 10032
USA

Brian E. Hew
Cancer Research Center of Hawaii
University of Hawaii at Manoa
Honolulu, HI 96813
USA

Markus Huber-Lang
Department of Traumatology, Hand-, Plastic-, and Reconstructive Surgery
University Hospital of Ulm
89075 Ulm
Germany

Deanna James
Department of Periodontics/Oral Health and Systemic Disease
University of Louisville School of Dentistry
Louisville, KY 40202
USA

Grace H. Kim
Department of Neurological Surgery
Columbia University
New York, NY 10032
USA

Andreas Klos
Department of Medical Microbiology
Medical School Hannover
30625 Hannover
Germany

Ricardo J. Komotar
Department of Neurological Surgery
Columbia University
New York, NY 10032
USA

Neel K. Krishna
Department of Microbiology and Molecular Cell Biology
Eastern Virginia Medical School
Norfolk, VA 23507
USA

Magda Kucia
Stem Cell Institute
James Graham Brown Cancer Center
University of Louisville
Louisville, KY 40202
USA

John D. Lambris
Department of Pathology and Laboratory Medicine
University of Pennsylvania
Philadelphia, PA 19104
USA

Susan M. Lea
Sir William Dunn School of Pathology
University of Oxford
South Parks Road
Oxford, OX1 3RE
UK

June Q. Lee
Cancer Research Center
University of Hawaii at Manoa
Honolulu, HI 96813
USA

Lawrence L.K. Leung
Department of Medicine, Division of Hematology
Palo Alto Health Care System
Stanford University School of Medicine and Veterans Administration
Stanford, CA 94305
USA

Jing Liang
Department of Neurology
University of Central South China University
P.R. China

Shuang Liang
Department of Periodontics/Oral Health and Systemic Disease
University of Louisville School of Dentistry
Louisville, KY 40202
USA

William J. Mack
Department of Neurological Surgery
Columbia University
New York, NY 10032
USA

Robin MacLaren
Centre de Recherche Hopital Laval
Université Laval
Québec, QC G1V 4G5
Canada

Linda Mathsson
Unit of Clinical Immunology
Uppsala University
Uppsala
Sweden

Megan R. McMullen
Department of Pathobiology
Cleveland Clinic
Cleveland, OH 44195
USA

Edward M. Medof
Department of Pathology
Case Western Reserve University
Cleveland, OH 44106
USA

Maxwell B. Merkow
Department of Neurological Surgery
Columbia University
New York, NY
USA

J. Mocco
Department of Neurological Surgery
Columbia University
New York, NY 10032
USA

Tom Eirik Mollnes
Institute of Immunology
University of Oslo
Oslo
Norway

Timothy Myles
Department of Medicine, Division of Hematology
Palo Alto Health Care System
Stanford University School of Medicine and Veterans Administration
Stanford, CA 94305
USA

Laura E. Nagy
Department of Pathobiology & Gastroenterology
Cleveland Clinic
Cleveland, OH 44195
USA

James Newhouse
Maui High Performance Computing Center
Kihei, HI 96753
USA

Bo Nilsson
Unit of Clinical Immunology
Uppsala University
Uppsala
Sweden

Kristina Nilsson-Ekdahl
Unit of Clinical Immunology
Uppsala University
Uppsala
Sweden

Toshihiko Nishimura
Department of Medicine, Division of Hematology
Palo Alto Health Care System
Stanford University School of Medicine and Veterans Administration
Stanford, CA 94305
USA

So-ichiro Nishiyama
Department of Microbiology
Aichi-Gakuin University School of Dentistry
Nagoya 464-8650
Japan

Peter G. Noakes
School of Biomedical Sciences
University of Queensland
St Lucia, QLD 4072
Australia

Kousaku Ohinata
Division of Food Science and Biotechnology
Graduate School of Agriculture
Kyoto University
Kyoto 611-0011
Japan

Marc L. Otten
Department of Neurological Surgery
Columbia University
New York, NY 10032
USA

Ellinor I.B. Peerschke
Department of Pathology, Center for Clinical Laboratories
The Mount Sinai School of Medicine
New York, NY 10029
USA

Michele T. Pritchard
Department of Pathobiology
Cleveland Clinic
Cleveland, OH 44195
USA

Janina Ratajczak
James Graham Brown Cancer Center
Stem Cell Institute
University of Louisville
Louisville, KY 40202
USA

Mariusz Z. Ratajczak
James Graham Brown Cancer Center
Stem Cell Institute
University of Louisville
Louisville, KY 40202
USA

Ryan Reca
James Graham Brown Cancer Center
Stem Cell Institute
University of Louisville
Louisville, KY 40202
USA

Daniel Ricklin
Department of Pathology and Laboratory Medicine
University of Pennsylvania
Philadelphia, PA 19104
USA

Daniel Rittirsch
Department of Pathology
University of Michigan Medical School
Ann Arbor, MI 48109
USA

Johan Rönnelid
Unit of Clinical Immunology
Uppsala University
Uppsala
Sweden

Shamsher S. Saini
Department of Microbiology and Immunology
University of Texas Medical Branch
Galveston, TX 77555
USA

J. Vidya Sarma
Department of Pathology
University of Michigan Medical School
Ann Arbor, MI 48109
USA

Muhamad-Ali K. Shakhatreh
Department of Periodontics/Oral Health and Systemic Disease
University of Louisville School of Dentistry
Louisville, KY 40202
USA

Gregory L. Stahl
Brigham and Women's Hospital
Center for Experimental Therapeutics and Reperfusion Injury
Harvard University
Boston, MA 02115
USA

Robert M. Starke
Department of Neurological Surgery
Columbia University
New York, NY 10032
USA

Abram Stavitsky
Department of Molecular Biology and Microbiology
Case Western Reserve University
Cleveland, OH 44106
USA

Michael E. Sughrue
Department of Neurological Surgery
Columbia University
New York, NY 10032
USA

Steve M. Taylor
School of Biomedical Sciences
University of Queensland
St Lucia, QLD 4072
Australia

Erdem Tüzün
Department of Neurology
University of Instambul
Istnabul
Turkey

Carl-Wilhelm Vogel
Cancer Research Center of Hawaii
University of Hawaii at Manoa
Honolulu, HI 96813
USA

Wu Wan
James Graham Brown Cancer Center
Stem Cell Institute
University of Louisville
Louisville, KY 40202
USA

Ming Wang
Department of Periodontics/Oral Health and Systemic Disease
University of Louisville School of Dentistry
Louisville, KY 40202
USA

Peter A. Ward
Department of Pathology
University of Michigan Medical School
Ann Arbor, MI 48109
USA

Trent M. Woodruff
School of Biomedical Sciences
University of Queensland
St Lucia, QLD 4072
Australia

Marcin Wysoczynski
James Graham Brown Cancer Center
Stem Cell Institute
University of Louisville
Louisville, KY 40202
USA

Jun Yan
Tumor Immunobiology Program of the James Graham Brown Cancer Center
Department of Medicine
University of Louisville School of Medicine
Louisville, KY 40202
USA

Ruowan Yan
Tumor Immunobiology Program of the James Graham Brown Cancer Center
Department of Medicine
University of Louisville School of Medicine
Louisville, KY 40202
USA

Huan Yang
Department of Neurology
University of Central South China University
P.R. China

Wei Yin
Department of Mechanical and Aerospace Engineering
Oklahoma State University
Stillwater, OK 74078
USA

Masaaki Yoshikawa
Division of Food Science and Biotechnology
Graduate School of Agriculture
Kyoto University
Kyoto 611-0011
Japan

Fuminobu Yoshimura
Department of Microbiology
Aichi-Gakuin University School of Dentistry
Nagoya 464-8650
Japan

Xiao-Lian Zhang
Hubei Province Key Laboratory of Allergy and Immune-related Diseases,
The State Key Laboratory of Virology
Wuhan University School of Medicine
Wuhan 430071.00
P.R. China

Ewa K. Zuba-Surma
James Graham Brown Cancer Center
Stem Cell Institute
University of Louisville
Louisville, KY 40202
USA

1. Adipokines and the Immune System: An Adipocentric View

Robin MacLaren[1,2], Wei Cui[1,2], and Katherine Cianflone[1,2]

[1]Department of Experimental Medicine, McGill University, Montreal, Canada
[2]Centre de Recherche Hôpital Laval, Université Laval, Québec, Canada

Abstract. There is increasing evidence of close interactions between the adipose and the immune systems. Adipocytes secrete multiple factors, including adipokines such as leptin and adiponectin that have both pro- and anti-inflammatory effects, and influence diseases involving the immune system. Further, adipose tissue also secretes various chemokines and cytokines, derived from either the adipocytes themselves, or the neighbouring cells including both resident and infiltrating macrophages. This close physical and paracrine interaction results in reciprocal actions of adipocytes, preadipocytes and macrophages within the microenvironment of the adipose tissue. Adipose tissue is a source of Acylation Stimulating Protein (ASP)/C3adesArg which interacts with the receptor C5L2 to stimulate triglyceride synthesis and glucose transport. C5L2, present on adipocytes, preadipocytes, macrophages, and numerous other myeloid and non-myeloid cells is also postulated to be a decoy receptor for C5a in immune cells. Several reviews within the past year have recently examined the role of C5L2 in C5a-mediated physiology. The present mini-review is an adipocentric view with emphasis on the role of ASP and C5L2 in lipid metabolism. C5L2 may play a role in mediating, on one hand, ASP stimulation of triglyceride synthesis in adipose, and, on the other hand, a role as mediator of C5a immune function. Both roles remain controversial, and will only be resolved with further studies.

1 Introduction

Adipose tissue, long considered as only an energy storehouse, has, in the last 15 years, expanded its role to an endocrine tissue. Adipose tissue secretes numerous factors that impact energy metabolism, energy storage and general metabolism, including immune functions. This recent shift in perspective has also revealed a number of previously unrecognized links between adipose tissue, liver and the immune system, including interaction of hormones/receptors from one system to the other, and a long list of common enzymes. While on the surface this may appear surprising, a common ancestor in evolution may underlie these similarities: in *Drosophila melanogaster*, the common fruit fly, liver, adipose and the haematopoietic system are all organized into a single functional unit: the "fat body". Thus, evolution may explain the overlapping biological profile of these organs, and the close link between immune and metabolic systems (Hotamisligil 2006).

J.D. Lambris (ed.), *Current Topics in Complement II.*
doi: 10:1007/978-0-387-78952-1_1, © Springer Science+Business Media, LLC 2008

2 Endocrine Functions of Adipose Tissue in the Immune System

It is now well recognized that adipose tissue secretes many autocrine, paracrine, and endocrine factors (adipokines) that have wide ranging effects in metabolism including satiety, regulation of energy expenditure and energy storage. However in addition to roles directly related to metabolism, adipokines are also involved in regulation of the following: immune function; cytokine induction, macrophage activation, neutrophil activation, recruitment of macrophage and T cells to adipose tissue (Lago et al. 2007; Guzik et al. 2006), and inflammatory diseases such as rheumatoid arthritis, osteoarthritis, sepsis, atherosclerosis, insulin resistance and obesity (Lago et al. 2007; Trayhurn and Wood 2004). The list of these adipokines is now extensive and growing every day. Readers are referred to the following detailed reviews (Lago et al. 2007; Bouloumie et al. 2005; Trayhurn and Wood 2004, 2005; Fain 2006). The most recent adipokines identified include examples such as visfatin, apelin, vaspin, hepcidin, and osteopontin (Beltowski 2006; Lago et al. 2007; Nomiyama et al. 2007). While some of the adipokines appear to be produced exclusively in adipose tissue, many of them are secreted elsewhere, including by myeloid cells and tissues. A brief list of those with immune associated functions is provided in Table 1.

Among those adipokines which are produced exclusively by adipose tissue and have been demonstrated to be involved in immune function are leptin and adiponectin. Leptin was among the first adipokines to be identified and is best known for its powerful role in satiety (Pelleymounter et al. 1995; Saladin et al. 1995). The absence of leptin in mice or humans results in early onset morbid obesity (Pelleymounter et al. 1995). The central role of leptin in satiety, hormone regulation, food intake and energy expenditure also led to effects on overall metabolism, especially insulin sensitivity (review Uysal et al. 1997). Beyond this, leptin plays key roles in reproduction, vascular function, regulation of bone mass and other targets in kidney, bowel, pancreas and muscle (reviewed by Guzik 2006).

Recently, the role of leptin in immune function has been explored in some detail. Leptin stimulates cytokine induction including secretion of IL-2 and factors influencing chemotaxis and macrophage activation. Leptin influences natural killer cytotoxicity, and affects neutrophil activation and naïve T cell proliferation. Leptin increases maturation and survival of thymic cells, and stimulates $T_{helper}1$ while inhibiting $T_{helper}2$ cells. As a consequence of the many varied functions of leptin, the links to disease are also broad. Leptin is suggested to play a role not only in insulin resistance and diabetes, but also in hepatitis, osteoarthritis, intestinal inflammation, sepsis, encephalomyelitis and rheumatoid arthritis, all diseases linked to abnormalities in immune function (Lago et al. 2007). Overall, leptin appears to provide a pro-inflammatory role in many aspects of immunity and inflammation (see recent review by Matarese et al. 2007). Adiponectin, a large molecular weight protein, is found circulating at high concentrations (relative to other adipokines, 5–10 μg/mL) and is present in multimeric forms (Koerner et al. 2005). Adiponectin interacts with cell surface receptors to stimulate fatty acid oxidation (Berg et al. 2001) and gene regulation (CD36 and PPARγ are just a few examples) (Tomas et al. 2002). In obesity, adiponectin decreases, and this lack of adiponectin contributes to changes in insulin sensitivity, dyslipidemia, cholesterol uptake and atherosclerotic plaque formation, endothelial damage, adhesion molecule and cytokine production, and inflammation, as reviewed elsewhere (Meier and Gressner 2004; Tilg and Moschen 2006). Therefore, altered adiponectin is linked not only to obesity, but also to atherosclerosis, diabetes, cardiovascular disease, liver injury and rheumatoid arthritis, with the general view that adiponectin is an anti-inflammatory molecule (Meier and Gressner 2004; Tilg and Moschen 2006).

In addition to leptin and adiponectin, a number of other adipokines have been linked to immune and inflammatory function, including visfatin, resistin, osteopontin, and the precursor to ASP, C3a and C3 (as discussed below). Overall, evidence supports that it is not a single adipokine, but adipokine dysregulation as a whole that is involved in insulin resistance, inflammatory processes and vascular and tissue inflammation. Further, activation of the innate immune system and inflammation are now considered key components in the development and progression of diabetes (Tataranni and Ortega 2005).

The link between pro-inflammatory factors, insulin resistance and adipose function was first proposed by Hotamisligil and Spiegelman (1994), when it was demonstrated that TNFα interfered with insulin signalling in adipose tissue. In spite of obesity, usually associated with insulin resistance, TNFα knockout mice remain insulin sensitive (Uysal et al. 1997). Further studies demonstrated that adipose tissue is a source of the source of TNFα . While adipocytes were initially suggested to be the main cytokine secreting cell (Funahashi et al. 1999), it is now well established that non-adipocyte cells in the stromal vascular fraction, such as macrophages, are an important source of many of these pro- and anti-inflammatory factors, chemokines, growth factors and proteases (see Table 1 and review (Bouloumie et al. 2005)). Many of these factors are involved in the adipose tissue microenvironment crosstalk between adipocytes, preadipocytes, resident macrophages and infiltrating macrophages.

3 Adipose Tissue Crosstalk: Adipocytes and Macrophages

The presence of macrophages in adipose tissue was first recognized in obese adipose tissue by Weisberg et al. (2003) and was recognized to be involved in insulin resistance (Wellen and Hotamisligil 2003; Xu et al. 2003). Lumeng et al. examined in detail the macrophage morphology and cell surface markers in lean and obese mice following high fat diet feeding (Lumeng et al. 2007). Their study demonstrated that the adipose tissue macrophage (ATM) population in obese mice is different from lean mice. In diet induced obese mice, the macrophages had characteristics of being "classically activated," including increased expression of pro-inflammatory genes and co-expression of F4/80$^+$ CD11c$^+$ markers in the gonadal

Table 1 Adipokines in metabolic and immune systems: This table is intended to give a brief overview of the overlap of these adipokines between metabolic and inflammatory roles but it is in no way a complete description of all functions and/or mechanisms. For more details on cardiovascular, satiety, inflammatory, and metabolic functions of these adipokines, please see reviews referenced in text.

	Leptin	**Adiponectin**	**Resistin**	**Visfatin**	**IL-6**	**TNF-α**
Associated diseases with hormone abnormalities	Obesity Insulin resistance Sepsis Rheumatoid arthritis Intestinal inflammation Thymic atrophy	Obesity Insulin resistance Diabetes Atherosclerosis Rheumatoid arthritis	Obesity Diabetes	Obesity Insulin resistance Sepsis	Inflammation Obesity sepsis Paget disease of bone Osteoporosis T-cell lymphoma	Obesity Insulin resistance Sepsis Cerebral malaria Multiple sclerosis Cancer Psoriasis Rheumatoid arthritis
Stimuli that alters hormone secretion from AdT	TNFα IL-6 Insulin Glucose Estrogen Angiotensin II	TNF-α Insulin IGF-1 Oleate Fasting / Satiety TZD – rosiglitizone	Inflammation TNFα LPS IL-6, IL-1, IL-12 Hyperglycemia Gonadal hormones TZD	TNF-α Insulin Cortisol TZD – rosiglitizone IL-1β	IL-1 TNFα	Adiponectin Gamma-interferon
Metabolic response to hormone	Satiety signal Regulation of insulin sensitivity Energy expenditure	Anti-diabetic Increase muscle FA oxidation Decrease hepatic glucose production pancreatic β-cell apoptosis protection	Regulation of insulin sensitivity	Reduced hepatocyte glucose release Increased peripheral glucose uptake	Macrophage recruitment to adipose tissue	Insulin resistance
Immune response to hormone	Chemotaxis Macrophage activation Natural killer cytotoxicity $T_{helper}1$ stimulation, $T_{helper}2$ inhibition Thymic T cell maturation/survival Neutrophil activation Links nutritional status to immune system	Anti-inflammatory Decrease T cell activation & proliferation Inhibit TNF-α & IL-6 release Increase IL-10 Decrease IFN-γ Vaso-protective	Pro-inflammatory Activate TNF-α & IL-6 release Increase leukocyte infliltration to tissues (including adipose tissue)	Delay neutrophil apoptosis	B-cell stimulation & differentiation	Induces ROS (reactive oxygen species)
Immune cells with receptor to hormone	Polymophonuclear leukocytes Monocytes Macrophage Lymphocytes	Monocytes Macrophage B & T cells NK cells	Macrophage monocytes			Stimulated T and B lyphocytes Most cells of myeloid origin

adipose tissue stromal vascular fraction (SVF). By contrast, adipose tissue from non-obese mice had less total macrophage content (based on percentage of cells positive for F4/80^{+} cells) and those macrophage had characteristics of being "alternatively activated," including increased expression of the anti-inflammatory IL-10 gene (Lumeng et al. 2007).

In agreement with the theory that both number and phenotype of macrophages changes with obesity, Cancello et al. (2005) examined subcutaneous adipose tissue from morbidly obese humans before and 3 months after bypass surgery when rapid weight loss occurs. Interestingly, when the subjects were obese, the infiltrating macrophages presented in crown-like structures surrounding mature adipocytes in a pattern similar to that seen after injury. Further studies in a variety of mouse models have consistently shown increased macrophage infiltration with classically activated phenotypes (pro-inflammatory) with obesity. Correlations exist with adipocyte cell size (Bouloumie et al. 2005; Weisberg et al. 2003) and adiposity (Bouloumie et al. 2005; Weisberg et al. 2003; Xu et al. 2003), although there is some debate as to whether plasma triglycerides are an independent factor (Cancello et al. 2006, 2007). The macrophages themselves predominantly localize around dead adipocytes (Neels and Olefsky 2006) (Cinti et al. 2005). Based on these (and other) experimental data, it is suggested that adipose tissue expansion during weight gain leads to recruitment of macrophages through a variety of signals. These signals include secretion of chemoattractant factors (as discussed below) as well as hypoxia.

4 Adipose Tissue Micro-Environment

Within the adipose tissue milieu, there are several ongoing processes (Fig. 1): (i) adipocytes release factors that cause chemotaxis and infiltration of circulating macrophages; (ii) macrophages then influence neighbouring adipocyte function; (iii) adipocytes, in turn, simultaneously influence macrophage function; (iv) both macrophages and adipocytes secrete factors that influence other tissues and systems, which in turn impact on the adipose tissue milieu (i.e. CNS control of satiety via leptin); and finally (v) preadipocytes contribute through cytokine secretion and direct effects on adipocyte function (Bouloumie et al. 2005).

4.1 Adipocyte-Mediated Recruitment, Differentiation and Macrophage Differentiation

With increased size of adipocytes, there is increased local secretion of most adipokines (a decreased secretion of adiponectin is the exception). The mechanism by which adipocytes recruit macrophages and influence their phenotype is currently being studied extensively. Several adipokines have been identified that recruit macrophages (MCP-1, osteopontin, TNFα, IL6, leptin). Enlarged adipocytes are more likely to become hypoxic and then necrotic, which signal macrophage recruitment. In fact, as the adipocyte cell grows (increases in lipid content), mRNA of several pro-inflammatory cytokines (TNFα, MCP-1, MIP-1α, and SAA3) increases disproportionately, potentially due to their production by other cells within the adipose tissue, particularly macrophages which also increase disproportionately with morbid obesity (Coenen et al. 2007). There are well recognized differences between subcutaneous and visceral adipose tissue depots in

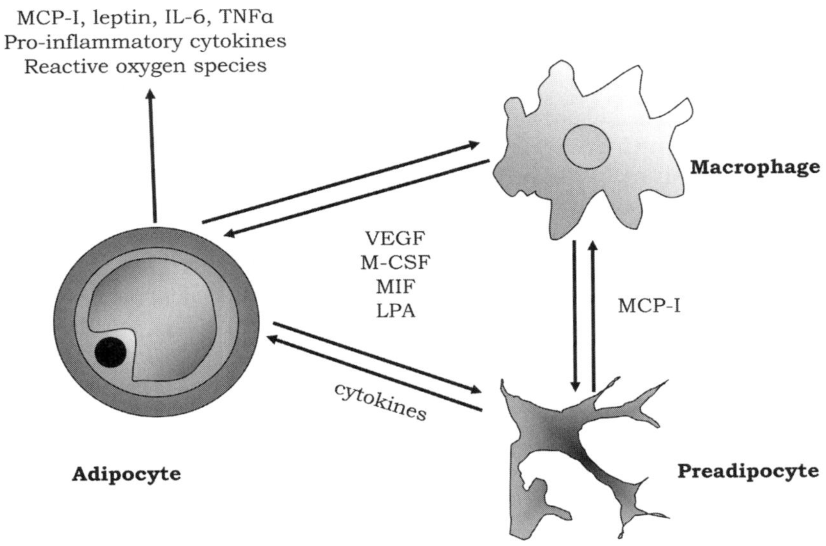

Fig. 1 Interactions between adipocytes, preadipocytes, macrophages and secreted factors within the adipose tissue micro-environment

mice and humans, and increased visceral adipose tissue mass is associated with increased inflammatory state (Lefebvre et al. 1998; Montague et al. 1998). In agreement with this, visceral adipose tissue has been demonstrated to have increased macrophage content as well (Tchoukalova et al. 2004; Weisberg et al. 2003).

4.2 Effects of Macrophages on Adipocyte Function

Following differentiation of the macrophages within the micro-environment of the adipose tissue, macrophages secrete M-CSF (macrophage colony-stimulating factor), VEGF, MIF and LPA (Bouloumie et al. 2005; Cancello and Clement 2006). There is also increased production of various pro-inflammatory cytokines and reactive oxygen species (Pausova 2006). The effects of macrophages on adipocyte function can be demonstrated directly in both co-culture and conditioned media experiments. As shown by Lumeng et al. (2007), co-culture of adipocytes with macrophages results in impaired insulin-stimulated glucose uptake, a key insulin action in adipocytes, via inhibition of GLUT4 transporter translocation to the cell surface. That this is likely mediated via soluble factors derived from macrophages is demonstrated by culture of adipocytes with macrophage conditioned media, which results in a similar insulin inhibitory effect (Lumeng et al. 2007).

4.3 Effects of Adipocytes on Macrophage Function

The converse action, adipocyte effects on macrophage function, is also evident. As demonstrated by Lumeng et al. (2007) direct co-culture of macrophages with adipocytes visually alters macrophage morphology. After several days in culture, the macrophages take on an elongated phenotype with long cellular processes. Increased intracellular lipid accumulation is also marked in the macrophages.

4.4 Contribution of Preadipocytes to Altered Adipocyte and Macrophage Function

Within the environment of the adipose tissue, there exists a pool of preadipocytes, which are proliferating cells capable of differentiation to mature lipid-laden adipocytes. These preadipocytes are also responsive to inflammatory signals. A recent study by Chung et al. (2006) demonstrated that LPS (lipopolysaccharide) stimulates inflammatory cytokine gene expression in the stromal vascular preadipocytes obtained from primary cultures of newly differentiated human adipocytes. LPS-induced inflammation in preadipocytes, acting via TLR2 and TLR4 receptors in preadipocytes as well as adipocytes, suppresses PPARγ activity, a key fat cell modulator, and reduces insulin responsiveness in adipocytes (Chung et al. 2006). The MCP-1 secreted by preadipocytes also enhanced macrophage recruitment and infiltration. Further, preadipocyte cells incubated in macrophage conditioned medium do not differentiate into adipocytes to the same extent. In fact, they express characteristics which are similar to macrophages, underlying the interrelationships within adipose tissue (Chung et al. 2006).

4.5 Pros and Cons of Macrophage Infiltration

The overwhelming interpretation of previous studies (as presented above) has suggested a negative effect of macrophage infiltration. Is all macrophage infiltration within the adipose tissue milieu necessarily associated with negative consequences? A study published recently by Odegaard et al. (2007) presents an interesting conundrum. As macrophage infiltration and production of various pro-inflammatory cytokines has been associated with insulin resistance, a macrophage specific knockout of PPARγ was constructed (Mac-PPARγKO). These mice are less susceptible to infection by *Leishmani a major* as evaluated by footpad swelling and footpad necrosis. In the absence of macrophage PPARγ, alternative activation of infiltrating macrophages is reduced, and indeed there is an overall reduction of macrophages detected by histological analysis in adipose tissue (Odegaard et al. 2007). Surprisingly, however, while this would have been expected to improve insulin sensitivity, it does not. In fact, the Mac-PPARγKO mice are less insulin sensitive with delayed glucose clearance following a glucose

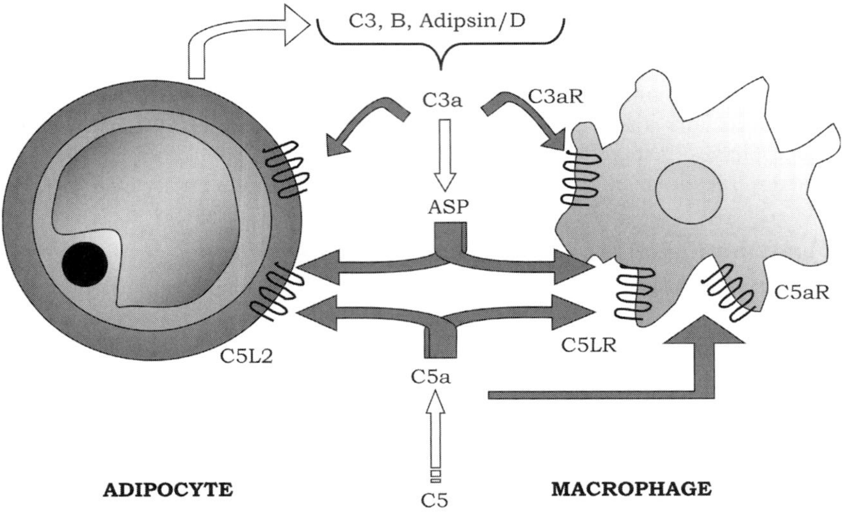

Fig. 2 Potential interactions between ASP/C3adR and C5a with C5L2, C5aR and C3aR in adipocytes and macrophages within the adipose tissue microenvironment

load or insulin challenge, and markedly increased plasma insulin levels. Clearly, the role of macrophages in adipose tissue infiltration is complex.

Further, the presence of macrophages in adipose tissue may decrease preadipocyte differentiation, thus restraining adipocyte tissue growth. Large fat-laden adipocytes are thought to be more insulin resistant (Pausova 2006). It may be that the initial infiltration of macrophages is intended to be compensatory or regulatory, and that the continued expansion of adipose tissue (due to over-consumption) initializes a vicious circle creating constant low grade inflammation and the accompanying negative consequences that are associated including associations with insulin resistance and atherosclerosis.

4.6 Effect of Weight Loss on Adipose Tissue Function

With weight gain, there is an increase both in adipocyte number (hyperplasia) as well as adipocyte size (hypertrophy) (Drolet et al. 2007). Many studies have demonstrated that with weight loss, there is a reduction in adipocyte tissue mass (DiGirolamo 1991). Recent data also suggests that weight loss may also be associated with decreased macrophage infiltration with bariatric surgery-induced weight loss (Cancello and Clement 2005). This study found a significant decrease in macrophage number with weight loss, correlating with BMI (body mass index) and mean adipocyte size. In addition, weight loss induced a phenotypical change in the ATM population with a significant increase in the anti-inflammatory alternatively activated macrophages after weight loss, as evidenced by increased secretion of IL-10 and a switch from M1-M2. To the author's knowledge, no studies on ATM changes with weight loss from diet and exercise have been completed, although a general reduction in inflammatory markers with weight loss (Cottam et al. 2004) suggests that ATM infiltration would likely decrease as well.

5 The Role of C5L2: Adipose vs. Immune

The recent identification of C5L2 (aka GPR77) as a potential ligand for both C5a, a powerful inflammatory factor, and ASP (C3adesArg), a stimulator of triglyceride synthesis in adipocytes, has added one addition link to the interaction between the adipose and the immune system. Although there are only a limited number of papers published since the first identification of C5L2 as a potential G-protein coupled receptor in 2000 (a total of ≈ 30, six of which are review articles), controversy remains regarding putative C5L2 ligands, C5L2 signalling capacity, and the physiological role of C5L2 in either adipose or immune systems (Table 2 and Figure 2). Controversy within the adipose-immune axis is not new. Other examples include the role of CD36 that has been proposed to be a fatty acid transporter in adipocytes (Harmon and Abumrad 1993) as well as a scavenger receptor in macrophages (Savill et al. 1991; Talle et al. 1983). The recent recognition of visfatin as an adipose tissue adipokine (Beltowski 2006), has partially overshadowed its previously identified function as a cytokine to differentiate B-cells and inhibit apoptosis of neutrophils in sepsis (reviewed by Pilz 2007) while known as PBEF (pre-B cell colony-enhancing factor). The roles of TNFα in lipid metabolism (Price et al. 1986) and then insulin resistance (Stephens and Pekala 1991) in addition to its known roles in immunity was also initially met with resistance, but is now well accepted, providing yet another example.

5.1 Metabolism, C3, and Acylation Stimulating Protein (ASP)

An additional example of the links between immune and metabolic systems is the presence of the alternative complement pathway in adipose tissue. Adipsin was initially characterized many years ago in mice by Spiegelman and colleagues as an adipose specific factor tightly linked to differentiation of preadipocytes to adipocytes (Wilkison et al. 1990). In fact, adipose is the major source of adipsin/factor D (Choy et al. 1992). Surprisingly, adipsin was later identified as the mouse homologue of human factor D (Choy et al. 1992), which led to the recognition of the adipose alternative complement pathway. In the alternative complement pathway adipocytes, upon differentiation, produce factors C3, B and adipsin leading to the production of C3adesArg (C3adR) (Choy and Spiegelman 1996, Choy et al. 1992).

However, prior to the identification of the adipose alternative complement pathway resulting in C3adR, the serum derived protein acylation stimulating protein (ASP) was characterized by its lipogenic functions (Cianflone et al. 1989).

ASP was later identified as C3adR, produced upon differentiation by adipose tissue (Baldo et al. 1993; Cianflone et al. 1994). Due to its lipogenic functions it is not surprising that both C3 secretion and ASP production increase in adipocytes following treatment with either insulin or dietary lipids (chylomicrons) (Maslowska et al. 1997; Scantlebury et al. 1998).

Interestingly, C3 KO mice, which are deficient in ASP, display delayed triglyceride and fatty acid clearance from the circulation following a fat meal, with a reduced capacity for fat storage in their adipose tissue (review (Cianflone et al. 2003; Xia et al. 2004)). Acute injection of ASP/C3adR normalizes the postprandial fat clearance. The alterations in fat clearance are also accompanied by hyperphagia and increased energy expenditure (review (Cianflone et al. 2003; Xia et al. 2004)). Finally, numerous studies have linked alterations in C3 as well as ASP with obesity, diabetes, cardiovascular disease (review (Cianflone et al. 2003; Muscari et al. 2007)). Taken together, this evidence from mice and humans highlights the ASP pathway as an interesting target for obesity research.

5.2 Identification of C5L2 and Ligand Binding in Transfected Cells

C5L2 (aka GPR77) was first identified independently by two groups as a putative orphan G-protein coupled receptor (GPCR) (Chen et al. 2007; Ohno et al. 2000). Based on the similarities demonstrated with the C3a receptor (C3aR) and C5a receptor (C5aR), this led to studies on the potential of anaphylatoxins to act as ligands, including C5a, C4a and C3a and the related desArg (dR) forms including ASP (C3adR). A relatively small number of published papers have actually directly examined ligand binding, primarily using cells transiently or stably expressing C5L2 (such as HEK and RBL)(Cain and Monk 2002; Johswich et al. 2006; Kalant et al. 2003a, 2005; Okinaga et al. 2003; Otto et al. 2004a; Scola et al. 2007). With respect to C5a and C5adR binding, there is agreement within the studies published that C5a is a high affinity ligand to C5L2 (Kd 0.5–10 nM), while C5adR has lower, but still robust affinity.

By contrast, there are fewer studies which have examined C3a and ASP/C3adR binding to C5L2, and the results are discordant. Studies by Klos and colleagues and by Gerard and colleagues both demonstrated an absence of C3a and C3adR/ASP binding (Johswich et al. 2006; Okinaga et al. 2003). By contrast, studies conducted in two additional laboratories, Monk and colleagues and Cianflone and colleagues, (Cain and Monk 2002; Kalant et al. 2003a, 2005) did demonstrate significant binding of ASP/C3adR and C3a, with affinities in the range of 50–200 nM. Comparisons between studies are problematic due to the extensive differences in many parameters including cell type utilized (HEK vs. RBL cells), receptor expression (transient vs. stably transfected), ligand purification (serum-derived vs. recombinant vs. protease generated), ligand labelling (iodinated vs. fluorescent), and binding methodologies (suspension cells vs. attached cells). Only Monk and

Table 2 Role of C5L2 and ASP (C3adesArg)
Summary is presented with explanations provided in the present text with reference to recent reviews (Ward 2007), (Monk et al. 2007), (Johswich et al. 2007), (Barrington et al. 2001; Cianflone et al. 2003; Kalant et al. 2003b). *CVD* cardiovascular disease; *EE* energy expenditure; *OM* omental; *SC* subcutaneous; *TG* triglyceride; *TZD* thiazolidinedione (diabetic medication)

	Adipocentric	**Immunocentric**
C5L2		
Production	Adipocytes (SC & OM), preadipocytes	Macrophages, neutrophils, HL-60, HeLa, astrocytes
Regulation	↑ Differentiation, insulin, TZD ↓ TNFα	↑ dbcAMP, noradrenaline ↓ TNFα ↑↓ IFNγ
Ligand binding	ASP(C3adR), C3a	C5a
Function	↑ TG synthesis & glucose transport	Decoy for C5a receptor
KO phenotype	Delayed postprandial TG clearance, hyperphagia, altered energy expenditure	↑or↓responses to C5a, negatively regulates inflammatory response in vivo
Disease	Hyperlipidemia	Septic shock
C3 & cleavage products		
Production	Adipocytes produce C3, B & adipsin (D), generate ASP	Macrophages produce C3, B and adipsin (D)
Regulation	↑ Dietary lipoproteins, insulin	↑ IL-1, IL-6, TNFα, estrogen ↓ IFNγ, glucocorticoids
Receptor binding	ASP binds only C5L2	C3a binds C5L2 & C3aR,
Function	↑ TG synthesis & glucose uptake	Host protection & inflammation
KO phenotype	Delayed postprandial TG clearance, hyperphagia, altered energy expenditure	Decreased inflammation, increased susceptibility bacterial infection (CLP), reduced TNFα release
Disease	↑ Obesity, CVD, diabetes ↓ weight loss, exercise, anorexia	Infection, humoral immune response, endotoxic shock

colleagues used recombinant derived C3a and C3adR (Kalant et al. 2003a). Klos and colleagues have pointed out the possibility of misinterpreting non-specific ^{125}I-C3a and ^{125}I-ASP binding to plastic as artefactual receptor binding (Johswich et al. 2006), and certainly we and others recognize the ‘stickiness’ of both of these proteins. On the other hand, this explanation cannot be used to explain the specific binding of fluorescently-labelled ASP to stably transfected C5L2-HEK cells in suspension, used in pre-screening and cell-sorting of these cells (Kalant et al.

2003a, 2005). An example of Fluos-ASP binding in comparison to anti-C5L2 antibody binding to sorted stably transfected C5L2-HEK cells is shown in Fig. 3.

The proposed affinities for C5a (with either C5aR or C5L2) are much stronger than the measured affinities of ASP with C5L2 (approximately 25- to 50-fold different). However, the physiological levels of serum ASP (C3adR) are also proportionally much greater than those of C5a: normal levels ASP = 25–100 nM (Maslowska et al. 1999), normal levels of C5a 1–2 nM (Huber-Lang et al. 2005), and accordingly it would be expected that both the binding affinity as well as the effective stimulatory dose would differ proportionally, with more ASP required (and available physiologically) for stimulation. These issues will likely be resolved with additional studies.

5.3 Signalling of C5L2 in Transfected Cells

As with ligand binding, the potential for direct signalling of the C5L2 receptor in C5L2 transfected cells remains controversial. Both Okinaga et al. (2003) and Kalant et al. (2005) have demonstrated phosphorylation of C5L2 subsequent to ligand treatment (C5a and ASP/C3adR, respectively), leading to β-arrestin translocation and internalization (Kalant et al. 2005), a marker of GPCR activation (Luttrell and Lefkowitz 2002). On the other hand, treatment with C5a was not found to stimulate ERK1/2 phosphorylation or intracellular Ca^{2+} fluxes (Okinaga et al. 2003), β hexosaminidase release (Kalant et al. 2003a) or stimulation of triglyceride synthesis and glucose transport (Kalant et al. 2005) in C5L2 transfected cells. By contrast, ASP/C3adR and C3a did stimulate triglyceride synthesis and glucose transport in stably transfected C5L2 cells (Kalant et al. 2005). It should be pointed out that the studies on ASP responsiveness were carried out in C5L2 stably transfected cells that had been screened and sorted on the basis of Fluos-ASP binding (Fig. 3). Again, as stated above, these issues remain controversial and will likely be resolved with additional studies.

5.4 A DRY Motif is not Required for Signalling

Based on the absence of a DRY (Asp-Arg-Tyr) motif in C5L2, which has been shown to be important in cell signalling in the C5aR, it has been suggested that C5L2 is obligately uncoupled from G-protein signalling (Johswich and Klos 2007; Okinaga et al. 2003). This is incorrect. While the motif is highly conserved in the rhodopsin family of GPCRs, particularly the aminergic GPCRs (Chung et al. 2002), directed mutations do not necessarily result in loss of function when the DRY motif is present and the DRY motif is not always present in functionally signalling GPCRs. There are several examples of mutations in an endogenous DRY motif that do not abolish cell signalling. (i) Mutation of the Arg in the α2A (Chung et al. 2002) and β2 (Seibold et al. 1998) adrenergic receptors show that the Arg is not required for G-protein activation by these GPCRs. (ii) In the CB2 cannabinoid receptor, mutation of the homologous Arg only partially reduced activity (Rhee et al. 2000), while (iii) Arg mutation in the type 1A angiotensin II receptor abolished coupling to some G-proteins but not others (Ohyama et al. 2002). On the other hand, some studies have shown a gain of constitutive activity with Arg substitutions (Chen et al. 2001; Fanelli et al. 1999; Scheer et al. 2000; Seibold et al. 1998). Further, the “conserved” DRY motif can present in various forms, including: DRS, NRY, ERY, DTW and RHG, preserving two, one or none of the “DRY” amino acids (Arora et al. 1997; Conner et al. 2007; Gruijthuijsen et al. 2004; Rosenkilde et al. 2005; Rovati et al. 2007; Scheer et al. 2000). In C5L2, the motif presents as a DLC motif in the human and DLF in mouse C5L2 (Kalant et al. 2005). Interestingly, the secretin family of GPCRs have a YL motif predicted to be in the same position as the first two amino acids of the rhodopsin family DRY motif. Mutation of the leucine in the vasoactive intestinal peptide 1 (VPAC1) receptor led to a pronounced impairment of G protein activation (Tams et al. 2001). The use of motifs to identify potentially important sequences in GPCRs is a valuable tool for analysis but, with exceptions as common as similarities, should be treated as a guideline rather than a rule. As with ligand binding, this issue of signalling motifs within C5L2 remains controversial.

5.5 Endogenous Expression of C5L2 and Functionality

C5L2 is widely expressed in various tissues including adipose, spleen, bone marrow, liver, intestine, lung, heart, brain, placenta and ovary (Chen et al. 2007; Gao et al. 2005; Gavrilyuk et al. 2005; Kalant et al. 2005; Ohno et al. 2000). Various cells, of both myeloid and non-myeloid origin, including human and mouse adipocytes and preadipocytes, neutrophils, immature but not mature dendritic cells, HL-60, HeLa cells and U937 macrophages express C5L2 (Huber-Lang et al. 2005; Johswich et al. 2006; Kalant et al. 2005; MacLaren et al. 2007; Ohno et al. 2000).

Examinations of ligand receptor responses in endogenously expressing cells are, however, few in number, and limited to functional responses, as no direct interaction with the C5L2 receptor has been demonstrated in these cells. With respect to C5a, ligand effects that could be directly attributed to the C5L2 receptor are complicated by the simultaneous expression of C5aR is these cells. Johswich et al. demonstrated an effect of C5a on Ca^{2+} fluxes in U937 cells and HL-60 cells, which was completely blocked in the presence of the C5aR inhibitor AcF[OPdChaAR] (Johswich et al. 2006), which does not prevent C5a binding to C5L2 (Otto et al. 2004b). While one explanation of this effect is that C5L2 is not a functional receptor, other explanations could include: (i) C5a activation of C5L2 does not result in increases in intracellular Ca^{2+} flux, or (ii) C5a activation of C5L2 requires C5aR (as discussed below in in vivo models).

By contrast, C3adR/ASP does not bind either the C3aR or the C5aR, and to date, C5L2 is the only identified ASP receptor (Kalant et al. 2003a, 2005). ASP, as

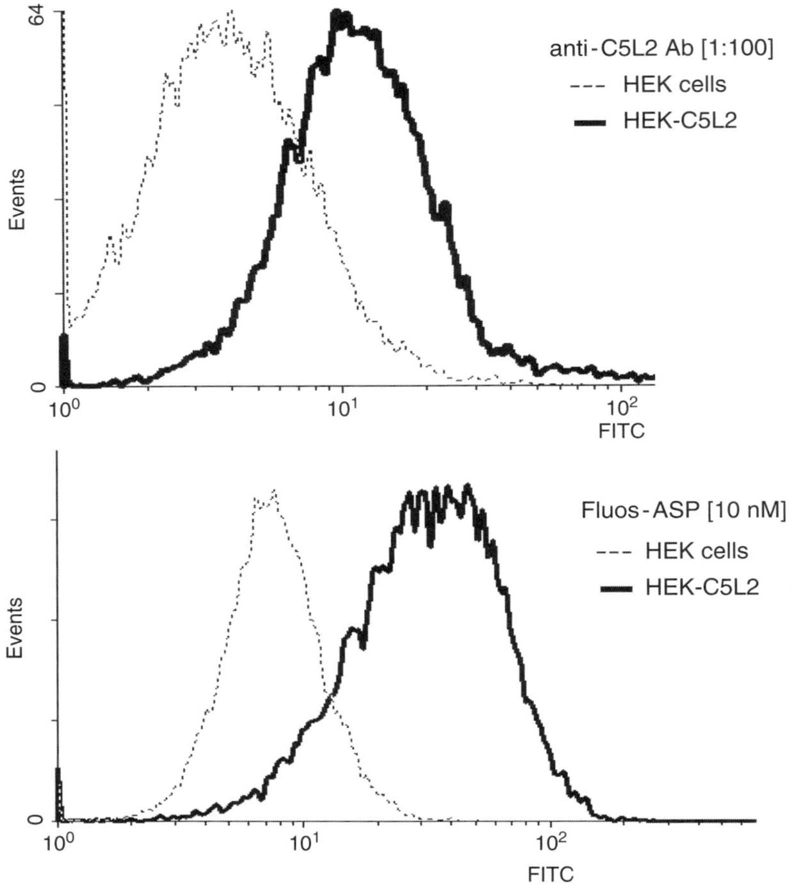

Fig. 3 Anti-C5L2 and Fluos-ASP binding to stably-transfected and sorted C5L2-HEK cells vs. HEK cells

well as C3a, stimulates triglyceride synthesis and glucose transport in preadipocytes and adipocytes (Kalant et al. 2003a, 2005), an effect that can be reduced using antisense and siRNA to C5L2 (Kalant et al. 2005), or blocked through antibody neutralization of C5L2, an effect comparable to anti-ASP neutralization (Cui et al. 2007). In preadipocytes, the ASP stimulation of lipogenesis is mediated via a PI3 kinase/phospholipase C/protein kinase C mechanism, an effect that is blocked using inhibitors to these pathways (Maslowska et al. 2006).

5.6 In Vivo Role of C5L2 in Mice and Humans

Notwithstanding the controversial results obtained in transfected cells, a role of C5L2 in both immune function and lipid metabolism is supported by evaluation of C5L2 expression in cells and the few studies published in mice. The ability of C5L2 to bind C5a without apparent activation has led to the suggestion that C5L2 may play a role as a decoy receptor, and therefore regulate the levels of C5a and C5adR availability (detailed in reviews (Johswich and Klos 2007; Monk et al. 2007; Ward 2008). An alternate role for C5L2 has also been proposed, based on a recent publication by Chen et al., suggesting that C5L2 is critical for the biological activities of C5a and C3a (Chen et al. 2007).

With respect to lipid metabolism, C5L2 expression in 3T3 adipocytes, a commonly used cell model, is increased with differentiation of adipocytes (MacLaren et al. 2007). Further, insulin and dexamethasone increased C5L2 mRNA, C5L2 cell surface expression and ASP binding, while TNFα decreased all three parameters (MacLaren et al. 2007). Rosiglitazone, a thiazolidinedione used to increase insulin sensitivity in diabetic patients, also increased C5L2 mRNA and ASP binding in adipocytes. In mice, C5L2 KO mice have reduced adipose tissue triglyceride synthesis and delayed postprandial lipid clearance, but are hyperphagic with increased muscle and heart fatty acid oxidation (Paglialunga et al. 2007). Interestingly, this is a similar phenotype to the C3 KO mice (Cianflone et al. 2003; Xia et al. 2004). A similar phenotype of delayed postprandial lipid clearance and altered lipid metabolism was obtained in wildtype mice injected for 10 days with either anti-C5L2 or anti-ASP neutralizing antibodies (Cui et al. 2007). Finally, in humans a C5L2 mutation has been associated with alterations in lipid metabolism (Marcil et al. 2006).

On the other hand, additional receptors for C3a/C3adR may yet exist. Using C5L2 KO mice, Honczarenko et al. demonstrated that the C5L2 receptor is not involved in C3a/C3adR mediated enhancement of bone marrow hematopoietic cell migration to CXCL12 (Honczarenko et al. 2005a). Honczarenko et al. (2005b), using C3aR KO mice, similarly suggested the presence of an alternate C3a/C3adR receptor based on their evaluation of responsiveness of hematopoietic cells to an SDF-1 gradient. As C3adR does not bind C3aR, the mechanism involved remains unidentified.

6 Summary

It is now generally well accepted that there is clear evidence of interaction between the adipose and the immune system. Activation of the innate immune system and a systemic low grade inflammatory state are integral parts of the response to obesity leading to insulin resistance and diabetes, and indeed, are used more and more for diagnosis of potential risk and treatment of risk factors. Vascular and tissue inflammation, endothelial dysfunction and cardiovascular disease are further complications. Notwithstanding the important role that C3a, C3aR, C5a, C5aR and potentially C5L2 play in the immune system, the implications of these and other immune factors in adipose tissue metabolism cannot be ignored. How the interaction between ASP and C5L2 vs. C5a and C5aR evolves remains to be seen.

Acknowledgments

This study was supported by grants from CIHR and NSERC to KC. K. Cianflone holds a Canada Research Chair in Adipose Tissue. R. MacLaren holds a trainee scholarship from NSERC.

References

Arora, K.K., Cheng, Z. and Catt, K.J. (1997) Mutations of the conserved DRS motif in the second intracellular loop of the gonadotropin-releasing hormone receptor affect expression, activation, and internalization. *Mol. Endocrinol.* 11, 1203–1212

Baldo, A., Sniderman, A.D., St Luce, S., Avramoglu, R.K., Maslowska, M., Hoang, B., Monge, J.C., Bell, A., Mulay, S. and Cianflone, K. (1993) The adipsin-acylation stimulating protein system and regulation of intracellular triglyceride synthesis. *J. Clin. Invest.* 92, 1543–1547

Barrington, R., Zhang, M., Fischer, M. and Carroll, M.C. (2001) The role of complement in inflammation and adaptive immunity. *Immunol. Rev.* 180, 5–15

Beltowski, J. (2006) Apelin and visfatin: unique "beneficial" adipokines upregulated in obesity? *Med. Sci. Monit.* 12, RA112–RA119

Berg, A.H., Combs, T.P., Du, X., Brownlee, M. and Scherer, P.E. (2001) The adipocyte-secreted protein Acrp30 enhances hepatic insulin action. *Nat. Med.* 7, 947–953

Bouloumie, A., Curat, C.A., Sengenes, C., Lolmede, K., Miranville, A. and Busse, R. (2005) Role of macrophage tissue infiltration in metabolic diseases. *Curr. Opin. Clin. Nutr. Metab Care.* 8, 347–354

Cain, S.A. and Monk, P.N. (2002) The orphan receptor C5L2 has high affinity binding sites for complement fragments C5a and C5a des-Arg(74). *J. Biol. Chem.* 277, 7165–7169

Cancello, R. and Clement, K. (2006) Is obesity an inflammatory illness? Role of low-grade inflammation and macrophage infiltration in human white adipose tissue. *BJOG* 113, 1141–1147

Cancello, R., Henegar, C., Viguerie, N., Taleb, S., Poitou, C., Rouault, C., Coupaye, M., Pelloux, V., Hugol, D., Bouillot, J.L., Bouloumie, A., Barbatelli, G., Cinti, S., Svensson, P.A., Barsh, G.S., Zucker, J.D., Basdevant, A., Langin, D. and Clement, K. (2005) Reduction of macrophage infiltration and chemoattractant gene expression changes in white adipose tissue of morbidly obese subjects after surgery-induced weight loss. *Diabetes* 54, 2277–2286

Cancello, R., Tordjman, J., Poitou, C., Guilhem, G., Bouillot, J.L., Hugol, D., Coussieu, C., Basdevant, A., Bar, H.A., Bedossa, P., Guerre-Millo, M. and Clement, K. (2006) Increased infiltration of macrophages in omental adipose tissue is associated with marked hepatic lesions in morbid human obesity. *Diabetes* 55, 1554–1561

Chen, A., Gao, Z.G., Barak, D., Liang, B.T. and Jacobson, K.A. (2001) Constitutive activation of A(3) adenosine receptors by site-directed mutagenesis. *Biochem. Biophys. Res. Commun.* 284, 596–601

Chen, N.J., Mirtsos, C., Suh, D., Lu, Y.C., Lin, W.J., McKerlie, C., Lee, T., Baribault, H., Tian, H. and Yeh, W.C. (2007) C5L2 is critical for the biological activities of the anaphylatoxins C5a and C3a. *Nature* 446, 203–207

Choy, L.N. and Spiegelman, B.M. (1996) Regulation of alternative pathway activation and C3a production by adipose cells. *Obes. Res.* 4, 521–532

Choy, L.N., Rosen, B.S. and Spiegelman, B.M. (1992) Adipsin and an endogenous pathway of complement from adipose cells. *J. Biol. Chem.* 267, 12736–12741

Chung, D.A., Wade, S.M., Fowler, C.B., Woods, D.D., Abada, P.B., Mosberg, H.I. and Neubig, R.R. (2002) Mutagenesis and peptide analysis of the DRY motif in the alpha2A adrenergic receptor: evidence for alternate mechanisms in G protein-coupled receptors. *Biochem. Biophys. Res. Commun.* 293, 1233–1241

Chung, S., Lapoint, K., Martinez, K., Kennedy, A., Boysen, S.M. and McIntosh, M.K. (2006) Preadipocytes mediate lipopolysaccharide-induced inflammation and insulin resistance in primary cultures of newly differentiated human adipocytes. *Endocrinology* 147, 5340–5351

Cianflone, K.M., Sniderman, A.D., Walsh, M.J., Vu, H.T., Gagnon, J. and Rodriguez, M.A. (1989) Purification and characterization of acylation stimulating protein. *J. Biol. Chem.* 264, 426–430

Cianflone, K., Roncari, D.A., Maslowska, M., Baldo, A., Forden, J. and Sniderman, A.D. (1994) Adipsin/acylation stimulating protein system in human adipocytes: regulation of triacylglycerol synthesis. *Biochemistry* 33, 9489–9495

Cianflone, K., Xia, Z. and Chen, L.Y. (2003) Critical review of acylation-stimulating protein physiology in humans and rodents. *Biochim. Biophys. Acta.* 1609, 127–143

Cinti, S., Mitchell, G., Barbatelli, G., Murano, I., Ceresi, E., Faloia, E., Wang, S., Fortier, M., Greenberg, A.S. and Obin, M.S. (2005) Adipocyte death defines macrophage localization and function in adipose tissue of obese mice and humans. *J. Lipid Res.* 46, 2347–2355

Coenen, K.R., Gruen, M.L., Chait, A. and Hasty, A.H. (2007) Diet-induced increases in adiposity, but not plasma lipids, promote macrophage infiltration into white adipose tissue. *Diabetes* 56, 564–573

Conner, A.C., Simms, J., Barwell, J., Wheatley, M. and Poyner, D.R. (2007) Ligand binding and activation of the CGRP receptor. *Biochem. Soc. Trans.* 35, 729–732

Cottam, D.R., Mattar, S.G., Barinas-Mitchell, E., Eid, G., Kuller, L., Kelley, D.E. and Schauer, P.R. (2004) The chronic inflammatory hypothesis for the morbidity associated with morbid obesity: implications and effects of weight loss. *Obes. Surg.* 14, 589–600

Cui, W., Paglialunga, S., Kalant, D., Lu, H., Roy, C., Laplante, M., Deshaies, Y. and Cianflone, K. (2007) Acylation stimulating protein/C5L2 neutralizing antibodies alter triglyceride metabolism in vitro and in vivo. *Am. J. Physiol Endocrinol. Metab.* 293, E1482–E1491

DiGirolamo, M. (1991) Cellular, metabolic, and clinical consequences of adipose mass enlargement in obesity. *Nutrition* 7, 287–289

Drolet, R., Richard, C., Sniderman, A.D., Mailloux, J., Fortier, M., Huot, C., Rheaume, C. and Tchernof, A. (2007) Hypertrophy and hyperplasia of abdominal adipose tissues in women. *Int. J. Obes. (Lond).* 32(2), 283–291

Fain, J.N. (2006) Release of interleukins and other inflammatory cytokines by human adipose tissue is enhanced in obesity and primarily due to the nonfat cells. *Vitam. Horm.* 74, 443–477

Fanelli, F., Barbier, P., Zanchetta, D., de Benedetti, P.G. and Chini, B. (1999) Activation mechanism of human oxytocin receptor: a combined study of experimental and computer-simulated mutagenesis. *Mol. Pharmacol.* 56, 214–225

Funahashi, T., Nakamura, T., Shimomura, I., Maeda, K., Kuriyama, H., Takahashi, M., Arita, Y., Kihara, S. and Matsuzawa, Y. (1999) Role of adipocytokines on the pathogenesis of atherosclerosis in visceral obesity. *Intern. Med.* 38, 202–206

Gao, H., Neff, T.A., Guo, R.F., Speyer, C.L., Sarma, J.V., Tomlins, S., Man, Y., Riedemann, N.C., Hoesel, L.M., Younkin, E., Zetoune, F.S. and Ward, P.A. (2005) Evidence for a functional role of the second C5a receptor C5L2. *FASEB J.* 19, 1003–1005

Gavrilyuk, V., Kalinin, S., Hilbush, B.S., Middlecamp, A., McGuire, S., Pelligrino, D., Weinberg, G. and Feinstein, D.L. (2005) Identification of complement 5a-like receptor (C5L2) from astrocytes: characterization of anti-inflammatory properties. *J. Neurochem.* 92, 1140–1149

Gruijthuijsen, Y.K., Beuken, E.V., Smit, M.J., Leurs, R., Bruggeman, C.A. and Vink, C. (2004) Mutational analysis of the R33-encoded G protein-coupled receptor of rat cytomegalovirus: identification of amino acid residues critical for cellular localization and ligand-independent signalling. *J. Gen. Virol.* 85, 897–909

Guzik, T.J., Mangalat, D. and Korbut, R. (2006) Adipocytokines – novel link between inflammation and vascular function? *J. Physiol Pharmacol.* 57, 505–528

Harmon, C.M. and Abumrad, N.A. (1993) Binding of sulfosuccinimidyl fatty acids to adipocyte membrane proteins: isolation and amino-terminal sequence of an 88-kD protein implicated in transport of long-chain fatty acids. *J. Membr. Biol.* 133, 43–49

Honczarenko, M., Lu, B., Nicholson-Weller, A., Gerard, N.P., Silberstein, L.E. and Gerard, C. (2005a) C5L2 receptor is not involved in C3a / C3a-desArg-mediated enhancement of bone marrow hematopoietic cell migration to CXCL12. *Leukemia* 19, 1682–1683

Honczarenko, M., Ratajczak, M., Nicholson-Weller, A. and Silberstein, L. (2005b) Complement C3a enhances CXCL12 (SDF-1)-mediated chemotaxis of bone marrow hematopoietic cells independently of C3a receptor. *J Immunol.* 175, 3698–3706

Hotamisligil, G.S. (2006) Inflammation and metabolic disorders. *Nature* 444, 860–867

Hotamisligil, G.S. and Spiegelman, B.M. (1994) Tumor necrosis factor alpha: a key component of the obesity-diabetes link. *Diabetes* 43, 1271–1278

Huber-Lang, M., Sarma, J.V., Rittirsch, D., Schreiber, H., Weiss, M., Flierl, M., Younkin, E., Schneider, M., Suger-Wiedeck, H., Gebhard, F., McClintock, S.D., Neff, T., Zetoune, F., Bruckner, U., Guo, R.F., Monk, P.N. and Ward, P.A. (2005) Changes in the novel orphan, C5a receptor (C5L2), during experimental sepsis and sepsis in humans. *J. Immunol.* 174, 1104–1110

Johswich, K. and Klos, A. (2007) C5L2 – an anti-inflammatory molecule or a receptor for acylation stimulating protein (C3a-desArg)? *Adv. Exp. Med. Biol.* 598, 159–180

Johswich, K., Martin, M., Thalmann, J., Rheinheimer, C., Monk, P.N. and Klos, A. (2006) Ligand specificity of the anaphylatoxin C5L2 receptor and its regulation on myeloid and epithelial cell lines. *J. Biol. Chem.* 281, 39088–39095

Kalant, D., Cain, S.A., Maslowska, M., Sniderman, A.D., Cianflone, K. and Monk, P.N. (2003a) The chemoattractant receptor-like protein C5L2 binds the C3a des-Arg77/ acylation-stimulating protein. *J. Biol. Chem.* 278, 11123–11129

Kalant, D., Maslowska, M., Scantlebury, T., Wang, H. and Cianflone, K. (2003b) Control of lipogenesis in adipose tissue and the role of acylation stimulating protein. *Can. J. Diabetes* 27, 154–171

Kalant, D., MacLaren, R., Cui, W., Samanta, R., Monk, P.N., Laporte, S.A. and Cianflone, K. (2005) C5L2 is a functional receptor for acylation-stimulating protein. *J. Biol. Chem.* 280, 23936–23944

Koerner, A., Kratzsch, J. and Kiess, W. (2005) Adipocytokines: leptin – the classical, resistin – the controversical, adiponectin – the promising, and more to come. *Best. Pract. Res. Clin. Endocrinol. Metab.* 19, 525–546

Lago, F., Dieguez, C., Gomez-Reino, J. and Gualillo, O. (2007) The emerging role of adipokines as mediators of inflammation and immune responses. *Cytokine Growth Factor Rev.* 18, 313–325

Lefebvre, A.M., Laville, M., Vega, N., Riou, J.P., van Gaal, L., Auwerx, J. and Vidal, H. (1998) Depot-specific differences in adipose tissue gene expression in lean and obese subjects. *Diabetes* 47, 98–103

Lumeng, C.N., Bodzin, J.L. and Saltiel, A.R. (2007) Obesity induces a phenotypic switch in adipose tissue macrophage polarization. *J. Clin. Invest.* 117, 175–184

Luttrell, L.M. and Lefkowitz, R.J. (2002) The role of beta-arrestins in the termination and transduction of G-protein-coupled receptor signals. *J. Cell Sci.* 115, 455–465

MacLaren, R., Kalant, D. and Cianflone, K. (2007) The ASP receptor C5L2 is regulated by metabolic hormones associated with insulin resistance. *Biochem. Cell Biol.* 85, 11–21

Marcil, M., Vu, H., Cui, W., Dastani, Z., Engert, J.C., Gaudet, D., Castro-Cabezas, M., Sniderman, A.D., Genest, J., Jr. and Cianflone, K. (2006) Identification of a novel C5L2 variant (S323I) in a French Canadian family with familial combined hyperlipemia. *Arterioscler. Thromb. Vasc. Biol.* 26, 1619–1625

Maslowska, M., Scantlebury, T., Germinario, R. and Cianflone, K. (1997) Acute in vitro production of acylation stimulating protein in differentiated human adipocytes. *J. Lipid Res.* 38, 1–11

Maslowska, M., Vu, H., Phelis, S., Sniderman, A.D., Rhode, B.M., Blank, D. and Cianflone, K. (1999) Plasma acylation stimulating protein, adipsin and lipids in non-obese and obese populations. *Eur. J. Clin. Invest.* 29, 679–686

Maslowska, M., Legakis, H., Assadi, F. and Cianflone, K. (2006) Targeting the signaling pathway of acylation stimulating protein. *J. Lipid Res.* 47, 643–652

Matarese, G., Leiter, E.H. and La Cava, A. (2007) Leptin in autoimmunity: many questions, some answers. *Tissue Antigens* 70, 87–95

Meier, U. and Gressner, A.M. (2004) Endocrine regulation of energy metabolism: review of pathobiochemical and clinical chemical aspects of leptin, ghrelin, adiponectin, and resistin. *Clin. Chem.* 50, 1511–1525

Monk, P.N., Scola, A.M., Madala, P. and Fairlie, D.P. (2007) Function, structure and therapeutic potential of complement C5a receptors. *Br. J. Pharmacol.* 152, 429–448

Montague, C.T., Prins, J.B., Sanders, L., Zhang, J., Sewter, C.P., Digby, J., Byrne, C.D. and O'Rahilly, S. (1998) Depot-related gene expression in human subcutaneous and omental adipocytes. *Diabetes* 47, 1384–1391

Muscari, A., Antonelli, S., Bianchi, G., Cavrini, G., Dapporto, S., Ligabue, A., Ludovico, C., Magalotti, D., Poggiopollini, G. and Zoli, M. (2007) Serum C3 is a stronger inflammatory marker of insulin resistance than C-reactive protein, leukocyte count, and erythrocyte sedimentation rate: comparison study in an elderly population. *Diabetes Care* 30, 2362–2368

Neels, J.G. and Olefsky, J.M. (2006) Inflamed fat: what starts the fire? *J. Clin. Invest.* 116, 33–35

Nomiyama, T., Perez-Tilve, D., Ogawa, D., Gizard, F., Zhao, Y., Heywood, E.B., Jones, K.L., Kawamori, R., Cassis, L.A., Tschop, M.H. and Bruemmer, D. (2007) Osteopontin mediates obesity-induced adipose tissue macrophage infiltration and insulin resistance in mice. *J. Clin. Invest.* 117, 2877–2888

Odegaard, J.I., Ricardo-Gonzalez, R.R., Goforth, M.H., Morel, C.R., Subramanian, V., Mukundan, L., Eagle, A.R., Vats, D., Brombacher, F., Ferrante, A.W. and Chawla, A. (2007) Macrophage-specific PPARgamma controls alternative activation and improves insulin resistance. *Nature* 447, 1116–1120

Ohno, M., Hirata, T., Enomoto, M., Araki, T., Ishimaru, H. and Takahashi, T.A. (2000) A putative chemoattractant receptor, C5L2, is expressed in granulocyte and immature dendritic cells, but not in mature dendritic cells. *Mol. Immunol.* 37, 407–412

Ohyama, K., Yamano, Y., Sano, T., Nakagomi, Y., Wada, M. and Inagami, T. (2002) Role of the conserved DRY motif on G protein activation of rat angiotensin II receptor type 1A. *Biochem. Biophys. Res. Commun.* 292, 362–367

Okinaga, S., Slattery, D., Humbles, A., Zsengeller, Z., Morteau, O., Kinrade, M.B., Brodbeck, R.M., Krause, J.E., Choe, H.R., Gerard, N.P. and Gerard, C. (2003) C5L2, a nonsignaling C5A binding protein. *Biochemistry* 42, 9406–9415

Otto, M., Hawlisch, H., Monk, P.N., Muller, M., Klos, A., Karp, C.L. and Kohl, J. (2004a) C5a mutants are potent antagonists of the C5a receptor (CD88) and of C5L2: position 69 is the locus that determines agonism or antagonism. *J. Biol. Chem.* 279, 142–151

Otto, M., Hawlisch, H., Monk, P.N., Muller, M., Klos, A., Karp, C.L. and Kohl, J. (2004b) C5a mutants are potent antagonists of the C5a receptor (CD88) and of C5L2: position 69 is the locus that determines agonism or antagonism. *J. Biol. Chem.* 279, 142–151

Paglialunga, S., Schrauwen, P., Roy, C., Moonen-Kornips, E., Lu, H., Hesselink, M.K., Deshaies, Y., Richard, D. and Cianflone, K. (2007) Reduced adipose tissue triglyceride synthesis and increased muscle fatty acid oxidation in C5L2 knockout mice. *J. Endocrinol.* 194, 293–304

Pausova, Z. (2006) From big fat cells to high blood pressure: a pathway to obesity-associated hypertension. *Curr. Opin. Nephrol. Hypertens.* 15, 173–178

Pelleymounter, M.A., Cullen, M.J., Baker, M.B., Hecht, R., Winters, D., Boone, T. and Collins, F. (1995) Effects of the obese gene product on body weight regulation in ob/ob mice. *Science* 269, 540–543

Pilz, S., Mangge, H., Obermayer-Pietsch, B. and Marz, W. (2007) Visfatin/pre-B-cell colony-enhancing factor: a protein with various suggested functions. *J. Endocrinol. Invest.* 30, 138–144

Price, S.R., Olivecrona, T. and Pekala, P.H. (1986) Regulation of lipoprotein lipase synthesis by recombinant tumor necrosis factor – the primary regulatory role of the hormone in 3T3-L1 adipocytes. *Arch. Biochem. Biophys.* 251, 738–746

Rhee, M.H., Nevo, I., Levy, R. and Vogel, Z. (2000) Role of the highly conserved Asp-Arg-Tyr motif in signal transduction of the CB2 cannabinoid receptor. *FEBS Lett.* 466, 300–304

Rosenkilde, M.M., Kledal, T.N. and Schwartz, T.W. (2005) High constitutive activity of a virus-encoded seven transmembrane receptor in the absence of the conserved DRY motif (Asp-Arg-Tyr) in transmembrane helix 3. *Mol. Pharmacol.* 68, 11–19

Rovati, G.E., Capra, V. and Neubig, R.R. (2007) The highly conserved DRY motif of class A G protein-coupled receptors: beyond the ground state. *Mol. Pharmacol.* 71, 959–964

Saladin, R., De Vos, P., Guerre-Millo, M., Leturque, A., Girard, J., Staels, B. and Auwerx, J. (1995) Transient increase in obese gene expression after food intake or insulin administration. *Nature* 377, 527–529

Savill, J., Hogg, N. and Haslett, C. (1991) Macrophage vitronectin receptor, CD36, and thrombospondin cooperate in recognition of neutrophils undergoing programmed cell death. *Chest* 99, 6S–7S

Scantlebury, T., Maslowska, M. and Cianflone, K. (1998) Chylomicron-specific enhancement of acylation stimulating protein and precursor protein C3 production in differentiated human adipocytes. *J. Biol. Chem.* 273, 20903–20909

Scheer, A., Costa, T., Fanelli, F., de Benedetti, P.G., Mhaouty-Kodja, S., Abuin, L., Nenniger-Tosato, M. and Cotecchia, S. (2000) Mutational analysis of the highly conserved arginine within the Glu/Asp-Arg-Tyr motif of the alpha(1b)-adrenergic receptor: effects on receptor isomerization and activation. *Mol. Pharmacol.* 57, 219–231

Scola, A.M., Higginbottom, A., Partridge, L.J., Reid, R.C., Woodruff, T., Taylor, S.M., Fairlie, D.P. and Monk, P.N. (2007) The role of the N-terminal domain of the complement fragment receptor C5L2 in ligand binding. *J. Biol. Chem.* 282, 3664–3671

Seibold, A., Dagarag, M. and Birnbaumer, M. (1998) Mutations of the DRY motif that preserve beta 2-adrenoceptor coupling. *Recept. Channels* 5, 375–385

Stephens, J.M. and Pekala, P.H. (1991) Transcriptional repression of the GLUT4 and C/EBP genes in 3T3-L1 adipocytes by tumor necrosis factor-alpha. *J. Biol. Chem.* 266, 21839–21845

Talle, M.A., Rao, P.E., Westberg, E., Allegar, N., Makowski, M., Mittler, R.S. and Goldstein, G. (1983) Patterns of antigenic expression on human monocytes as defined by monoclonal antibodies. *Cell Immunol.* 78, 83–99

Tams, J.W., Knudsen, S.M. and Fahrenkrug, J. (2001) Characterization of a G protein coupling "YL" motif of the human VPAC1 receptor, equivalent to the first two amino acids in the "DRY" motif of the rhodopsin family. *J. Mol. Neurosci.* 17, 325–330

Tataranni, P.A. and Ortega, E. (2005) A burning question: does an adipokine-induced activation of the immune system mediate the effect of overnutrition on type 2 diabetes? *Diabetes* 54, 917–927

Tchoukalova, Y.D., Sarr, M.G. and Jensen, M.D. (2004) Measuring committed preadipocytes in human adipose tissue from severely obese patients by using adipocyte fatty acid binding protein. *Am. J. Physiol. Regul. Integr. Comp Physiol.* 287, R1132–R1140

Tilg, H. and Moschen, A.R. (2006) Adipocytokines: mediators linking adipose tissue, inflammation and immunity. *Nat. Rev. Immunol.* 6, 772–783

Tomas, E., Tsao, T.S., Saha, A.K., Murrey, H.E., Zhang, C.C., Itani, S.I., Lodish, H.F. and Ruderman, N.B. (2002) Enhanced muscle fat oxidation and glucose transport by ACRP30 globular domain: acetyl-CoA carboxylase inhibition and AMP-activated protein kinase activation. *Proc. Natl. Acad. Sci.U S A* 99, 16309–16313

Trayhurn, P. and Wood, I.S. (2004) Adipokines: inflammation and the pleiotropic role of white adipose tissue. *Br. J. Nutr.* 92, 347–355

Trayhurn, P. and Wood, I.S. (2005) Signalling role of adipose tissue: adipokines and inflammation in obesity. *Biochem. Soc. Trans.* 33, 1078–1081

Uysal, K.T., Wiesbrock, S.M., Marino, M.W. and Hotamisligil, G.S. (1997) Protection from obesity-induced insulin resistance in mice lacking TNF-alpha function. *Nature* 389, 610–614

Ward, P.A. (2008) Role of the complement in experimental sepsis. *J. Leukoc. Biol.* 83(3), 467–470

Weisberg, S.P., McCann, D., Desai, M., Rosenbaum, M., Leibel, R.L. and Ferrante, A.W., Jr. (2003) Obesity is associated with macrophage accumulation in adipose tissue. *J. Clin. Invest.* 112, 1796–1808

Wellen, K.E. and Hotamisligil, G.S. (2003) Obesity-induced inflammatory changes in adipose tissue. *J. Clin. Invest.* 112, 1785–1788

Wilkison, W.O., Min, H.Y., Claffey, K.P., Satterberg, B.L. and Spiegelman, B.M. (1990) Control of the adipsin gene in adipocyte differentiation. Identification of distinct nuclear factors binding to single- and double-stranded DNA. *J. Biol. Chem.* 265, 477–482

Xia, Z., Stanhope, K.L., Digitale, E., Simion, O.M., Chen, L., Havel, P. and Cianflone, K. (2004) Acylation-stimulating protein (ASP)/complement C3adesArg deficiency results in increased energy expenditure in mice. *J. Biol. Chem.* 279, 4051–4057

Xu, H., Barnes, G.T., Yang, Q., Tan, G., Yang, D., Chou, C.J., Sole, J., Nichols, A., Ross, J.S., Tartaglia, L.A. and Chen, H. (2003) Chronic inflammation in fat plays a crucial role in the development of obesity-related insulin resistance. *J. Clin .Invest.* 112, 1821–1830

2. The Role of Complement in Stroke Therapy

Ricardo J. Komotar, Grace H. Kim, Marc L. Otten, Benjamin Hassid, J. Mocco, Michael E. Sughrue, Robert M. Starke, William J. Mack, Andrew F. Ducruet, Maxwell B. Merkow, Matthew C. Garrett, and E. Sander Connolly

Department of Neurological Surgery, Columbia University, New York, NY, USA

Abstract. Cerebral ischemia and reperfusion initiate an inflammatory process which results in secondary neuronal damage. Immunomodulatory agents represent a promising means of salavaging viable tissue following stroke. The complement cascade is a potent mediator of inflammation which is activated following cerebral ischemia. Complement is deposited on apoptotic neurons which likely leads to injury in adjacent viable cells. Studies suggest that blocking the complement cascade during the early phases of infarct evolution may result in decreased penumbral tissue infarction and limit the extent of brain injury. Additionally, other elements of the complement cascade may play a critical role in cell survival. In this paper, we review the role of the complement cascade in neuronal damage following ischemic injury and emphasize possible therapeutic targets.

1 Introduction

Stroke is a leading cause of morbidity and mortality in the U.S. (Dirnagl et al. 1999). This condition severely impacts quality of life, and its burden on health care resources is staggering (Dirnagl et al. 1999; Fischer et al. 1995; Gladstone et al. 2002). To further complicate the issue, there exists a paucity of effective stroke therapies, the majority of which are not options for approximately 97–98% of patients presenting with acute ischemic stroke (Multicentre Acute Stroke Trial – Italy (MAST-I) Group 1995). To this end, research efforts have focused on developing therapeutic avenues that ameliorate ischemic brain injury and can be administered safely at delayed time points.

Arterial occlusion initiates an inflammatory cascade with spatial and temporal variation that can be exploited for therapeutic benefit. In particular, although the ischemic core suffers irreversible damage, the more peripheral portions (penumbra) continue to be viable and represent salvageable tissue. Although establishing reperfusion is clearly an integral component to stroke management, manipulation of the inflammatory processes involved in cerebral ischemia will be critical to achieve maximal therapeutic benefit. In addition to the relatively delayed nature of neuroinflammatory pathophysiology, the inflammatory cascade offers numerous potential targets subject to a broad arsenal of immunomodulatory agents. In this paper, we review the complement cascade, a potent inflammatory mediator involved with neuronal damage following ischemic injury.

J.D. Lambris (ed.), *Current Topics in Complement II.*

doi: 10:1007/978-0-387-78952-1_2,

2 Inflammation Following Cerebral Ischemia

Reperfusion of ischemic parenchyma activates pro-inflammatory effectors, thereby exacerbating cerebral injury. As demonstrated in both animals and humans, inflammation leads to the release of proteases and free radicals by infiltrating neutrophils and increased leukocyte adhesion leading to capillary plugging, both of which are deleterious and expand the zone of irreversible injury (Bednar et al. 1991, 1997; Dawson et al. 1996; del Zoppo et al. 1991; Dirnagl et al. 1999; Matsuo et al. 1994; Shimakura et al. 2000; Small et al. 1999; Sughrue and Connolly 2004; Yano et al. 2003). A major component of the inflammatory response involves the complement cascade. Complement is deposited on apoptotic neurons following cerebral ischemia (Cowell et al. 2003a; Huang et al. 1999; Schafer et al. 2000). While the majority of systemic complement is produced hepatically, extrahepatic cells are also capable of manufacturing complement proteins (Morgan and Gasque 1996, 1997). Although the origin of neuronal complement deposition remains unclear, various brain cells have been demonstrated to produce mRNA for the entire complement cascade in vitro (Gasque et al. 1992, 1993; van Beek et al. 2003). Moreover, ischemic environments appear to up-regulate the expression of complement proteins, notably C1q (Schafer et al. 2000). Given the loss of blood-brain-barrier integrity that occurs after stroke, it is also likely that some portion of complement deposited in the brain enters with systemic circulation (Belayev et al. 1996). Regardless, given the diffusible nature of complement cascade byproducts, a reasonable hypothesize is that apoptotic cells injure adjacent viable cells by promoting inflammation via complement activation (Chakraborti et al. 2000).

3 Complement Mediated Ischemia/Reprefusion Injury

Complement has been shown to play a role in ischemia/reperfusion injury throughout the body. The majority of this work has focused on myocardial ischemia, where complement deposition occurs within 3 h of onset, which may be related to the focal loss of regulators of complement activation (RCA) in the infarct region (Vakeva et al. 1994). In addition, animals genetically deficient in complement (Kilgore et al. 1998), as well as those treated with soluble complement inhibitor-1 (sCR1) (Lazar et al. 1998; Weisman et al. 1990), C1-esterase inhibitor (Buerke et al. 1995), and C5 activation blocking antibodies (Vakeva et al. 1998) have attenuated post-ischemic myocardial injury. The hypothesis of complement mediated ischemia/reperfusion injury has been further strengthened by similar benefits observed in renal (Duffield et al. 2001), intestinal (Stahl et al. 2003), and limb (Weiser et al. 1996) ischemia. Drawing conclusions from experiments relying on complement inhibition, however, is problematic due in part to their lack of specificity. For instance, recent work utilizing C1-esterase inhibitor has demonstrated that the protection mediated by this inhibitor does not involve C1q but instead results from inhibition of cell recruitment and inflammation (De Simoni et al. 2004; Storini et al. 2005).

The complement cascade is activated in a host of neurological diseases (Gasque et al. 1992, 1993, 1996, 2000, 2002; Morgan and Gasque 1996, 1997; Morgan et al. 1997; Rogers et al. 1996; van Beek et al. 2003). While not the primary inciting event, complement up-regulation in the acute stage of neurologic disease likely exacerbates injury. More specifically, complement has been demonstrated to have pro-inflammatory effects in dementia secondary to Alzheimer's disease (Eikelenboom et al. 1989; Head et al. 2001; Matsuoka et al. 2001; McGeer et al. 1989), and Pick's Disease (Gasque et al. 2000; Singhrao et al. 1996; Yasuhara et al. 1994), while transgenic mice producing a soluble form of the murine complement inhibitor Crry, were shown to develop less profound brain injury secondary to head trauma (Rancan et al. 2003).

Neurons, in vitro, are capable of spontaneously activating the complement cascade (Gasque et al. 1996, 2000; Singhrao et al. 2000). Singhrao et al. (2000) demonstrated this phenomenon, as human primary neurons cultured in the presence of complement activate the entire complement cascade without the presence of external stimuli, eventually leading to neuronal lysis by MAC. This may be the result of a relative deficiency in Decay activating factor (DAF, CD55) and Membrane Cofactor Protein (MCP, CD46). Diminished levels of these factors render neurons susceptible to spontaneous complement activation (Singhrao et al. 2000). Although these data suggest that neurons may be susceptible to lysis by complement, it is important to note that certain soluble complement inhibitors, such as C4-binding protein and Factor H, are mainly produced by astrocytes and likely lacking in a pure culture of neurons (D'Ambrosio et al. 2001).

4 Complement Mediated Cell Clearance

Complement deposition on the surface of pathogens greatly increases the rate at which these opsonized pathogens are phagocytosed and destroyed (Mevorach 1999). Studies have also demonstrated that early complement components such as C1q, C3, and C4 play a critical role in facilitating the clearance of apoptotic cells (Fishelson et al. 2001; Mevorach 1999, 2000, 2003; Mevorach et al. 1998). This may explain why genetic complement deficiencies strongly predispose individuals to develop autoimmune diseases, such as lupus, which are generally considered to result in part from the failure to clear apoptotic cells that spill cellular debris after acute tissue injury (Elward and Gasque 2003; Fadok 1999; Fadok and Henson 1998; Fadok et al. 1998a,b; Fishelson et al. 2001; Mevorach 2003).

It has been demonstrated that phagocytosis of apoptotic cells is not a biologically neutral event, but rather has profound effects on cytokine expression (Fadok 1999; Fadok and Chimini 2001; Fishelson et al. 2001; Savill 2000). In experimental models, for example, macrophage ingestion of apoptotic cells reduces expression of pro-inflammatory cytokines such as TNF-alpha and IL-1, and increases expression of anti-inflammatory cytokines such as TGF-β (Fadok 1999; Knepper-Nicolai et al. 1998; Ren et al. 2001; Savill 1997). This phenomenon has also been demonstrated to occur in human microglia, with the interesting caveat that growth factors such as NGF are also released (De Simone et al. 2003). In another study, mice doubly transgenic for human amyloid precursor peptide (hAPP) and sCrry, experienced significantly more extensive Alzheimer's-type neurodegeneration compared to mice only transgenic for hAPP (Wyss-Coray et al. 2002). While the exact mechanism for this observation is currently unclear, it is hypothesized that this increase in neuronal loss is secondary to failed amyloid plaque clearance in the complement-inhibited mice (Wyss-Coray et al. 2002). In short, complement inhibition appears to exacerbate certain disease processes as a result of decreased clearance of cellular debris.

5 Implications of Complement Inhibition

Work involving non-specific pharmacologic complement inhibitors, such as cobra venom factor (Cowell et al. 2003b; Figueroa et al. 2005; Vasthare et al. 1998), and C1-esterase inhibitor (Akita et al. 2001; De Simoni et al. 2003), supports the hypothesis that complement activation contributes to cerebral ischemia/reperfusion injury. Moreover, studies involving treatment with a specific complement inhibitor, soluble complement receptor-1 (sCR1), led to significant reductions in neutrophil and platelet aggregation, as well as significantly improved neurological function following experimental stroke. Cerebral infarct volume was also reduced, although this difference was not statistically significant (Huang et al. 1999). These results were more marked when a sialylated form of the sCR1 molecule, which both inhibits complement activation and blocks neutrophil adhesion via P/E-selectin, was administered (Huang et al. 1999). Taken together, these data strongly suggest that blocking the complement cascade during the early phases of infarct evolution may be able to decrease recruitment of penumbral tissue into the infarct and thereby limit the extent of brain injury following stroke.

Despite systemic complement inhibition, however, sCR1 treatment did not improve outcome in non-human primate stroke models (Mocco et al. 2006a). While the reasons for failure to translate remain unclear, further dissection of the complement cascade revealed that complete C3 inhibition was critical for neuroprotection (Mocco et al. 2006b). The protective effects of C3 blockade appear to be mediated by C3a and associated with diminished PMN infiltration. Moreover, C3 depletion and/or blockade lead not only to improved outcome and decreased PMN infiltration, but also to reduced P-selectin transcription and microvascular thrombosis (Atkinson et al. 2006).

6 Complement and Neurogenesis/Neurorecovery

Recent studies have supported the role of complement in basal and ischemia-induced neurogenesis, as C3a and C5a are highly expressed on neural progenitor cells (Rahpeymai et al. 2006). In addition, C3 deficient and C3aR deficient mice, as well as wild type mice treated with C3aR antagonists for ten days showed reduced basal neurogenesis. Moreover, C3 deficient mice had decreased ischemia-induced neurogenesis and demonstrated no benefit in outcome following permanent MCA occlusion.

While the anaphylotoxins C3a and C5a are best known for their profound pro-inflammatory effects, more recently it has been appreciated that these molecules play critical roles in promoting cell survival, differentiation, and regeneration following injury (Markiewski et al. 2004; Ratajczak et al. 2004; Strey et al. 2003). For example, it has been shown that C3 –/– knockout mice subjected to either toxic injury or partial hepatectomy, demonstrate marked impairment of liver regeneration secondary to their failure to produce C3a and C5a (Markiewski et al. 2004; Strey et al. 2003). C3a and C5a are thought to prime cells for rapid tissue regeneration/repair by increasing their responsiveness to specific growth factors (Markiewski et al. 2004; Strey et al. 2003). Similarly, work by Ratajczak and colleagues have implicated C3a in the engraftment of hematopoetic stem cells into bone marrow stroma (Ratajczak et al. 2004). They demonstrated that C3 –/– knockout mice irradiated and then administered bone marrow mononuclear cells demonstrate a profound impairment in the ability of these stem cells to form functional colony forming units when compared to wild-type controls. This effect was hypothesized to result from a combination of two C3 mediated mechanisms. First, C3 activation releases C3a, which, acting in a large part through its catabolite C3ades-Arg, binds the C5L2 receptor, potentiating the stem cell chemotactic response to the growth factor stromal-derived factor-1. Second, C3b deposited in the stroma is broken down to iC3b, mediating stem cell-stroma attachment via the iC3b-CR3 interaction.

While the precise effects of C3a and C5a on neurorecovery remain unclear, there is evidence that these molecules have non-inflammatory effects in the CNS and may effect neuronal survival following stress. Mice genetically deficient in C5 demonstrate markedly increased susceptibility to glutamate excitotoxic neuronal death following an intrahippocampal injection of kainate (Pasinetti et al. 1996). Additionally, more recent in vitro work has demonstrated that pretreatment of neuronal cultures with C5a reduces induced neuronal cell death, in part due to MAP-kinase dependent inhibition of Caspase activation (Mukherjee and Pasinetti 2000, 2001; Osaka et al. 1999).

Thus, there exists a potential conflict in the most recent findings of complement-related cerebral ischemia research: whereas C3 inhibition has been shown to reduce infarct volume in some stroke models, C3 deficiency has been shown to hinder neurogenesis. It will be critical for future studies to determine the etiology for these differences, as timing, dosing, and the type of experimental model implemented may play important roles.

7 Steps Towards Clinical Translation

Since anti-C3 strategies appear to be protective via diminished anaphylatoxin dependent neutrophil trafficking and improved microvascular patency within the penumbra, C3 blockade may not be beneficial in a setting where reperfusion is not significant. In support of this hypothesis, data not yet published from our laboratory has found C3a receptor antagonism protects mice from transient, but not permanent cerebral ischemia. Although transient and permanent MCA occlusion resulted in similar degrees of endothelial ICAM-a expression and granulocyte infiltration into post-ichemic cortex, C3aR antagonism markedly diminished this phenomenon only in cases involving reperfusion. These findings suggest that translational efforts to bring anti-complement therapies to the bedside in patients suffering acute ischemic stroke should concentrate on administration in the setting of known reperfusion, for example following proven recanalization with IV TPA or endovascular methods.

Along these lines, clinical studies regarding anti-inflammatory stroke therapies have been discouraging thus far. For example, the Enlimomab (anti-ICAM-1) antibody Acute Stroke Trial was halted following enrollment of 625 patients, as Enlimomab treated patients demonstrated worse outcomes than placebo treated patients (Furuya et al. 2001). Subsequent investigation revealed, however, that the mouse antibody utilized in this study activated the complement cascade when incubated with blood from healthy human volunteers (Vuorte et al. 1999).

While complete inhibition of complement may appear therapeutic following stroke, recent advances indicate that the complement cascade is multi-potent and critical to ensuring proper mammalian physiology. Therefore, a thorough understanding of each component is essential. While C3a has been shown to mediate ischemia/reperfusion injury in the brain by recruiting and activating neutrophils, the exact mechanisms by which the C5a receptor mediates its effects are still unknown. Future studies are necessary to elucidate the mechanism by which the C5a receptor is upregulated in the brain following ischemia/reperfusion injury and the specific targets through which it mediates its effects. In addition, efforts to validate post-ischemic efficacy should be pursued.

8 Conclusion

Anti-complement strategies appear to hold great promise for the development of stroke therapeutics. Yet caution should be exercised, for if any knowledge has been gained from the many failed attempts at translating stroke therapies to the bedside, it is that cavalier application of under-elucidated therapies lead to wasted resources and the potential for poor patient outcomes. Since it is clear that the complement cascade is a complex and intricate system with widely varied effects, it will be critical for future studies to determine the specific roles of the individual complement components and the mechanisms by which each component affects cerebral injury in the context of stroke.

References

Akita N., Nakase H., Kanemoto Y., Kaido T., Nishioka T., Sakaki T. (2001) [The effect of C1 esterase inhibitor on ischemia: reperfusion injury in the rat brain]. No to Shinkei – Brain and Nerve 53:641–644

Atkinson C., Zhu H., Qiao F., Varela J.C., Yu J., Song H., Kindy M.S., Tomlinson S. (2006) Complement-dependent P-selectin expression and injury following ischemic stroke. *Journal of Immunology* 177:7266–7274

Bednar M.M., Raymond S., McAuliffe T., Lodge P.A., Gross C.E. (1991) The role of neutrophils and platelets in a rabbit model of thromboembolic stroke. *Stroke* 22:44–50

Bednar M.M., Gross C.E., Howard D.B., Lynn M. (1997) Neutrophil activation in acute human central nervous system injury. *Neurological Research* 19:588–592

Belayev L., Busto R., Zhao W., Ginsberg M.D. (1996) Quantitative evaluation of blood-brain barrier permeability following middle cerebral artery occlusion in rats. *Brain Research* 739:88–96

Buerke M., Murohara T., Lefer A.M. (1995) Cardioprotective effects of a C1 esterase inhibitor in myocardial ischemia and reperfusion. *Circulation* 91:393–402

Chakraborti T., Mandal A., Mandal M., Das S., Chakraborti S. (2000) Complement activation in heart diseases. Role of oxidants. *Cellular Signalling* 12:607–617

Cowell R.M., Plane J.M., Silverstein F.S. (2003a) Complement activation contributes to hypoxic-ischemic brain injury in neonatal rats. *Journal of Neuroscience* 23:9459–9468

Cowell R.M., Plane J.M., Silverstein F.S. (2003b) Complement activation contributes to hypoxic-ischemic brain injury in neonatal rats. *Journal of Neuroscience* 23:9459–9468

D'Ambrosio A.L., Pinsky D.J., Connolly E.S. (2001) The role of the complement cascade in ischemia/reperfusion injury: implications for neuroprotection. *Molecular Medicine* 7:367–382

Dawson D.A., Ruetzler C.A., Carlos T.M., Kochanek P.M., Hallenbeck J.M. (1996) Polymorphonuclear leukocytes and microcirculatory perfusion in acute stroke in the SHR. *Keio Journal of Medicine* 45:248–252; discussion 252–243

De Simone R., Ajmone-Cat M.A., Tirassa P., Minghetti L. (2003) Apoptotic PC12 cells exposing phosphatidylserine promote the production of anti-inflammatory and neuroprotective molecules by microglial cells. *Journal of Neuropathology and Experimental Neurology* 62:208–216

De Simoni M.G., Storini C., Barba M., Catapano L., Arabia A.M., Rossi E., Bergamaschini L. (2003) Neuroprotection by complement (C1) inhibitor in mouse transient brain ischemia. *Journal of Cerebral Blood Flow and Metabolism* 23:232–239

De Simoni M.G., Rossi E., Storini C., Pizzimenti S., Echart C., Bergamaschini L. (2004) The powerful neuroprotective action of C1-inhibitor on brain ischemia-reperfusion injury does not require C1q. *American Journal of Pathology* 164:1857–1863

del Zoppo G.J., Schmid-Schonbein G.W., Mori E., Copeland B.R., Chang C.M. (1991) Polymorphonuclear leukocytes occlude capillaries following middle cerebral artery occlusion and reperfusion in baboons. *Stroke* 22:1276–1283

Dirnagl U., Iadecola C., Moskowitz M.A. (1999) Pathobiology of ischaemic stroke: an integrated view. *Trends in Neurosciences* 22:391–397

Duffield J.S., Ware C.F., Ryffel B., Savill J. (2001) Suppression by apoptotic cells defines tumor necrosis factor-mediated induction of glomerular mesangial cell apoptosis by activated macrophages. *American Journal of Pathology* 159:1397–1404

Eikelenboom P., Hack C.E., Rozemuller J.M., Stam F.C. (1989) Complement activation in amyloid plaques in Alzheimer's dementia. *Virchows Archiv. B. Cell Pathology* 56:259–262

Elward K., Gasque P. (2003) "Eat me" and "don't eat me" signals govern the innate immune response and tissue repair in the CNS: emphasis on the critical role of the complement system. *Molecular Immunology* 40:85–94

Fadok V.A. (1999) Clearance: the last and often forgotten stage of apoptosis. *Journal of Mammary Gland Biology & Neoplasia* 4:203–211

Fadok V.A., Chimini G. (2001) The phagocytosis of apoptotic cells. *Seminars in Immunology* 13:365–372
Fadok V.A., Henson P.M. (1998) Apoptosis: getting rid of the bodies. *Current Biology* 8:R693–R695
Fadok V.A., Bratton D.L., Frasch S.C., Warner M.L., Henson P.M. (1998a) The role of phosphatidylserine in recognition of apoptotic cells by phagocytes [see comment]. *Cell Death and Differentiation* 5:551–562
Fadok V.A., Warner M.L., Bratton D.L., Henson P.M. (1998b) CD36 is required for phagocytosis of apoptotic cells by human macrophages that use either a phosphatidylserine receptor or the vitronectin receptor (alpha v beta 3). *Journal of Immunology* 161:6250–6257
Figueroa E., Gordon L.E., Feldhoff P.W., Lassiter H.A. (2005) The administration of cobra venom factor reduces post-ischemic cerebral injury in adult and neonatal rats. *Neuroscience Letters* 380:48–53
Fischer B., Schmoll H., Riederer P., Bauer J., Platt D., Popa-Wagner A. (1995) Complement C1q and C3 mRNA expression in the frontal cortex of Alzheimer's patients. *Journal of Molecular Medicine* 73:465–471
Fishelson Z., Attali G., Mevorach D. (2001) Complement and apoptosis. *Molecular Immunology* 38:207–219
Furuya K., Takeda H., Azhar S., McCarron R.M., Chen Y., Ruetzler C.A., Wolcott K.M., DeGraba T.J., Rothlein R., Hugli T.E., del Zoppo G.J., Hallenbeck J.M. (2001) Examination of several potential mechanisms for the negative outcome in a clinical stroke trial of enlimomab, a murine anti-human intercellular adhesion molecule-1 antibody: a bedside-to-bench study. *Stroke* 32:2665–2674
Gasque P., Julen N., Ischenko A.M., Picot C., Mauger C., Chauzy C., Ripoche J., Fontaine M. (1992) Expression of complement components of the alternative pathway by glioma cell lines. *Journal of Immunology* 149:1381–1387
Gasque P., Ischenko A., Legoedec J., Mauger C., Schouft M.T., Fontaine M. (1993) Expression of the complement classical pathway by human glioma in culture. A model for complement expression by nerve cells. *Journal of Biological Chemistry* 268:25068–25074
Gasque P., Thomas A., Fontaine M., Morgan B.P. (1996) Complement activation on human neuroblastoma cell lines in vitro: route of activation and expression of functional complement regulatory proteins. *Journal of Neuroimmunology* 66:29–40
Gasque P., Dean Y.D., McGreal E.P., VanBeek J., Morgan B.P. (2000) Complement components of the innate immune system in health and disease in the CNS. *Immunopharmacology* 49:171–186
Gasque P., Neal J.W., Singhrao S.K., McGreal E.P., Dean Y.D., Van B.J., Morgan B.P. (2002) Roles of the complement system in human neurodegenerative disorders: pro-inflammatory and tissue remodeling activities. *Molecular Neurobiology* 25:1–17
Gladstone D.J., Black S.E., Hakim A.M. (2002) Toward wisdom from failure: lessons from neuroprotective stroke trials and new therapeutic directions. *Stroke* 33:2123–2136
Head E., Azizeh B.Y., Lott I.T., Tenner A.J., Cotman C.W., Cribbs D.H. (2001) Complement association with neurons and beta-amyloid deposition in the brains of aged individuals with Down Syndrome. *Neurobiology of Disease* 8:252–265
Huang J., Kim L.J., Mealey R., Marsh H.C., Jr., Zhang Y., Tenner A.J., Connolly E.S., Jr., Pinsky D.J. (1999) Neuronal protection in stroke by an sLex-glycosylated complement inhibitory protein. *Science* 285:595–599

Kilgore K.S., Park J.L., Tanhehco E.J., Booth E.A., Marks R.M., Lucchesi B.R. (1998) Attenuation of interleukin-8 expression in C6-deficient rabbits after myocardial ischemia/reperfusion. *Journal of Molecular and Cellular Cardiology* 30:75–85

Knepper-Nicolai B., Savill J., Brown S.B. (1998) Constitutive apoptosis in human neutrophils requires 99synergy between calpains and the proteasome downstream of caspases. *Journal of Biological Chemistry* 273:30530–30536

Lazar H.L., Hamasaki T., Bao Y., Rivers S., Bernard S.A., Shemin R.J. (1998) Soluble complement receptor type I limits damage during revascularization of ischemic myocardium. *Annals of Thoracic Surgery* 65:973–977

Markiewski M.M., Mastellos D., Tudoran R., DeAngelis R.A., Strey C.W., Franchini S., Wetsel R.A., Erdei A., Lambris J.D. (2004) C3a and C3b activation products of the third component of complement (C3) are critical for normal liver recovery after toxic injury. *Journal of Immunology* 173:747–754

Matsuo Y., Onodera H., Shiga Y., Nakamura M., Ninomiya M., Kihara T., Kogure K. (1994) Correlation between myeloperoxidase-quantified neutrophil accumulation and ischemic brain injury in the rat. Effects of neutrophil depletion. *Stroke* 25:1469–1475

Matsuoka Y., Picciano M., Malester B., LaFrancois J., Zehr C., Daeschner J.M., Olschowka J.A., Fonseca M.I., O'Banion M.K., Tenner A.J., Lemere C.A., Duff K. (2001) Inflammatory responses to amyloidosis in a transgenic mouse model of Alzheimer's disease. *American Journal of Pathology* 158:1345–1354

McGeer P.L., Akiyama H., Itagaki S., McGeer E.G. (1989) Activation of the classical complement pathway in brain tissue of Alzheimer patients. *Neuroscience Letters* 107:341–346

Mevorach D. (1999) The immune response to apoptotic cells. *Annals of the New York Academy of Sciences* 887:191–198

Mevorach D. (2000) Opsonization of apoptotic cells. Implications for uptake and autoimmunity. *Annals of the New York Academy of Sciences* 926:226–235

Mevorach D. (2003) Systemic lupus erythematosus and apoptosis: a question of balance. *Clinical Reviews in Allergy and Immunology* 25:49–60

Mevorach D., Mascarenhas J.O., Gershov D., Elkon K.B. (1998) Complement-dependent clearance of apoptotic cells by human macrophages. *Journal of Experimental Medicine* 188:2313–2320

Mocco J., Mack W.J., Ducruet A.F., King R.G., Sughrue M.E., Coon A.L., Sosunov S.A., Sciacca R.R., Zhang Y., Marsh H.C., Jr., Pinsky D.J., Connolly E.S., Jr. (2006a) Preclinical evaluation of the neuroprotective effect of soluble complement receptor type 1 in a nonhuman primate model of reperfused stroke. *Journal of Neurosurgery* 105:595–601

Mocco J., Mack W.J., Ducruet A.F., Sosunov S.A., Sughrue M.E., Hassid B.G., Nair M.N., Laufer I., Komotar R.J., Claire M., Holland H., Pinsky D.J., Connolly E.S., Jr. (2006b) Complement component C3 mediates inflammatory injury following focal cerebral ischemia. *Circulation Research* 99:209–217

Morgan B.P., Gasque P. (1996) Expression of complement in the brain: role in health and disease. *Immunology Today* 17:461–466

Morgan B.P., Gasque P. (1997) Extrahepatic complement biosynthesis: where, when and why? *Clinical and Experimental Immunology* 107:1–7

Morgan B.P., Gasque P., Singhrao S., Piddlesden S.J. (1997) The role of complement in disorders of the nervous system. *Immunopharmacology* 38:43–50

Mukherjee P., Pasinetti G.M. (2000) The role of complement anaphylatoxin C5a in neurodegeneration: implications in Alzheimer's disease. *Journal of Neuroimmunology* 105:124–130

Mukherjee P., Pasinetti G.M. (2001) Complement anaphylatoxin C5a neuroprotects through mitogen-activated protein kinase-dependent inhibition of caspase 3. *Journal of Neurochemistry* 77:43–49

Osaka H., Mukherjee P., Aisen P.S., Pasinetti G.M. (1999) Complement-derived anaphylatoxin C5a protects against glutamate-mediated neurotoxicity. *Journal of Cellular Biochemistry* 73:303–311

Pasinetti G.M., Tocco G., Sakhi S., Musleh W.D., DeSimoni M.G., Mascarucci P., Schreiber S., Baudry M., Finch C.E. (1996) Hereditary deficiencies in complement C5 are associated with intensified neurodegenerative responses that implicate new roles for the C-system in neuronal and astrocytic functions. *Neurobiology of Disease* 3:197–204

Rahpeymai Y., Hietala M.A., Wilhelmsson U., Fotheringham A., Davies I., Nilsson A.K., Zwirner J., Wetsel R.A., Gerard C., Pekny M., Pekna M. (2006) Complement: a novel factor in basal and ischemia-induced neurogenesis. *Embo Journal* 25:1364–1374

Rancan M., Morganti-Kossmann M.C., Barnum S.R., Saft S., Schmidt O.I., Ertel W., Stahel P.F. (2003) Central nervous system-targeted complement inhibition mediates neuroprotection after closed head injury in transgenic mice. *Journal of Cerebral Blood Flow and Metabolism* 23:1070–1074

Multicentre Acute Stroke Trial – Italy (MAST-I) Group. (1995) Randomised controlled trial of streptokinase, aspirin, and combination of both in treatment of acute ischaemic stroke. *Lancet* 346:1509–1514

Ratajczak M.Z., Reca R., Wysoczynski M., Kucia M., Baran J.T., Allendorf D.J., Ratajczak J., Ross G.D. (2004) Transplantation studies in C3-deficient animals reveal a novel role of the third complement component (C3) in engraftment of bone marrow cells. *Leukemia* 18:1482–1490

Ren Y., Stuart L., Lindberg F.P., Rosenkranz A.R., Chen Y., Mayadas T.N., Savill J. (2001) Nonphlogistic clearance of late apoptotic neutrophils by macrophages: efficient phagocytosis independent of beta 2 integrins. *Journal of Immunology* 166:4743–4750

Rogers C.A., Gasque P., Piddlesden S.J., Okada N., Holers V.M., Morgan B.P. (1996) Expression and function of membrane regulators of complement on rat astrocytes in culture. *Immunology* 88:153–161

Savill J. (1997) Apoptosis in resolution of inflammation. *Journal of Leukocyte Biology* 61:375–380

Savill J. (2000) Apoptosis in resolution of inflammation. *Kidney and Blood Pressure Research* 23:173–174

Schafer M.K., Schwaeble W.J., Post C., Salvati P., Calabresi M., Sim R.B., Petry F., Loos M., Weihe E. (2000) Complement C1q is dramatically up-regulated in brain microglia in response to transient global cerebral ischemia. *Journal of Immunology* 164:5446–5452

Shimakura A., Kamanaka Y., Ikeda Y., Kondo K., Suzuki Y., Umemura K. (2000) Neutrophil elastase inhibition reduces cerebral ischemic damage in the middle cerebral artery occlusion. *Brain Research* 858:55–60

Singhrao S.K., Neal J.W., Gasque P., Morgan B.P., Newman G.R. (1996) Role of complement in the aetiology of Pick's disease? *Journal of Neuropathology and Experimental Neurology* 55:578–593

Singhrao S.K., Neal J.W., Rushmere N.K., Morgan B.P., Gasque P. (2000) Spontaneous classical pathway activation and deficiency of membrane regulators render human neurons susceptible to complement lysis. *American Journal of Pathology* 157:905–918

Small D.L., Morley P., Buchan A.M. (1999) Biology of ischemic cerebral cell death. *Progress in Cardiovascular Diseases* 42:185–207

Stahl G.L., Xu Y., Hao L., Miller M., Buras J.A., Fung M., Zhao H. (2003) Role for the alternative complement pathway in ischemia/reperfusion injury [see comment]. *American Journal of Pathology* 162:449–455

Storini C., Rossi E., Marrella V., Distaso M., Veerhuis R., Vergani C., Bergamaschini L., De Simoni M.G. (2005) C1-inhibitor protects against brain ischemia-reperfusion injury via inhibition of cell recruitment and inflammation. *Neurobiology of Diseases* 19:10–17

Strey C.W., Markiewski M., Mastellos D., Tudoran R., Spruce L.A., Greenbaum L.E., Lambris J.D. (2003) The proinflammatory mediators C3a and C5a are essential for liver regeneration. *Journal of Experimental Medicine* 198:913–923

Sughrue M.E., Connolly E.S., Jr. (2004) Effectively bridging the preclinical/clinical gap: the results of the ASTIN trial. *Stroke* 01.STR.0000121164.0000129117.0000121144

Vakeva A., Morgan B.P., Tikkanen I., Helin K., Laurila P., Meri S. (1994) Time course of complement activation and inhibitor expression after ischemic injury of rat myocardium. *American Journal of Pathology* 144:1357–1368

Vakeva A.P., Agah A., Rollins S.A., Matis L.A., Li L., Stahl G.L. (1998) Myocardial infarction and apoptosis after myocardial ischemia and reperfusion: role of the terminal complement components and inhibition by anti-C5 therapy. *Circulation* 97:2259–2267

van Beek J., Elward K., Gasque P. (2003) Activation of complement in the central nervous system: roles in neurodegeneration and neuroprotection. *Annals of the New York Academy of Sciences* 992:56–71

Vasthare U.S., Barone F.C., Sarau H.M., Rosenwasser R.H., DiMartino M., Young W.F., Tuma R.F. (1998) Complement depletion improves neurological function in cerebral ischemia. *Brain Research Bulletin* 45:413–419

Vuorte J., Lindsberg P.J., Kaste M., Meri S., Jansson S.E., Rothlein R., Repo H. (1999) Anti-ICAM-1 monoclonal antibody R6.5 (Enlimomab) promotes activation of neutrophils in whole blood. *Journal of Immunology* 162:2353–2357

Weiser M.R., Williams J.P., Moore F.D., Jr., Kobzik L., Ma M., Hechtman H.B., Carroll M.C. (1996) Reperfusion injury of ischemic skeletal muscle is mediated by natural antibody and complement. *Journal of Experimental Medicine* 183:2343–2348

Weisman H.F., Bartow T., Leppo M.K., Marsh H.C., Jr., Carson G.R., Concino M.F., Boyle M.P., Roux K.H., Weisfeldt M.L., Fearon D.T. (1990) Soluble human complement receptor type 1: in vivo inhibitor of complement suppressing post-ischemic myocardial inflammation and necrosis. *Science* 249:146–151

Wyss-Coray T., Yan F., Lin A.H., Lambris J.D., Alexander J.J., Quigg R.J., Masliah E. (2002) Prominent neurodegeneration and increased plaque formation in complement-inhibited Alzheimer's mice. *Proceedings of the National Academy of Sciences of the United States of America* 99:10837–10842

Yano T., Anraku S., Nakayama R., Ushijima K. (2003) Neuroprotective effect of urinary trypsin inhibitor against focal cerebral ischemia-reperfusion injury in rats. *Anesthesiology* 98:465–473

Yasuhara O., Aimi Y., McGeer E.G., McGeer P.L. (1994) Expression of the complement membrane attack complex and its inhibitors in Pick disease brain. *Brain Research* 652:346–349

3. Food Intake Regulation by Central Complement System

Kousaku Ohinata and Masaaki Yoshikawa

Division of Food Science and Biotechnology, Graduate School of Agriculture, Kyoto University, Gokasho Uji, Kyoto 611-0011, Japan

Abstract. Complement C3a and C5a are released from C3 and C5, respectively, on activation of the complement system and play an important role in immune response. C3a, C5a and their receptors have been revealed to be present in the central nervous system (CNS) as well as the peripheral immune system. We found that centrally administered C3a suppresses food intake, while C5a stimulates food intake, and their food intake regulation may be associated with the prostaglandin system. We propose that complement C3a and C5a are regulators not only of the immune system but also of the CNS.

1 Introduction

The complement system, consisting of a series of proteins from C1q to C9, is activated in response to pathogenic microorganism invasion, and plays an important role in immune response (Law and Reid 1995). Among these proteins, complement C3a (77 amino acid residues) and C5a (74 amino acid residues) are released from the amino terminus regions of C3 and C5, respectively, by enzymatic degradation during the complement system activation (Law and Reid 1995). C3a and C5a stimulate histamine secretion in mast cells, increase capillary vessel permeability and contract smooth muscle (Law and Reid 1995). C5a has chemotaxic activity and facilitates phagocytes mobilization (Law and Reid 1995). C3a and C5a are also called anaphylatoxins, since they induce anaphylactic-like reactions after their peripheral administration (Law and Reid 1995).

C3a and C5a receptors, seven transmembrane G-protein coupled receptors (Ames et al. 1996; Tornetta et al. 1997; Gerard and Gerard 1991; Boulay et al. 1991), were thought to be restricted to the peripheral immune system; however, it has been reported that these receptors are also present in the central nervous system (CNS), including glial cells and neurons (Nataf et al. 1999; Davoust et al. 1999; Gasque et al. 1997, 1998; O'Barr et al. 2001); however, little is known about central functions of C3a and C5a. In the current review, we focused on food intake regulation by C3a and C5a.

J.D. Lambris (ed.), *Current Topics in Complement II.*
doi: 10:1007/978-0-387-78952-1_3, © Springer Science+Business Media, LLC 2008

2 C3a Effect on Food Intake

2.1 Anorexigenic Action of Central C3a

It was reported that C3a administered into the perifornical hypothalamic region of non-fasted rats potentiated feeding stimulation by norepinephrine (Schupf et al. 1983); however, we found that intracerebroventricular (i.c.v.)-administered C3a suppressed food intake in fasted mice (Ohinata et al. 2002). C3a-des-Arg was inactive under this experimental condition (Ohinata et al. 2002), indicating that the anorexigenic effect was specific to C3a. We thus revealed that C3a agonist suppresses food intake after central administration.

It is known that acute infections trigger a host defense reaction with several physiological and behavioral changes, including anorexia (Langhans 2007). Lipopolysaccharide (LPS), a Gram-negative bacterial cell-wall used as a model of microbial infections, suppresses food intake after central and peripheral administration (Langhans 2007; Inui 2001). Intraperitoneally (i.p) administered LPS increased C3 mRNA levels in the median eminence and arcuate nucleus of the hypothalamus (Nadeau and Rivest 2001), which is an important brain site for food intake regulation. C3a receptor mRNA expression was also elevated in the mouse brain after i.p. injection of LPS (Nadeau and Rivest 2001). Although it is not clear whether the complement components originate from astrocytes and microglia in the brain parenchyma or systemic immune cells after leakage from the blood-brain barrier (BBB), the complement system might be involved in anorexia during an acute inflammatory response (Nadeau and Rivest 2001). Furthermore, anorexia under infection and inflammation might also be associated with stimulating anorexigenic cytokines such as tumor necrosis factor (TNF)-α, interleukin (IL)-1β and IL-6 derived from neurons and glial cells as well as peripheral lymphocytes and/or monocyte macrophages (Hopkins and Rothwell 1995; Inui 1999a). Interestingly, mice lacking C3a receptor had elevated plasma IL-1β concentrations and enhanced lethality to endotoxin LPS shock (Kildsgaard 2000), suggesting that C3a receptor acts as an anti-inflammatory receptor. It is also suggested that the central complement system might be activated during acute brain injury, such as ischemia and trauma, and chronic neurodegeneration, including Alzheimer's disease and Huntington's disease (Nadeau and Rivest 2001; van Beek et al. 2003; Gasque et al. 2000). It should be clarified how the complement system, including anorexigenic C3a, contributes to eating disorders in these diseases.

The other major source of C3a is white adipose tissue, which is well-known as not only an energy storage tissue but also an active endocrine organ (Cianflone et al. 2003). Adipocyte also produces and secretes leptin, which is an afferent signal from the periphery to the brain that regulates adipose tissue mass (Inui 1999b). The level of leptin is positively correlated with body fat mass (Inui 1999b). Leptin reduces appetite and increases energy expenditure via the CNS (Inui 1999b). Similarly to leptin, C3a decreases food intake after central administration (Ohinata et al. 2002). It is thought that after C3a secretion from white adipose tissue, the

C-terminal arginine of C3a was considerably removed to give C3a-des-Arg (acylation-stimulating protein, ASP) by carboxypeptidase. Plasma concentrations of C3a-des-Arg are elevated in human obesity (Sniderman et al. 1991). Fasting, chronic hypocaloric diets or gastric bypass surgery, which lead to decreased body weight, also increases plasma C3a-des-Arg levels (Sniderman et al. 1991). In contrast to leptin, C3a-des-Arg increases energy storage by stimulating triacylglycerol synthesis and increasing glucose transport independently of C3a receptor (Sniderman et al. 1991). C3a receptor agonists with resistance to peptidase, especially carboxypeptidase, might be used as anti-obesity drugs.

2.2 C3a Agonist Peptides Derived from Natural Proteins

A number of low molecular peptides with C3a agonist properties have also been found in the enzymatic digests of human, animal and plant proteins (Takahashi et al. 1996, 1997, 1998; Chiba et al. 1989), and we investigated whether these C3a agonist peptides change food intake after central or peripheral administration. We previously isolated C3a agonist peptides such as human albutensin A (Ala-Phe-Lys-Ala-Trp-Ala-Val-Ala-Arg) (Takahashi et al. 1998), bovine casoxin C (Tyr-Ile-Pro-Ile-Gln-Tyr-Val-Leu-Ser-Arg) (Chiba et al. 1989; Takahashi et al. 1997) and oryzaetensin (Gly-Tyr-Pro-Met-Tyr-Pro-Leu-Pro-Arg) (Takahashi et al. 1996) derived from serum albumin, κ casein and rice albumin, respectively, by guinea pig ileum-contraction assay. Primary structure at the carboxy termini of these peptides have homology to that of C3a, Leu-Gly-Leu-Ala-Arg, which is minimally essential for the C3a activity. Among these peptides, human albutensin A decreased food intake after i.c.v. and i.p. administration (Ohinata et al. 2002). Although albutensin A had low affinities for C5a and bombesin receptors as well as C3a receptor (Ohinata et al. 2002; Takahashi et al. 1998), the anorexigenic activity induced by albutensin A may be mediated by C3a receptor for the following reasons: 1) Albutensin A suppressed food intake in mice lacking bombesin receptor subtype 1, which mainly mediates bombesin-induced food intake suppression (Ohki-Hamazaki et al. 1999); 2) It has been reported that centrally administered C5a increased food intake in rats (Schupf and Williams 1987; Williams et al. 1985) and we confirmed the orexigenic effect of C5a after i.c.v. administration under our experimental conditions (Ohinata et al. 2002); 3) Albutensin A-des-Arg without affinity for C3a receptor did not show anorexigenic activity i.c.v. and i.p. administration. Taken together, food intake suppression of albutensin A may be mediated through C3a receptor.

[Trp^5]-oryzatensin(5–9) (Trp-Pro-Leu-Pro-Arg, WPLPR) is an agonist peptide for C3a receptor designed based on the C-terminal region of ileum-contracting peptide oryzatensin derived from rice protein (Jinsmaa et al. 2001). WPLPR suppressed food intake after i.c.v. and i.p administration (Ohinata et al. 2007). Orally administered WPLPR also suppressed food intake (Ohinata et al. 2007). WPLPR may be at least partly absorbed into the blood, probably because of its low molecular weight and resistance to digestive enzymes in the gastrointestinal tract

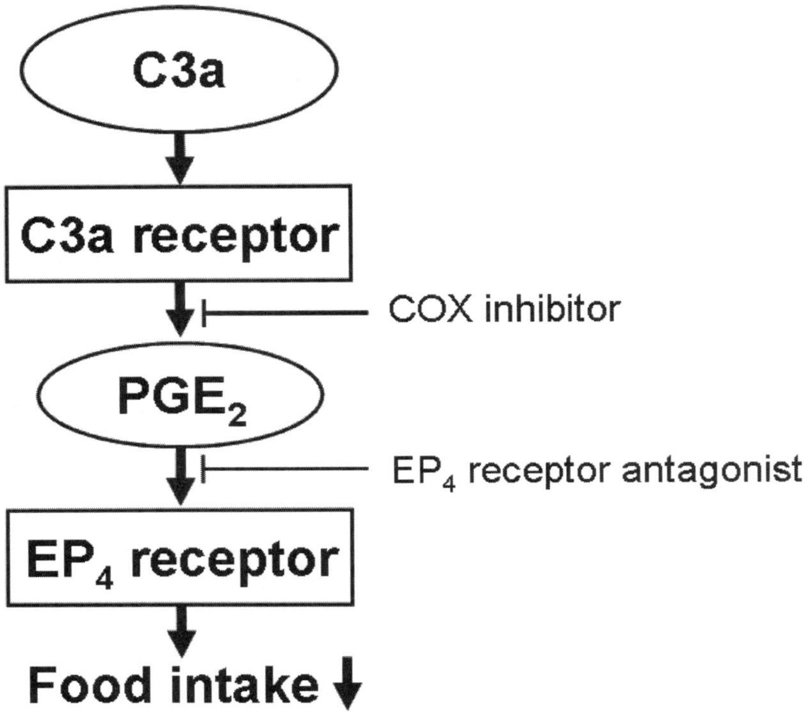

Fig. 1 Model of food intake regulation by a C3a agonist. A C3a agonist suppresses food intake via PGE_2 production and EP_4 receptor activation

owing to presence of two proline residues. Both C3a agonist peptides and C3a itself suppressed food intake after central administration in fasted mice. Further investigations will reveal whether peripherally administered WPLPR crosses the BBB or whether its signal is neurally transmitted to the CNS after peripheral C3a receptor activation.

Since anorexigenic peptides are known to often decrease the gastric emptying rate, we tested whether C3a agonist peptides delayed it. Centrally and i.p. administered human albutensin A and WPLPR delayed the gastric emptying rate at a higher dose than that required for food intake suppression (Ohinata et al. 2002, 2007). These results suggest that peripherally administered human albutensin A and WPLPR might suppress food intake via acting on the central C3a receptor rather than via inhibiting peripheral gastrointestinal motility.

2.3 Involvement of the PGE_2-EP_4 Receptor Pathway

We also investigated the mechanism underlying the anorexigenic activity of a C3a agonist peptide, WPLPR. Its anorexigenic activity was blocked by pretreatment with a cyclooxygenase (COX) inhibitor indomethacin (Ohinata et al. 2007). Among COX products, prostaglandin (PG) E_2 is known to suppress food intake. We found that anorexigenic activity of the C3a agonist peptide was completely inhibited by an antagonist for EP_4 receptor among four receptor subtypes for PGE_2 (Ohinata et al. 2007). These results suggest that the C3a agonist might suppress food intake via PGE_2 production followed by activation of EP_4 receptor (Fig. 1).

PGE_2, a bioactive lipid produced in the CNS of mammals, including humans, has a variety of physiologically and pathophysiologically central actions on wakefulness, fever, pain response and food intake (Hayaishi 1991; Ushikubi et al. 1998; Huang et al. 2003; Horton 1964; Levine and Morley 1981). PGE_2 exerts its actions through four different types of G-protein-coupled seven-transmembrane receptors, known as EP_{1-4} (Narumiya et al. 1999; Narumiya and FitzGerald 2001). Recently, it has been revealed that EP_3 and EP_4 mediate the febrile response and wakefulness of PGE_2, respectively (Ushikubi et al. 1998; Huang et al. 2003). We found that an EP_4 agonist, ONO-AE1-329, decreased food intake after i.c.v. administration among four highly selective EP_1-EP_4 agonists (Ohinata et al. 2006). The anorexigenic action of ONO-AE1-329 and PGE_2 was blocked by an EP_4 antagonist ONO-AE3-208 (Ohinata et al. 2006). These results suggest that EP_4 activation in the CNS suppressed food intake.

It was reported that PGE_2 also inhibited gastrointestinal motility (Van Miert et al. 1983). An EP_4 agonist suppressed the gastric emptying rate at a higher dose than required for food intake suppression (Ohinata et al. 2006). These results were consistent with C3a agonist peptides such as human albutensin A and WPLPR suppressing gastric empting at a higher dose than that required for food intake suppression (Ohinata et al. 2002 and 2007).

Since PGE_2 is reported to elevate blood glucose levels (Yatomi et al. 1987), we examined whether an EP_4 agonist could change blood glucose levels, which might also be associated with food intake regulation. An EP_4 agonist at a dose of 10 nmol/mouse elevated fasted blood glucose levels after central administration in mice (Ohinata et al. 2006). I.c.v.-administered human albutensin A also increased blood glucose levels (Ohinata et al. 2002). These suppressed food intake at a lower dose than that required to elevate blood glucose, suggesting that hyperglycemia might be independent of anorexigenic activities.

The EP_4 receptor is widely distributed throughout the whole body and its mRNA is also expressed in almost all mouse tissue (Narumiya et al. 1999). In the brain, EP_4 mRNA is expressed in the hypothalamic area, which plays important roles in the regulation of food intake (Narumiya et al. 1999). In the hypothalamus, EP_4 mRNA was abundantly localized in the paraventricular nucleus (PVN) and the supraoptic nucleus in rats (Zhang and Rivest 1999). PGE_2 is produced from arachidonic acid by COX followed by PGE synthase, and acts near its production site (Narumiya et al. 1999; Matsuoka et al. 2003). It was reported that COX and PGE synthase are constitutively present in the PVN (Matsuoka et al. 2003). The relationships between the PGE_2-EP_4 system and hypothalamic neuropeptides regulating food intake should be clarified. To our knowledge, a C3a agonist

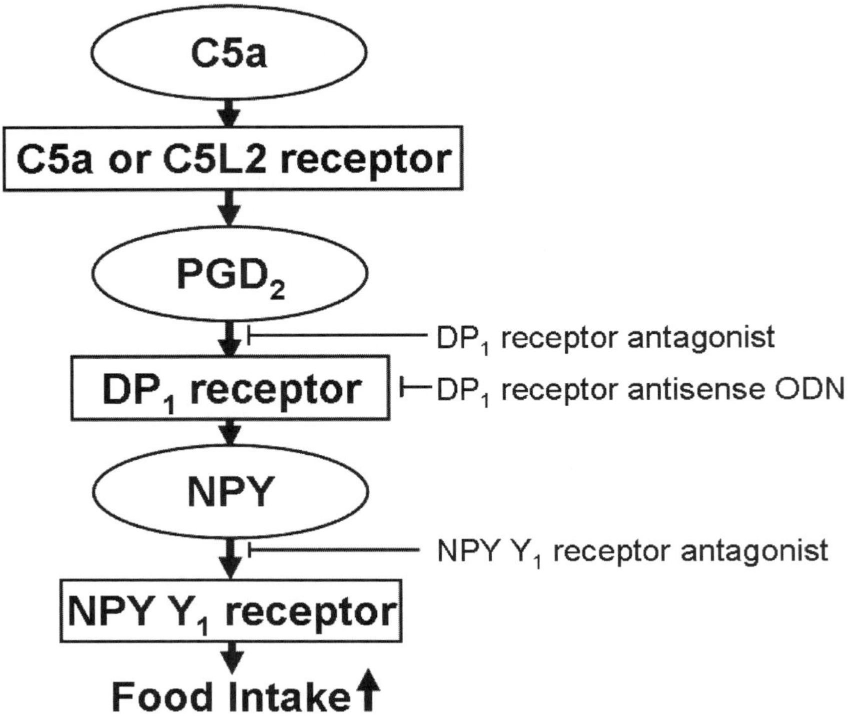

Fig. 2 Model of food intake regulation by C5a. C5a stimulates food intake via PGD_2 production and DP_1 receptor activation coupled to the NPY system

peptide, WPLPR, is the first example of an anorexigenic peptide to activate the PGE_2-EP_4 pathway.

Mediators of C3a agonists in ileum contraction are thought to be histamine and acetylcholine as well as PGE_2 (Takahashi et al. 1996). Histamine and PGE_2 are anorexigenic (Levine and Morley 1981; Ohinata et al. 2006; Ookuma et al. 1993; Sakata 1995); however, the anorexigenic action of WPLPR was not blocked by pretreatment of an antagonist for histamine H_1 receptor (Ohinata et al. 2007), which mediated the anorexigenic activity of histamine (Ookuma et al. 1993; Sakata 1995; Wada et al. 1991), suggesting that the suppression of food intake by WPLPR is independent of the histamine system. In addition, an antagonist for cholecystokinin1 (CCK_1) receptor (Moran and Kinzig 2004; Strader and Woods 2005; Marczak et al. 2006), which mediated the anorexigenic activity of a well-known anorexigenic peptide CCK, did not change the food intake suppression induced by WPLPR (Ohinata et al. 2007). Taken together, a C3a agonist, WPLPR, decreases food intake through PGE_2 production and EP_4 receptor activation but not through anorexigenic signaling by histamine H_1 and CCK_1 receptors.

3 C5a Effect on Food Intake

3.1 Orexigenic Action of Central C5a

It was previously reported that C5a injected into the perifornical region of the hypothalamus stimulated food intake in non-fasted rats (Schupf and Williams 1987; Williams et al. 1985). The increased food intake induced by C5a was inhibited by phentolamine, an α-adrenergic antagonist, suggesting that C5a activates an α-adrenergic receptor in the hypothalamus (Williams et al. 1985). C5a receptor, a seven transmembrane G-protein-coupled receptor, is found in neurons, astrocytes and microglia in the CNS, and is up-regulated during inflammatory conditions such as meningitis, brain trauma and multiple sclerosis (Nataf et al. 1999). C5L2 receptor, a recently identified G-protein-coupled receptor belonging to a subfamily of C3a, C5a and fMLP receptors that are related to the chemokine receptor family (Ohno et al. 2000), showed affinity for C5a, C3a, C5a-des-Arg and C3a-des-Arg, and is found in astrocytes of the CNS (Monk et al. 2007; Scola et al. 2007; Gavrilyuk et al. 2005). Centrally administered C5a stimulated food intake in mice (Ohinata et al. 2002); however, C5a-des-Arg, which had no affinity for C5a receptor, was inactive (unpublished data), suggesting that C5a receptor might mediate the orexigenic activity of C5a. On the other hand, C3a-des-Arg, which promotes positive energy balance by stimulating triacylglycerol synthesis via C5L2 receptor, was reported to increase food intake after i.c.v. administration in rats (Saleh et al. 2001), indicating involvement of C5L2 receptor in food intake stimulation. Further investigations will elucidate whether C5a stimulates food intake via C5a or C5L2 receptor.

3.2 Involvement of PGD_2-DP_1 Receptor

PGD_2 is a mediator of C5a in peripheral tissue such as rat Kupffer cells (Puschel et al. 1996). PGD_2, the most abundant PG produced in the CNS (Narumiya et al. 1982), is involved in various central actions including sleep induction, hypothermia and attenuation of the pain response (Hayaishi 1991, 2002; Urade and Hayaishi 1999; Eguchi et al. 1999). PGD_2 is produced endogenously from arachidonic acid, via PGH_2, by PGD synthase in the brain (Urade et al. 1993; Beuckmann et al. 2000), and prostaglandin DP_1 receptor for PGD_2 is also present in the CNS (Hirata et al. 1994; Mizoguchi et al. 2001; Oida et al. 1997). Recently, we found that central PGD_2 stimulates food intake via DP_1 receptor (Ohinata et al. 2008). We also found that the orexigenic activity of C5a was mediated by DP_1 receptor using a DP_1 receptor-selective antagonist or antisense oligodeoxynucleotide (unpublished data). NPY, a well-known orexigenic peptide that is abundant in the hypothalamus of the brain (Inui 1999b,c; Asakawa et al. 2001, Blomqvist and Herzog 1997), mediates the orexigenic actions of endogenous neuropeptides such as ghrelin and orexin. We found that the orexigenic activity of C5a is coupled to the neuropeptide Y (NPY) system (unpublished data), consistent with our report that PGD_2 stimulates food intake through activating NPY Y_1 receptor, downstream of DP_1 receptor (Ohinata et al. 2008). Taken together, C5a stimulates food intake via PGD_2 production and DP_1 receptor activation followed by activating the NPY system (Fig. 2). Although LPS is known to suppress food intake, it activates orexigenic C5a and its receptor after i.p. administration (Schupf and Williams 1987; Williams et al. 1985). The physiological significance of C5a in food intake regulation during normal and pathological conditions should be clarified in the future.

Acknowledgement

This work was supported in part by Grants-in-Aid for Scientific Research from the Japanese Society for the Promotion of Science to MY and KO, and a PROBRAIN grant from the Bio-oriented Technology Research Advancement Institution to MY.

References

Ames RS, Li Y, Sarau HM, Nuthulaganti P, Foley JJ, Ellis C, Zeng Z, Su K, Jurewicz AJ, Hertzberg RP, Bergsma DJ, Kumar C. (1996) Molecular cloning and characterization of the human anaphylatoxin C3a receptor. J Biol Chem. 271(34):20231–20234

Asakawa A, Inui A, Kaga T, Yuzuriha H, Nagata T, Ueno N, Makino S, Fujimiya M, Niijima A, Fujino MA, Kasuga M. (2001) Ghrelin is an appetite-stimulatory signal from stomach with structural resemblance to motilin. *Gastroenterology*. 120(2):337–345

Beuckmann CT, Lazarus M, Gerashchenko D, Mizoguchi A, Nomura S, Mohri I, Uesugi A, Kaneko T, Mizuno N, Hayaishi O, Urade Y. (2000) Cellular localization of lipocalin-type prostaglandin D synthase (beta-trace) in the central nervous system of the adult rat. *J Comp Neurol*. 428(1):62–78

Blomqvist AG, Herzog H. (1997) Y-receptor subtypes-how many more? *Trends Neurosci*. 20:294–298

Boulay F, Mery L, Tardif M, Brouchon L, Vignais P. (1991) Expression cloning of a receptor for C5a anaphylatoxin on differentiated HL-60 cells. *Biochemistry*. 30(12):2993–2999

Chiba H, Tani F, Yoshikawa M. (1989) Opioid antagonist peptides derived from kappa-casein. *J Dairy Res*. 56(3):363–366

Cianflone K, Xia Z, Chen LY. (2003) Critical review of acylation-stimulating protein physiology in humans and rodents. *Biochim Biophys Acta*. 1609(2):127–143

Davoust N, Jones J, Stahel PF, Ames RS, Barnum SR. (1999) Receptor for the C3a anaphylatoxin is expressed by neurons and glial cells. *Glia*. 26(3):201–211

Eguchi N, Minami T, Shirafuji N, Kanaoka Y, Tanaka T, Nagata A, Yoshida N, Urade Y, Ito S, Hayaishi O. (1999) Lack of tactile pain (allodynia) in lipocalin-type prostaglandin D synthase-deficient mice. *Proc Natl Acad Sci U S A*. 96(2):726–730

Gasque P, Singhrao SK, Neal JW, Gotze O, Morgan BP. (1997) Expression of the receptor for complement C5a (CD88) is up-regulated on reactive astrocytes, microglia, and endothelial cells in the inflamed human central nervous system. *Am J Pathol*. 150(1):31–41

Gasque P, Singhrao SK, Neal JW, Wang P, Sayah S, Fontaine M, Morgan BP. (1998) The receptor for complement anaphylatoxin C3a is expressed by myeloid cells and nonmyeloid cells in inflamed human central nervous system: analysis in multiple sclerosis and bacterial meningitis. *J Immunol.* 160(7):3543–3554

Gasque P, Dean YD, McGreal EP, VanBeek J, Morgan BP. (2000) Complement components of the innate immune system in health and disease in the CNS. *Immunopharmacology.* 49(1–2):171–186

Gavrilyuk V, Kalinin S, Hilbush BS, Middlecamp A, McGuire S, Pelligrino D, Weinberg G, Feinstein DL. (2005) Identification of complement 5a-like receptor (C5L2) from astrocytes: characterization of anti-inflammatory properties. *J Neurochem.* 92(5):1140–1149

Gerard NP, Gerard C. (1991) The chemotactic receptor for human C5a anaphylatoxin. *Nature.* 349(6310):614–617

Hayaishi O. (1991) Molecular mechanisms of sleep-wake regulation: roles of prostaglandins D_2 and E_2. *FASEB J.* 5(11):2575–2581

Hayaishi O. (2002) Molecular genetic studies on sleep-wake regulation, with special emphasis on the prostaglandin D_2 system. *J Appl Physiol.* 92(2):863–868

Hirata M, Kakizuka A, Aizawa M, Ushikubi F, Narumiya S. (1994) Molecular characterization of a mouse prostaglandin D receptor and functional expression of the cloned gene. *Proc Natl Acad Sci U S A.* 91(23):11192–11196

Hopkins SJ, Rothwell NJ. (1995) Cytokines and the nervous system. I: expression and recognition. *Trends Neurosci.* 18(2):83–88

Horton EW. (1964) Actions of prostaglandins E_1, E_2 and E_3 on the central nervous system. *Br J Pharmacol.* 22:189–192

Huang ZL, Sato Y, Mochizuki T, Okada T, Qu WM, Yamatodani A, Urade Y, Hayaishi O. (2003) Prostaglandin E_2 activates the histaminergic system via the EP_4 receptor to induce wakefulness in rats. *J Neurosci.* 23(14):5975–5983

Inui A. (1999a) Cancer anorexia-cachexia syndrome: are neuropeptides the key? *Cancer Res.* 59(18):4493–4501

Inui A. (1999b) Feeding and body-weight regulation by hypothalamic neuropeptides – mediation of the actions of leptin. *Trends Neurosci.* 22(2):62–67

Inui A. (1999c) Neuropeptide Y feeding receptors-are multiple subtypes involved? *Trends Pharmacol Sci.* 20:43–46

Inui A. (2001) Cytokines and sickness behavior: implications from knockout animal models. *Trends Immunol.* 22(9):469–473

Jinsmaa Y, Takenaka Y, Yoshikawa M. (2001) Designing of an orally active complement C3a agonist peptide with anti-analgesic and anti-amnesic activity. *Peptides.* 22(1):25–32

Kildsgaard J, Hollmann TJ, Matthews KW, Bian K, Murad F, Wetsel RA. (2000) Cutting edge: targeted disruption of the C3a receptor gene demonstrates a novel protective anti-inflammatory role for C3a in endotoxin-shock. *J Immunol.* 165(10):5406–5409

Langhans W. (2007) Signals generating anorexia during acute illness. *Proc Nutr Soc.* 66(3):321–330

Law SK, Reid KB. (1995) In: D Male, D Richwood (Eds.), *Complement*, Oxford University Press, New York

Levine AS, Morley JE. (1981) The effect of prostaglandins (PGE_2 and $PGF_{2}\alpha$) on food intake in rats. *Pharmacol Biochem Behav.* 15(5):735–738

Marczak ED, Ohinata K, Lipkowski AW, Yoshikawa M. (2006) Arg-Ile-Tyr (RIY) derived from rapeseed protein decreases food intake and gastric emptying after oral administration in mice. *Peptides.* 27(9):2065–2068

Matsuoka Y, Furuyashiki T, Bito H, Ushikubi F, Tanaka Y, Kobayashi T, Muro S, Satoh N, Kayahara T, Higashi M, Mizoguchi A, Shichi H, Fukuda Y, Nakao K, Narumiya S.

(2003) Impaired adrenocorticotropic hormone response to bacterial endotoxin in mice deficient in prostaglandin E receptor EP1 and EP3 subtypes. *Proc Natl Acad Sci U S A*. 100(7):4132–4137

Mizoguchi A, Eguchi N, Kimura K, Kiyohara Y, Qu WM, Huang ZL, Mochizuki T, Lazarus M, Kobayashi T, Kaneko T, Narumiya S, Urade Y, Hayaishi O. (2001) Dominant localization of prostaglandin D receptors on arachnoid trabecular cells in mouse basal forebrain and their involvement in the regulation of non-rapid eye movement sleep. *Proc Natl Acad Sci U S A*. 98(20):11674–11679

Monk PN, Scola AM, Madala P, Fairlie DP. (2007) Function, structure and therapeutic potential of complement C5a receptors. *Br J Pharmacol*. 152(4):429–448

Moran TH, Kinzig KP. (2004) Gastrointestinal satiety signals II. Cholecystokinin. *Am J Physiol Gastrointest Liver Physiol*. 286(2):G183–G188

Nadeau S, Rivest S. (2001) The complement system is an integrated part of the natural innate immune response in the brain. *FASEB J*. 15(8):1410–1412

Narumiya S, FitzGerald GA. (2001) Genetic and pharmacological analysis of prostanoid receptor function. *J Clin Invest*. 108(1):25–30

Narumiya S, Ogorochi T, Nakao K, Hayaishi O. (1982) Prostaglandin D_2 in rat brain, spinal cord and pituitary: basal level and regional distribution. *Life Sci*. 31(19):2093–2103

Narumiya S, Sugimoto Y, Ushikubi F. (1999) Prostanoid receptors: structures, properties, and functions. *Physiol Rev*. 79(4):1193–1226

Nataf S, Stahel PF, Davoust N, Barnum SR. (1999) Complement anaphylatoxin receptors on neurons: new tricks for old receptors? *Trends Neurosci*. 22(9):397–402

O'Barr SA, Caguioa J, Gruol D, Perkins G, Ember JA, Hugli T, Cooper NR. (2001) Neuronal expression of a functional receptor for the C5a complement activation fragment. *J Immunol*. 166(6):4154–4162

Ohinata K, Inui A, Asakawa A, Wada K, Wada E, Yoshikawa M. (2002) Albutensin A and complement C3a decrease food intake in mice. *Peptides*. 23(1):127–133

Ohinata K, Suetsugu K, Fujiwara Y, Yoshikawa M. (2006) Activation of prostaglandin E receptor EP_4 subtype suppresses food intake in mice. *Prostaglandins Other Lipid Mediat*. 81(1–2):31–36

Ohinata K, Suetsugu K, Fujiwara Y, Yoshikawa M. (2007) Suppression of food intake by a complement C3a agonist [Trp^5]-oryzatensin(5–9). *Peptides*. 28(3):602–606

Ohinata K, Takagi K, Biyajima K, Fujiwara Y, Fukumoto S, Eguchi N, Urade Y, Asakawa A, Fujimiya M, Inui A, Yoshikawa M. (2008) Central prostaglandin D_2 stimulates food intake via the neuropeptide Y system in mice. *FEBS Lett*. 582(5):679–684

Ohki-Hamazaki H, Sakai Y, Kamata K, Ogura H, Okuyama S, Watase K, Yamada K, Wada K. (1999) Functional properties of two bombesin-like peptide receptors revealed by the analysis of mice lacking neuromedin B receptor. *J Neurosci*. 19(3):948–954

Ohno M, Hirata T, Enomoto M, Araki T, Ishimaru H, Takahashi TA. (2000) A putative chemoattractant receptor, C5L2, is expressed in granulocyte and immature dendritic cells, but not in mature dendritic cells. *Mol Immunol*. 37(8):407–412

Oida H, Hirata M, Sugimoto Y, Ushikubi F, Ohishi H, Mizuno N, Ichikawa A, Narumiya S. (1997) Expression of messenger RNA for the prostaglandin D receptor in the leptomeninges of the mouse brain. *FEBS Lett*. 417(1):53–56

Ookuma K, Sakata T, Fukagawa K, Yoshimatsu H, Kurokawa M, Machidori H, Fujimoto K. (1993) Neuronal histamine in the hypothalamus suppresses food intake in rats. *Brain Res*. 628(1–2):235–242

Puschel GP, Nolte A, Schieferdecker HL, Rothermel E, Gotze O, Jungermann K. (1996) Inhibition of anaphylatoxin C3a- and C5a- but not nerve stimulation- or Noradrenaline-dependent increase in glucose output and reduction of flow in Kupffer cell-depleted perfused rat livers. *Hepatology*. 24(3):685–690

Sakata T. (1995) Histamine receptor and its regulation of energy metabolism. *Obes Res.* 3 Suppl 4:541S–548S

Saleh J, Blevins JE, Havel PJ Barrett JA, Gietzen DW, Cianflone K. (2001) Acylation stimulating protein (ASP) acute effects on postprandial lipemia and food intake in rodents. *Int J Obes Relat Metab Disord.* 25(5):705–713

Schupf N, Williams CA. (1987) Psychopharmacological activity of immune complexes in rat brain is complement dependent. *J Neuroimmunol.* 13(3):293–303

Schupf N, Williams CA, Hugli TE, Cox J. (1983) Psychopharmacological activity of anaphylatoxin C3a in rat hypothalamus. *J Neuroimmunol.* 5(3):305–316

Scola AM, Higginbottom A, Partridge LJ, Reid RC, Woodruff T, Taylor SM, Fairlie DP, Monk PN. (2007) The role of the N-terminal domain of the complement fragment receptor C5L2 in ligand binding. *J Biol Chem.* 282(6):3664–3671

Sniderman AD, Cianflone KM, Eckel RH. (1991) Levels of acylation stimulating protein in obese women before and after moderate weight loss. *Int J Obes.* 15(5):333–336

Strader AD, Woods SC. (2005) Gastrointestinal hormones and food intake. *Gastroenterology.* 128(1):175–191

Takahashi M, Moriguchi S, Ikeno M, Kono S, Ohata K, Usui H, Kurahashi K, Sasaki R, Yoshikawa M. (1996) Studies on the ileum-contracting mechanisms and identification as a complement C3a receptor agonist of oryzatensin, a bioactive peptide derived from rice albumin. *Peptides.* 17(1):5–12

Takahashi M, Moriguchi S, Suganuma H, Shiota A, Tani F, Usui H, Kurahashi K, Sasaki R, Yoshikawa M. (1997) Identification of casoxin C, an ileum-contracting peptide derived from bovine kappa-casein, as an agonist for C3a receptors. *Peptides.* 18(3):329–336

Takahashi M, Moriguti S, Minami T, Suganuma H, Shiota A, Takenaka Y, Tani F, Sasaki R, Yoshikawa M. (1998) Albutensin A, an ileum-contracting peptide derived from serum albumin, acts through both receptors for complement C3a, and C5a. *Peptide Science.* 5:29–35

Tornetta MA, Foley JJ, Sarau HM, Ames RS. (1997) The mouse anaphylatoxin C3a receptor: molecular cloning, genomic organization, and functional expression. *J Immunol.* 158(11):5277–5282

Urade Y, Hayaishi O. (1999) Prostaglandin D_2 and sleep regulation. *Biochim Biophys Acta.* 1436(3):606–615

Urade Y, Kitahama K, Ohishi H, Kaneko T, Mizuno N, Hayaishi O. (1993) Dominant expression of mRNA for prostaglandin D synthase in leptomeninges, choroid plexus, and oligodendrocytes of the adult rat brain. *Proc Natl Acad Sci U S A.* 90(19):9070–9074

Ushikubi F, Segi E, Sugimoto Y, Murata T, Matsuoka T, Kobayashi T, Hizaki H, Tuboi K, Katsuyama M, Ichikawa A, Tanaka T, Yoshida N, Narumiya S. (1998) Impaired febrile response in mice lacking the prostaglandin E receptor subtype EP3. *Nature.* 395(6699):281–284

van Beek J, Elward K, Gasque P. (2003) Activation of complement in the central nervous system: roles in neurodegeneration and neuroprotection. *Ann N Y Acad Sci.* 992:56–71

Van Miert AS, Van Duin CT, Woutersen-Van Nijnanten FM. (1983) Effect of intracerebroventricular injection of PGE_2 and 5HT on body temperature, heart rate and rumen motility of conscious goats. *Eur J Pharmacol.* 92(1–2):143–146

Wada H, Inagaki N, Itowi N, Yamatodani, A. (1991) Histaminergic neuron system in the brain: distribution and possible functions. *Brain Res Bull.* 27(3–4):367–370

Williams CA, Schupf N, Hugli TE. (1985) Anaphylatoxin C5a modulation of an alpha-adrenergic receptor system in the rat hypothalamus. *J Neuroimmunol.* 9(1–2):29–40

Yatomi A, Iguchi A, Yanagisawa S, Matsunaga H, Niki I, Sakamoto N. (1987) Prostaglandins affect the central nervous system to produce hyperglycemia in rats. *Endocrinology*. 121(1):36–41

Zhang J, Rivest S. (1999) Distribution, regulation and colocalization of the genes encoding the EP_2- and EP_4-PGE_2 receptors in the rat brain and neuronal responses to systemic inflammation. *Eur J Neurosci*. 11(8):2651–2668

4. A Pivotal Role of Activation of Complement Cascade (CC) in Mobilization of Hematopoietic Stem/Progenitor Cells (HSPC)

Mariusz Z. Ratajczak*, Marcin Wysoczynski, Ryan Reca, Wu Wan, Ewa K. Zuba-Surma, Magda Kucia, and Janina Ratajczak

James Graham Brown Cancer Center, Stem Cell Institute, University of Louisville, Louisville, KY, USA
*mzrata01@louisville.edu

Abstract. Complement cascade (CC) and innate immunity emerge as important and underappreciated modulators of trafficking of hematopoietic stem/progenitor cells (HSPC). Accordingly, we reported that (i) C becomes activated in bone marrow (BM) during G-CSF-induced mobilization by the classical immunoglobulin (Ig)-dependent pathway, and that (ii) C3 cleavage fragments increase the responsiveness of HSPC to an stromal derived factor-1 (SDF-1) gradient. Furthermore, our recent data in immunodeficient mice support the concept that the CC is a major factor modulating egress of HSPC from bone marrow (BM) into peripheral blood (PB). Thus, in light of these findings, mobilization of HSPC could be envisioned as part of an immune response that requires CC activation by the classical Ig-dependent and/or Ig-independent pathways. Hence modulation of CC activation could allow for the development of more efficient mobilization strategies in patients who are poor mobilizers of HSPC.

1 Introduction

Hematopoietic stem/progenitor cells (HSPC) circulate in peripheral blood (PB) under steady state conditions at very low levels in order to keep in balance a pool of stem cells in the bone marrow (BM) microenvironment that is distributed in bones located in different parts of the body. Thus PB could be envisioned as a "highway" by which HSPC are trafficking in the organism while relocating between hematopoietic niches (Lapidot et al. 2005; Quesenberry and Becker 1998; Ratajczak et al. 2006).

Similarly, during hematopoietic transplant, allogeneic donor- or autologous patient-derived HSPC are infused intravenously and circulate for a short period of time in PB and subsequently, from PB migrate (home) to the hematopoietic niches in BM, where they are supposed to engraft permanently (Lapidot et al. 2005). In contrast, during a reverse process called mobilization, HSPC egress from the BM into PB – as it is seen, for example, after the administration of certain drugs (e.g., granulocyte colony stimulating factor – G-CSF) (Papayannopoulou 2004), during

J.D. Lambris (ed.), *Current Topics in Complement II.*
doi: 10:1007/978-0-387-78952-1_4,

systemic inflammation (Ratajczak et al. 2006) or stress related to tissue organ injury (Reca et al. 2006; Ratajczak et al. 2004b).

We reported recently that complement cascade (CC) becomes activated in BM during (i) conditioning for hematopoietic transplantation by myeloablative radio-chemotherapy, (ii) G-CSF induced mobilization and (iii) several models of tissue organ/injury (Allendorf et al. 2003; Ratajczak et al. 2004a,b, 2006; Reca et al. 2003). These observations in toto suggest the existence of a close link between inflammation, tissue damage and stem cell trafficking. In this context, CC and innate immunity emerged as important and underappreciated modulators of this process (Ratajczak et al. 2006; Reca et al. 2006). The role of CC in mobilization of HSPC could be explained by (i) enhancement of responsiveness of HSPC to the gradient of the major BM-secreted chemoattractant for these cells that is α-chemokine stromal derived factor-1 (SDF-1), (ii) modulation of adhesiveness of HSPC in BM microenvironment, as well as (iii) an probably increase in permeability of bone marrow endothelium that is crucial for homing and egress/mobilization of HSPC respectively (Reca et al. 2007; Kucia et al. 2005; Ratajczak et al. 2007).

Accordingly, mounting evidence accumulated that SDF-1 – seven transmembrane span $G_{\alpha i}$ protein coupled receptor CXCR4 axis plays a crucial role in the retention and homing of HSPC in BM as well as in their egress (mobilization) from BM into PB (Lapidot et al. 2005; Kucia et al. 2005; Ratajczak et al. 2007). We reported recently that cleavage fragments of the third component (C3) of the CC – C3a and $_{desArg}$C3a play an important role in enhancing the homing responses of HSPC to an SDF-1 gradient (Allendorf et al. 2003; Ratajczak et al. 2004a,c, 2006; Reca et al. 2003). We also noticed that G-CSF, which is a more commonly used drug to induce mobilization, activates CC and this activation depends on the classical immunoglobulin (Ig)-dependent CC activation pathway (Reca et al. 2007). This suggests a potential link between Ig level, CC activation and mobilization of HSPC. Moreover, since patients suffering from severe combined immuno-deficiency (SCID) mobilize poorly (Sekhsaria et al. 1996), we hypothesized that this could be directly linked to the lack of CC activating Ig in these patients (Reca et al. 2007).

In this chapter, we will discuss this novel role of CC in regulating the trafficking of HSPC and present a concept that the mobilization of HSPC and their egress from BM occurs as part of an immune response (Ratajczak et al. 2006; Reca et al. 2007).

2 Mobilization of HSPC and Problem of Poor Mobilizers

The forced migration of HSPC from BM into PB is called mobilization (Papayannopoulou 2004). Mobilization could be envisioned as a reverse mechanism (mirror image) for stem cell homing (Fig. 1) and is important from a clinical point of view as a procedure that allows for the collection of HSPC for hematopoietic transplantation (Papayannopoulou 2004). As mentioned above,

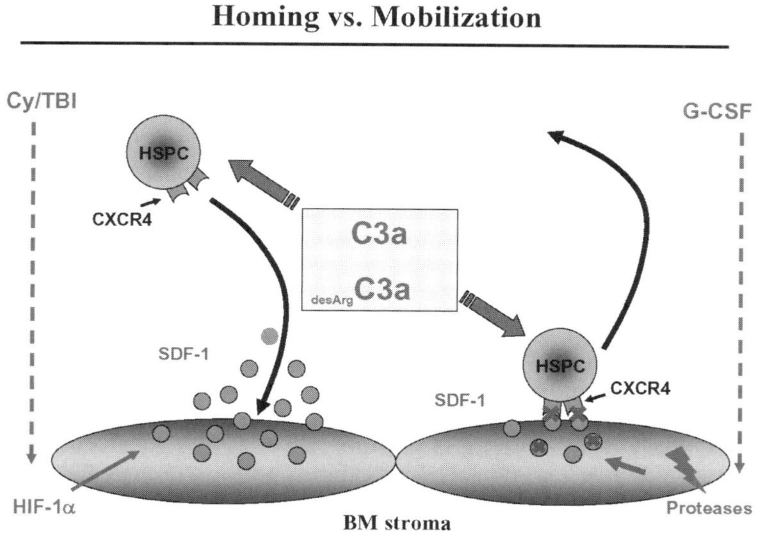

Fig. 1 Homing of HSPC into BM and their mobilization into PB. Homing of HSPC to BM after transplantation and mobilization of HSPC from BM into PB depends on the activity of the SDF-1-CXCR4 axis. Conditioning for transplantation by chemotherapy (e.g., cyclophosphamide - CY) and/or total body irradiation (TBI) increases via hypoxia inducible factor-1 (HIF-1a) the concentration of SDF-1 in the BM microenvironment. This results in chemoattraction of $CXCR4^+$ HSPC. During the reverse process, mobilization, a mobilizing agent (e.g., G-CSF) i) increases the concentration of proteases in the BM microenvironment, degrading SDF-1 and CXCR4 and ii) decreases SDF-1 expression at the mRNA level. This results in the release of $CXCR4^+$ HSPC into PB. Both these processes are modulated by C3 cleavage fragments (C3a, ${}_{desArg}C3a$ and iC3b) whose BM-concentration increases due to CC activation/cleavage both during conditioning for transplant by CY/TBI or stimulation of mobilization by G-CSF

mobilization can be performed on the patient itself (autologous transplantation) or on an MHC related or unrelated donor to the patient (allogeneic transplantation). The molecular mechanisms governing mobilization of HSPC are still not well understood. However, as shown in Fig. 1 attenuation of SDF-1–CXCR4 interaction between BM-secreted SDF-1 and HSPC-expressed CXCR4 plays a pivotal role in release of these cells from the BM into PB (Levesque et al. 2002, 2003).

Many agents have been described that may induce mobilization of HSPC (Papayannopoulou 2004). G-CSF most frequently employed in the clinic efficiently mobilizes HSPC after a few consecutive daily injections (Papayannopoulou 2004; Levesque et al. 2003). Other compounds such as polysaccharides (e.g., zymosan) mobilize HSPC within 1 h after a single injection (Papayannopoulou 2004). Mobilization could also be induced by chemokines (e.g., IL-8, Gro-β), growth factors (e.g., VEGF) and CXCR4 antagonists (e.g., AMD3100, T140), and it is additionally modulated by lipopolysaccharide (LPS) that is released by intestinal bacteria (Broxmeyer et al. 2005; Fukuda et al. 2007; Velders et al. 2004). Unfortunately, ~25% of patients do not respond efficiently to currently recommended mobilization protocols and are termed poor mobilizers. Therefore, shedding more light on the molecular mechanisms governing process of HSPC mobilization could allow us to develop more efficient mobilization strategies.

3 G-CSF-Induced Mobilization Triggers Activation of CC

The model of G-CSF-induced HSPC mobilization is the one best studied so far and evidence from several laboratories suggests that attenuation of the SDF-1–CXCR4 axis precedes egress of HSPC from the BM into PB (Levesque et al. 2002, 2003). It is proposed that after G-CSF infusion the BM turns into a proteolytic microenvironment (Fig. 1) that leads to enzymatic degradation of SDF-1 and enzymatic processing of the N-terminus of CXCR4 (Levesque et al. 2003). In the process of proteolysis of SDF-1 and CXCR4 various proteolytic enzymes are secreted by myeloid precursors and granulocytes in BM, e.g., elastase, cathepsin and metalloproteinases (MMPs) (Levesque et al. 2003). Furthermore, as recently reported, G-CSF may directly down-regulate the expression of SDF-1 at the mRNA level in BM (Semerad et al. 2005). The end result is a decrease in responsiveness of HSPC to an SDF-1 gradient, and their release into circulation.

We propose that G-CSF induced proteolytic processes in the BM that leads to "BM injury" also triggers the local activation of CC and C3 cleavage (Ratajczak et al. 2006). The mechanism of CC activation involves mostly the classical (Ig-dependent) but also to some degree the alternative (Ig-independent) CC activation pathways (Reca et al. 2007). The former is triggered after exposure of a neo-epitope in BM tissue damaged by secreted proteases. The neo-epitope binds naturally occurring antibodies that circulate in peripheral blood and via C1q triggers CC activation through the classical pathway (Allendorf et al. 2003; Reca et al. 2007). Furthermore, it is likely that C3 could also be cleaved/activated directly by mentioned above G-CSF-induced proteolytic proteases (Reca et al. 2007). Similarly, since G-CSF activates the coagulation system (Canales et al. 2002) and thrombin has recently been identified as a C5 activator (Huber-Lang et al. 2006), CC could be also activated in G-CSF mobilized patients as a result of activation of the coagulation cascade.

However, G-CSF induced mobilization of HSPC involves several mechanisms, we hypothesized that lack of neo-epitope binding naturally occurring Ig could be responsible for a reported poor G-CSF induced mobilization in immunodeficient patients (Reca et al. 2007; Sekhsaria et al. 1996).

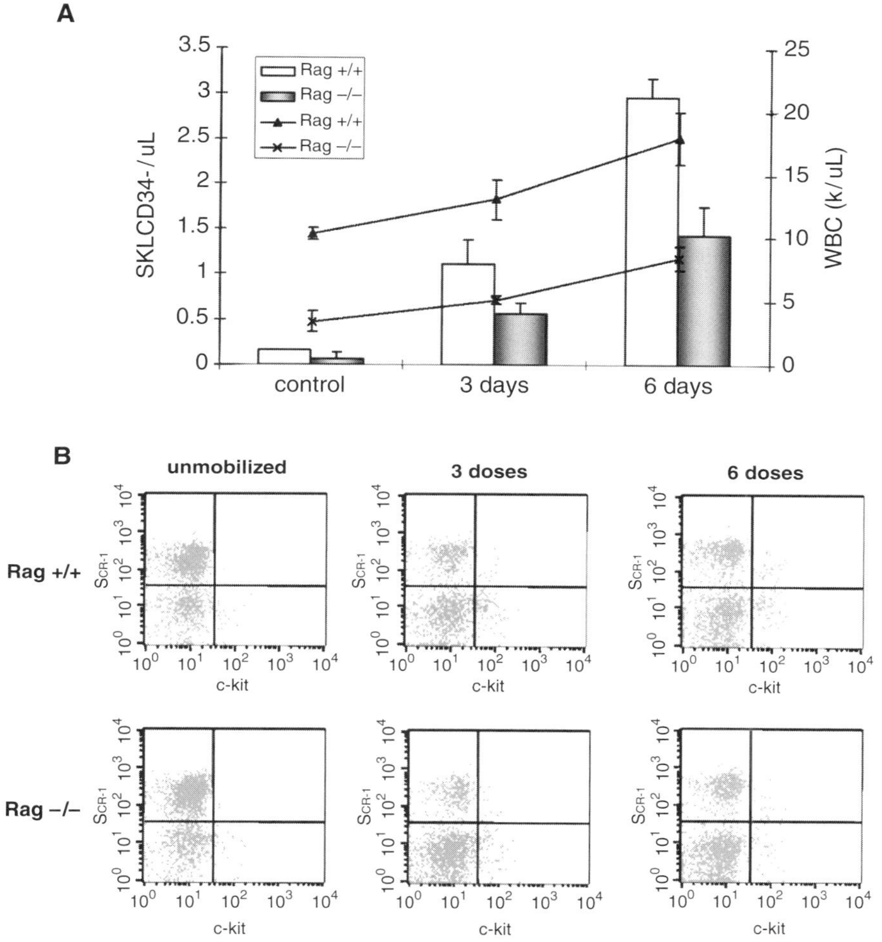

Fig. 2 G-CSF-induced mobilization is impaired RAG2$^{-/-}$ mice – as evidenced by a number of circulating primitive CD34^{-}Sca-1^{+}kit^{+}lin^{-} (CSKL) cells.

(a) Rag2$^{-/-}$ as well as age- and sex-matched wt (B6) mice were mobilized for 3 or 6 days with G-CSF (250 μg/kg s.c./day) ($n = 16$ animals/group). Number of circulating CSKL cells/μl of PB in B6 mice compared to wt mice. $P < 0.09$ as compared to wt mice mobilized for 3 days. * $p<0.0001$ as compared to wt mice mobilized for 6 days

(b) Lin^{-} PBMC were stain with anty-c-kit, anty-Sca-1 and anty-CD34. Representative dot-plots for CSKL analysis are shown

4 Experimental Evidence that Ig-Deficient RAG2, SCID and Jh Mice, but not T-Cell-Depleted Mice, Respond Poorly to G-CSF Induced Mobilization

Based on our preliminary data and fact that immunodefcient patients are poor mobilizers (Sekhsaria et al. 1996), we became interested on a potential role of Ig and their role in CC activation during mobilization process. To better address the role of CC activation in HSPC mobilization, we mobilized mice that lack Ig (RAG2, SCID and Jh) by G-CSF or zymosan, compounds that activate CC and cleave C3 by the classical Ig-dependent and the alternative Ig-independent pathways, respectively. In addition, we evaluated mobilization in C5-deficient animals.

Therefore, to test the role of Ig in triggering CC activation in G-CSF-induced mobilization of HSPC, immunodeficient RAG2 and SCID mice (lacking both B and T lymphocytes and Ig) and Jh mice (selectively Ig-deficient) were mobilized by G-CSF for 3 or 6 days. First, we observed that under normal steady-state conditions all of these mice which display an unaffected myeloid compartment and normal numbers of CFU-GM in BM, have a slightly lower number of circulating CFU-GM and a significantly lower number of white blood cells (WBC) in their PB compared to their wt littermates (Reca et al. 2007). The decrease in number of circulating WBC was related to a severe deficiency of circulating lymphocytes that in mice are a major fraction of nucleated cells in PB.

More importantly, we found that after administration of G-CSF, the immunodeficient RAG2 and SCID mice had significantly less (~four times) circulating clonogenic CFU-GM in their PB at days 3 and 6 of mobilization as compared to genotypically matched wt mice (Reca et al. 2007). Similarly, the number of primitive Sca-1^{+}kit^{+}lin^{-} (SKL) cells as well as CD34^{-} Sca-1^{+}kit^{+}lin^{-} (CSKL) in immunodeficient SCID and RAG2 mice mobilized by G-CSF was significantly lower than in wt littermates (Fig. 2). Interestingly, we also noticed that the level of mobilization of HSPC in these animals correlated positively with activation of MMP-9 but not MMP-2 (Fig. 3).

Since RAG2 and SCID mice lack also T cells, in order to exclude the possibility that T lymphocytes could account in part for the defective mobilization of these mice, we performed mobilization in wt T-cell-depleted animals. To achieve this, wt BALB/c mice were depleted of T-lymphocytes using a cocktail of antibodies (Abs) against CD4^{+} and CD8^{+} T lymphocytes. The efficiency of T-depletion after administration of Abs was confirmed by FACS analysis. Subsequently, wt control mice and T-cell- depleted mice were mobilized for 6 days by G-CSF. We did not find any effect of T-cell depletion on the number of circulating CFU-GM in PB, although the WBC (white blood cell) count in PB was slightly reduced due to Ab-mediated T cell depletion (Reca et al. 2007). This supports that lack of T-lymphocytes in immundeficient RAG2 and SCID mice was not responsible for poor G-CSF induced mobilization in these animals.

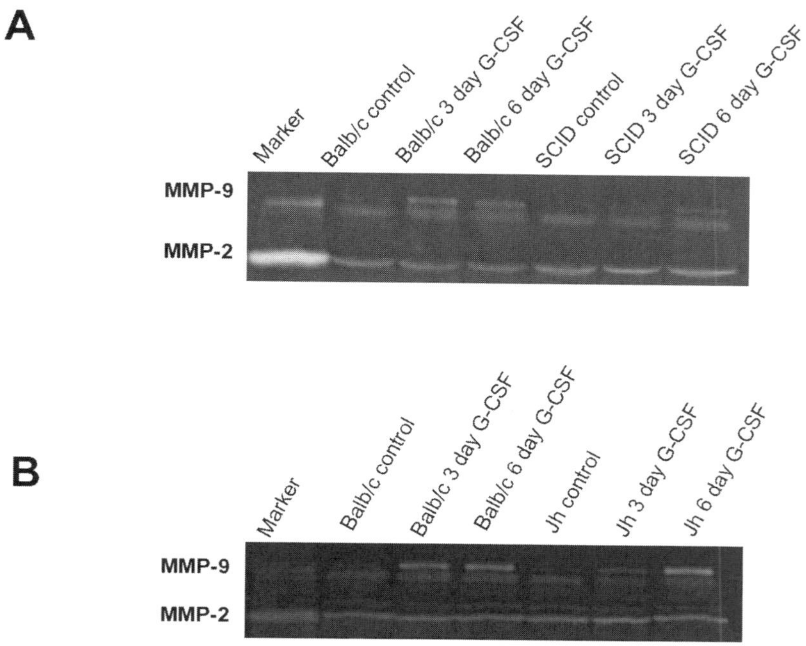

Fig. 3 Mobilization of HSPC in wt and immunodeficient mice as a function of MMP-9 activation.
(a) SCID as well as age- and sex-matched wt (Balb-c) mice were mobilized for 3 or 6 days with G-CSF (250 μg/kg s.c./day) (n = 16 animals/group). Activity of MMP-9 but not MMP-2 by zymography correlates with mobilization in these mice.
(b) Ig-deficient Jh mice as well as as well as age- and sex-matched wt (Balb-c) mice were mobilized for 3 or 6 days with G-CSF (250 μg/kg s.c./day) (*n* = 16 animals/group). Activity of MMP-9 but not MMP-2 by zymography correlates with mobilization in these mice.

Based on these observations in SCID and RAG2 mice, and lack o T cell involvement in mobilization process, we turned our attention to Jh mice that selectively lack Ig. We observed that the mobilization response to G-CSF of the selectively Ig-deficient mice (Jh) was lower than in wt mice (Reca et al. 2007). Jh mice similarly as immunodeficient RAG2 and SCID mice had a significant decrease (~ four times) in the numbers of circulating clonogenic CFU-GM in their PB at days 3 and 6 of mobilization as compared to genotypically matched wt mice. Similarly, all Ig-deficient Jh animals displayed also a lower number of mobilized PB white blood cell counts (Reca et al. 2007).

This observation, together with the fact that Ig deficiency correlates with poor G-CSF induced mobilization, supports the pivotal role of Ig in mobilization of HSPC. However, since some degree of mobilization still occurred in all Ig-deficient mice (SCID, RAG2, Jh), CC must become activated during G-CSF administration also by alternative mechanisms as well. In fact, C3a cleavage fragments remained still detectable in the serum of these animals (Reca et al. 2007).

5 Defective Mobilization in RAG2, SCID and Jh Mice is Restored by Purified Immunoglobulins Supports Further a Pivotal Role of Ig in this Process

To obtain proof that poor mobilization in RAG2, SCID and Jh mice could be explained by a lack of C-activating Ig, we reconstituted these mice with purified wt Ig and subsequently mobilized them for 6 consecutive days with a suboptimal (five times lower as in previous experiments) dose of G-CSF (50 μg/kg/day).

We noticed that the defective G-CSF mobilization observed in these mice was significantly improved if they were supplemented with wt Ig prior to G-CSF mobilization and corresponded with elevated PB white blood cell counts and increased number of CFU-GM circulating in PB (Reca et al. 2007). Of interest, Ig infusion enhanced mobilization not only in Ig-deficient (SCID, RAG2 and Jh) but also in wt animals, particularly in C57BL/6 mice, which are regarded as poor mobilizers (Papayannopoulou T 2004), however this effect was less pronounced when mobilization was forced by a high dose of G-CSF (250 μg/kg/day).

This indicates that wt Ig contains an admixture of naturally occurring antibodies that are able to recognize a neo-epitope and activate CC in BM that has been exposed to G-CSF (Allendorf et al. 2003). This specific requirement for wt Ig was subsequently supported in the experiments in which irrelevant purified monoclonal IgG or IgM antibodies were used and no increase in mobilization was observed (Reca et al. 2007). At the same time in our control experiments, the potential contamination of Ig preparations by endotoxin, which alone can mobilize HSPC (Velders et al. 2004), was excluded by showing an enhancing effect of Ig + suboptimal dose of G-CSF in C3He/J mice. In our control experiments we also noticed that the addition of Ig preparations alone without G-CSF did not mobilize mice.

6 Evidence that CC Activation and C3a Generation in Serum correlate with G-CSF Induced HSPC Mobilization

C3 activation leads to the generation of liquid (C3a and desArgC3a) as well as solid phase (iC3b) cleavage fragments (Fig. 4). We reported in the past that in G-CSF induced mobilization, C3 was cleaved and $C3a_{des\text{-}Arg}$ (liquid phase C3 cleavage product) became detectable by ELISA in serum (Allendorf et al. 2003; Ratajczak et al. 2004a,c, 2006; Reca et al. 2003). At the same time, the solid phase C3 cleavage product, iC3b, was deposited on BM-derived fibroblasts and endothelial cells as determined by FACS analysis of these cells stained with anti-iC3b antibodies (Ratajczak et al. 2004c).

To test directly whether C becomes activated or not during G-CSF mobilization, we at first tested the serum of wt, RAG2, SCID and Jh mice for the presence of $_{desArg}$C3a (soluble C3 cleavage fragment) before and after G-CSF mobilization. The serum of G-CSF-mobilized animals revealed impaired C activation/cleavage in Ig-deficient animals and its increase in animals supplemented with wt Ig (Reca et al. 2007). As expected, no CC activation/cleavage was detected in C3$^{-/-}$ mice (Reca et al. 2007). In the next step using FACS, we evaluated iC3b deposition on BM mononuclear cells (MNC) in Ig-deficient RAG2 mice mobilized by G-CSF. We found no increase in iC3b deposits in RAG2 animals mobilized by G-CSF compared to wt C57BL/6 mice, although an increase was detectable on BM MNC if the mice were supplemented with Ig during mobilization (Reca et al. 2007). This again supports a pivotal role of Ig in CC activation in G-CSF mobilized animals.

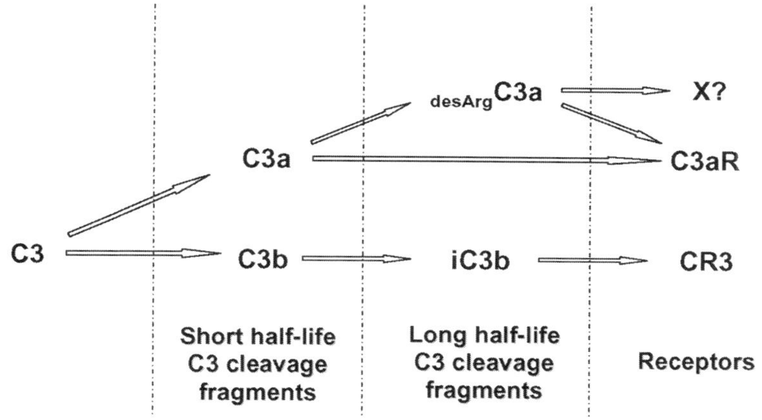

Fig. 4 Cleavage of C3 into biologically active fragments. Cleavage/activation of C3 in BM is initiated by a C3-convertase to generate fluid-phase C3a and stromal cell-bound C3b. Both C3a and C3b have a short half-life. Fluid C3a is rapidly degraded to $_{des\text{-}Arg}$C3a and bound C3b is proteolyzed to iC3b. While C3a mainly activates the C3aR, $_{des\text{-}Arg}$C3a binds to another unidentified receptor X? iC3b tethers HSPC by interacting with CR3. All these C3 cleavage fragments increase responsiveness of HSPC to an SDF-1 gradient

7 In Contrast to G-CSF-Mobilization, RAG2, SCID and Jh Mice Display Normal Zymosan-Induced Mobilization

Zymosan activates CC by the Ig-independent alternative pathway (Reca et al. 2007). To better understand the role of Ig and the classical pathway of CC activation in mobilization, we performed similar mobilization studies in RAG2, SCID and Jh mice employing zymosan as a mobilizing agent. These studies revealed that these immunodeficient animals respond to zymosan like their wt-matched controls, and mobilization of CFU-GM was again associated with CC activation/cleavage (Reca et al. 2007). Notably, as predicted $C3^{-/-}$ mice very poorly mobilize CFU-GM in response to zymosan. These results further support the evidence for a pivotal role of CC activation in the mobilization of HSPC. However, since zymosan activates CC by the alternative Ig-independent pathway, all Ig-deficient (RAG2, SCID and Jh) mice mobilized normally (Reca et al. 2007).

8 Impaired Mobilization in C5-Deficient Mice Supports a Pivotal Role for CC in Egress of HSPC from BM and their Mobilization into PB

In previous studies, we reported that C3 cleavage fragments increase responsiveness of HSPC to an SDF-1 gradient, and postulated that C3a and $C3a_{desArg}$ provide a "last line of defense" to protect HSPC from uncontrolled egress from BM during mobilization (Allendorf et al. 2003; Ratajczak et al. 2004a,c, 2006; Reca et al. 2003, 2007). Our current data, however, show that activation of the CC cascade downstream from C3 is crucial to execute both G-CSF- and zymosan-induced egress of HSPC.

Accordingly, since both classical and alternative pathways of CC activation merge at C3, subsequently leading to the activation of C5, to better address the role of C in mobilization, we performed mobilization studies in C5-deficient mice. We found that in contrast to C3 deficient mice that, as we reported, are easy mobilizers (Ratajczak et al. 2004a) the C5 deficient mice were severely suppressed in response to both G-CSF (~five times) and zymosan (~three times) (Reca et al. 2007). The different effect of C3 and C5 cleavage fragments on mobilization of HSPC suggests that the CC cascade modulates egress of HSPC in both a negative (C3) and positive (C5) manner and our recent observations that C5-deficient and Ig-deficient mice, which do not activate CC by the classical pathway, are poor mobilizers support this notion.

However, this observation does not, at first glance, corroborate our previous report on the mobilization of $C3^{-/-}$ and $C3aR^{-/-}$ mice, which we reported to be easy mobilizers (Ratajczak et al. 2004a). This apparent discrepancy can be explained by the fact that the various CC cleavage fragments have different effects on mobilization. Based on the previous and current data a more complex picture of the role of CC in HSPC mobilization has now emerged. Accordingly, we hypothesize that mobilization/retention of HSPC in BM is regulated differentially at various levels of the CC activation cascade (Fig. 5), and that the CC system may contain internal checks and balances that modulate trafficking of HSPC. We believe that the soluble (C3a, $C3a_{desArg}$), and solid (iC3b) phases of C3 cleavage are primarily involved in retention of HSPC in BM. In this context the BM microenvironment, by expressing C3a and $C3a_{desArg}$, is increasing the responsiveness of HSPC to an SDF-1 gradient as a "last line of defense" against an uncontrolled egress of HSPC from BM into PB (Ratajczak et al. 2004a,c; Reca et al. 2003). Similarly, iC3b deposited in the BM microenvironment tethers HSPC and increases their retention

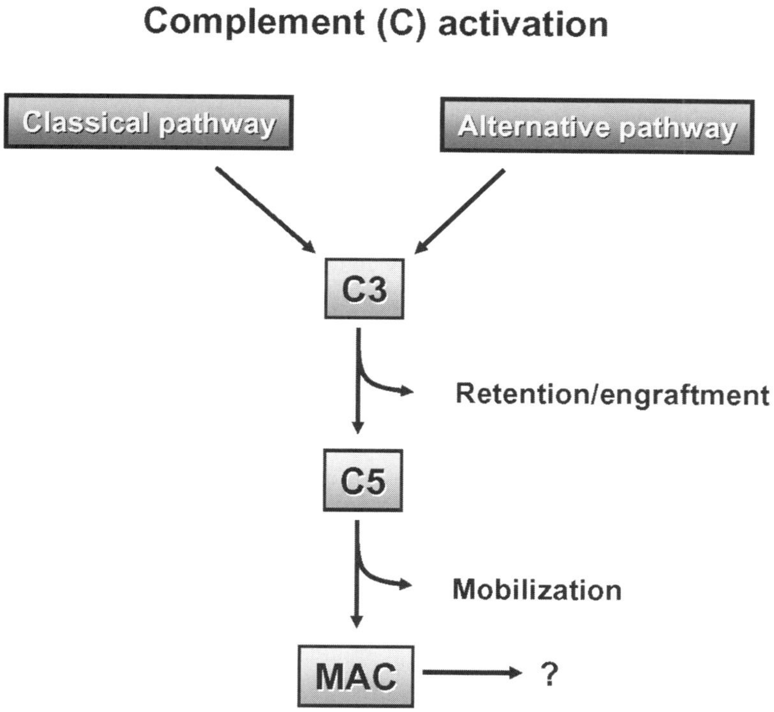

Fig. 5 Complement cascade activation products differentially modulate stem cell trafficking. Upon activation of the CC system C3 and C5 cleavage products play opposite roles in the retention or mobilization of cells, respectively. While liquid phase C3 cleavage fragments (C3a and $C3a_{desArg}$) enhance responsiveness of HSPC to an SDF-1 gradient, solid phase cleavage fragment iC3b deposited onto surrounding surfaces helps to retain HSPC in their niche. In contrast as we hypothesize, activation of C5 promotes mobilization of HSPC by increasing the permeability of the BM endothelium and the recruitment/degranulation of granulocytes. Thus we propose that an activated CC can affect stem cell mobilization in a negative (C3) or a positive (C5) way. The potential contribution of other CC cascade proteins (e.g., membrane attack complex – MAC) requires further study

in BM (Ratajczak et al. 2004a). This explains why C3- or C3aR-deficient mice that lack "this last line of defense" are easy mobilizers.

On the other hand, C3 activation is obligatory for activation of downstream CC proteins including C5 and we demonstrate here that lack of C5 activation is associated with impaired mobilization of HSPC. An explanation of this phenomenon might be that C5 cleavage fragments (C5a and $C5a_{desArg}$) known as potent anaphylatoxins increase the vascular permeability of BM vasculature as well as activate granulocytes to release proteolytic enzymes (Reca et al. 2007) which could potentially activate CC by alternative mechanisms and we are currently testing this hypothesis. Thus, this recent observation further confirms the role of the CC cascade as a pivotal modulator of stem cell mobilization and a molecular explanation of the phenomenon which C5 cleavage products play in egress of HSPC from BM needs further studies.

Overall, our data supports the idea that CC is involved in crucial mechanisms responsible for the egress of HSPC from the BM. On the other hand, since the generation of two less potent anaphylatoxins (C3a and C4a) is not impaired in C5-deficient mice, this explains why some HSPC are still mobilized even in the total absence of C5a. Furthermore, it is also likely that some other downstream C proteins (e.g., C6 or the membrane attack complex) may be required for the egress of HSPC from BM. This issue will be addressed by performing mobilization in C6 deficient mice as well as in mice deficient both in C3 and C5.

9 Conclusions

In conclusion, our data support the concept that the activation of CC is a major factor modulating egress of HSPC from BM into PB. We found that (i) G-CSF- but not zymosan-induced mobilization, was severely reduced in RAG2, SCID and Jh mice, (ii) impaired G-CSF-induced mobilization was restored after infusion of purified wild type Ig, and (iii) mobilization was severely reduced in C5-deficient mice. Thus, in light of our recent findings, mobilization of HSPC could be envisioned as part of an immune response that requires CC activation by the classical Ig-dependent and/or Ig-independent pathways. Hence modulation of CC activation could allow for the development of more efficient mobilization strategies in patients who are poor mobilizers of HSPC. Finally, competitive repopulating studies using mobilized PBMNC performed in murine model will address the effect of CC on mobilization of the most primitive long-term repopulating hematopoiesis stem cells.

Acknowledgments

This work was supported by an NIH grant R01 DK074720-01 to MZR.

References

Allendorf, D. J., Baran, J. T., Dyke, C. W., Ratajczak, M. Z. and Ross, G. D. (2003). Unified Science & Technology for Reducing Biological Threats & Countering Terrorism. University of New Mexico: Albuquerque, NM, USA

Broxmeyer, H. E., Orschell, C. M., Clapp, D. W., Hangoc, G., Cooper, S., Plett, P. A., Liles, W. C., Li, X., Graham-Evans, B., Campbell, T. B., Calandra, G., Bridger, G., Dale, D. C. and Srour, E. F. (2005). Rapid mobilization of murine and human hematopoietic stem and progenitor cells with AMD3100, a CXCR4 antagonist. *J Exp Med* 201, 1307–1318

Canales, M. A., Arrieta, R., Gomez-Rioja, R., Diez, J., Jimenez-Yuste, V. and Hernandez-Navarro, F. (2002). Induction of a hypercoagulability state and endothelial cell activation by granulocyte colony-stimulating factor in peripheral blood stem cell donors. *J Hematother Stem Cell Res* 11, 675–681

Fukuda, S., Bian, H., King, A. G. and Pelus, L. M. (2007). The chemokine GRObeta mobilizes early hematopoietic stem cells characterized by enhanced homing and engraftment. *Blood* 110, 860–869

Huber-Lang, M., Sarma, J. V., Zetoune, F. S., Rittirsch, D., Neff, T. A., McGuire, S. R., Lambris, J. D., Warner, R. L., Flierl, M. A., Hoesel, L. M., Gebhard, F., Younger, J. G., Drouin, S. M., Wetsel, R. A. and Ward, P. A. (2006). Generation of C5a in the absence of C3: a new complement activation pathway. *Nat Med* 12, 682–687

Kucia, M., Reca, R., Miekus, K., Wanzeck, J., Wojakowski, W., Janowska-Wieczorek, A., Ratajczak, J. and Ratajczak, M. Z. (2005). Trafficking of normal stem cells and metastasis of cancer stem cells involve similar mechanisms: pivotal role of the SDF-1-CXCR4 axis. *Stem Cells* 23, 879–894

Lapidot, T., Dar, A. and Kollet, O. (2005). How do stem cells find their way home? *Blood* 106, 1901–1910

Levesque, J. P., Hendy, J., Takamatsu, Y., Williams, B., Winkler, I. G. and Simmons, P. J. (2002). Mobilization by either cyclophosphamide or granulocyte colony-stimulating factor transforms the bone marrow into a highly proteolytic environment. *Exp Hematol* 30, 440–449

Levesque, J. P., Hendy, J., Takamatsu, Y., Simmons, P. J. and Bendall, L. J. (2003). Disruption of the CXCR4/CXCL12 chemotactic interaction during hematopoietic stem cell mobilization induced by GCSF or cyclophosphamide. *J Clin Invest* 111, 187–196

Papayannopoulou, T. (2004). Current mechanistic scenarios in hematopoietic stem/progenitor cell mobilization. *Blood 103*, 1580–1585

Quesenberry, P. J. and Becker, P. S. (1998). Stem cell homing: rolling, crawling, and nesting. *Proc Natl Acad Sci U S A* 95, 15155–15157

Ratajczak, J., Reca, R., Kucia, M., Majka, M., Allendorf, D. J., Baran, J. T., Janowska-Wieczorek, A., Wetsel, R. A., Ross, G. D. and Ratajczak, M. Z. (2004a). Mobilization studies in mice deficient in either C3 or C3a receptor (C3aR) reveal a novel role for complement in retention of hematopoietic stem/progenitor cells in bone marrow. *Blood* 103, 2071–2078

Ratajczak, M. Z., Kucia, M., Reca, R., Majka, M., Janowska-Wieczorek, A. and Ratajczak, J. (2004b). Stem cell plasticity revisited: CXCR4-positive cells expressing mRNA for early muscle, liver and neural cells 'hide out' in the bone marrow. *Leukemia* 18, 29–40

Ratajczak, M. Z., Reca, R., Wysoczynski, M., Kucia, M., Baran, J. T., Allendorf, D. J., Ratajczak, J. and Ross, G. D. (2004c). Transplantation studies in C3-deficient animals reveal a novel role of the third complement component (C3) in engraftment of bone marrow cells. *Leukemia* 18, 1482–1490

Ratajczak, M. Z., Reca, R., Wysoczynski, M., Yan, J. and Ratajczak, J. (2006). Modulation of the SDF-1-CXCR4 axis by the third complement component (C3) – implications for trafficking of CXCR4+ stem cells. *Exp Hematol* 34, 986–995

Ratajczak, M. Z., Machalinski, B., Wojakowski, W., Ratajczak, J. and Kucia M. (2007). A hypothesis for an embryonic origin of pluripotent Oct-4(+) stem cells in adult bone marrow and other tissues. *Leukemia* 21, 860–867

Reca, R., Mastellos, D., Majka, M., Marquez, L., Ratajczak, J., Franchini, S., Glodek, A., Honczarenko, M., Spruce, L. A., Janowska-Wieczorek, A., Lambris, J. D. and Ratajczak, M. Z. (2003). Functional receptor for C3a anaphylatoxin is expressed by normal hematopoietic stem/progenitor cells, and C3a enhances their homing-related responses to SDF-1. *Blood* 101, 3784–3793

Reca, R., Wysoczynski, M., Yan, J., Lambris, J. D. and Ratajczak, M. Z. (2006). The role of third complement component (C3) in homing of hematopoietic stem/progenitor cells into bone marrow. *Adv Exp Med Biol* 586, 35–51

Reca, R., Cramer, D., Yan, J., Laughlin, M. J., Janowska-Wieczorek, A., Ratajczak, J. and Ratajczak, M. Z. (2007). A novel role of complement in mobilization; Immunodeficient mice are poor G-Csf mobilizers because they lack complement-activating immunoglobulins. *Stem Cells* 25, 3093–3100

Semerad, C. L., Christopher, M. J., Liu, F., Short, B., Simmons, P. J., Winkler, I., Levesque, J. P., Chappel, J., Ross, F. P. and Link, D. C. (2005). G-CSF potently inhibits osteoblast activity and CXCL12 mRNA expression in the bone marrow. *Blood* 106, 3020–3027

Sekhsaria, S., Fleisher, T. A., Vowells, S., Brown, M., Miller, J., Gordon, I., Blaese, R. M., Dunbar, C. E., Leitman, S. and Malech, H. L. (1996). Granulocyte colony-stimulating factor recruitment of CD34+ progenitors to peripheral blood: impaired mobilization in chronic granulomatous disease and adenosine deaminase – deficient severe combined immunodeficiency disease patients. *Blood* 88, 1104–1112

Velders, G. A., van Os, R., Hagoort, H., Verzaal, P., Guiot, H. F., Lindley, I. J., Willemze, R., Opdenakker, G. and Fibbe, W. E. (2004). Reduced stem cell mobilization in mice receiving antibiotic modulation of the intestinal flora: involvement of endotoxins as cofactors in mobilization. *Blood* 103, 340–346

5. Regulation of Tissue Inflammation by Thrombin-Activatable Carboxypeptidase B (or TAFI)

Lawrence L.K. Leung, Toshihiko Nishimura, and Timothy Myles

Department of Medicine, Division of Hematology, Stanford University School of Medicine and Veterans Administration Palo Alto Health Care System, Stanford, CA, USA

Abstract. Thrombin-activatable procarboxypeptidase B (proCPB or thrombin-activatable fibrinolysis inhibitor or TAFI) is a plasma procarboxypeptidase that is activated by the thrombin-thrombomodulin complex on the vascular endothelial surface. The activated CPB removes the newly exposed carboxyl terminal lysines in the partially digested fibrin clot, diminishes tissue plasminogen activator and plasminogen binding, and protects the clot from premature lysis. We have recently shown that CPB is catalytically more efficient than plasma CPN, the major plasma anaphylatoxin inhibitor, in inhibiting bradykinin, activated complement C3a, C5a, and thrombin-cleaved osteopontin in vitro. Using a thrombin mutant (E229K) that has minimal procoagulant properties but retains the ability to activate protein C and proCPB in vivo, we showed that infusion of E229K thrombin into wild type mice reduced bradykinin-induced hypotension but it had no effect in proCPB-deficient mice, indicating that the beneficial effect of E229K thrombin is mediated through its activation of proCPB and not protein C. Similarly proCPB-deficient mice displayed enhanced pulmonary inflammation in a C5a-induced alveolitis model and E229K thrombin ameliorated the magnitude of alveolitis in wild type but not proCPB-deficient mice. Thus, our in vitro and in vivo data support the thesis that thrombin-activatable CPB has broad anti-inflammatory properties. By specific cleavage of the carboxyl terminal arginines from C3a, C5a, bradykinin and thrombin-cleaved osteopontin, it inactivates these active inflammatory mediators. Along with the activation of protein C, the activation of proCPB by the endothelial thrombin-thrombomodulin complex represents a homeostatic feedback mechanism in regulating thrombin's pro-inflammatory functions in vivo.

1 Introduction

Thrombin-activatable procarboxypeptidase B (proCPB or thrombin-activatable fibrinolysis inhibitor or TAFI) is a recently described plasma procarboxypeptidase that plays a role in modulating fibrin clot lysis. In this chapter, we reviewed recent studies and suggest that its physiological function is not limited to controlling fibrinolysis but may serve a much broader anti-inflammatory role in the regulation of thrombin and tissue inflammation.

J.D. Lambris (ed.), *Current Topics in Complement II.*
doi: 10:1007/978-0-387-78952-1_5, © Springer Science+Business Media, LLC 2008

2 Procoagulant and Anticoagulant Properties of Thrombin

The procoagulant activities of thrombin are well established: it is the final enzyme of the clotting cascade and responsible for forming the fibrin clot; in a positive feedback fashion, it also activates other clotting factors and accelerates the efficiency of the cascade; and it activates platelets via the cleavage of the protease activated receptors (PAR) to form platelet aggregates. Thus thrombin is the major physiological mediator for the formation of platelet-fibrin thrombus at the site of vascular injury (Coughlin 2005). It is noteworthy that the expression of the PAR receptors is not restricted to platelets and they also exist on endothelial cells, monocytes, and smooth muscle cells; activation of PARs in these cells generally leads to a proliferative and pro-inflammatory phenotype and contributes to thrombin's proinflammatory properties.

While the procoagulant properties of thrombin are well recognized, it is less well appreciated that thrombin also has anticoagulant properties (Esmon 2003). This is mediated by its binding to thrombomodulin (TM), an integral membrane protein found on the vascular endothelial cell surface. Binding of thrombin to TM acts like a molecular switch – thrombin is not inhibited as an enzyme, but its substrate specificity is completely altered. When bound to TM, thrombin no longer clots fibrinogen or activates platelets, but instead acquires the ability to activate protein C, a pro-enzyme (zymogen) in plasma, efficiently (by >1,000-fold). Activated protein C cleaves and inactivates activated clotting factors Va and VIIIa, two key cofactors in the clotting cascade, and dampens thrombin generation. It also has intrinsic anti-inflammatory properties and is efficacious in the treatment of severe sepsis (Esmon 2005; Bernard et al. 2001).

3 Dissociation of Thrombin's Procoagulant and Anticoagulant Properties

It is remarkable that the procoagulant and anticoagulant properties of thrombin can be completely dissociated by a single mutation, with the discovery of E229K thrombin (Leung and Gibbs 1997). To define the structure-function relationships of thrombin, a library of 53 thrombin mutants was created; based on the crystal structure of thrombin and using the alanine scanning mutagenesis approach, we systematically replaced all the surface exposed hydrophilic amino acid residues on thrombin with alanine. The thrombin mutants were then screened empirically for those that retained protein C activation but had diminished clotting function. Thrombin mutant E229A was identified (Gibbs et al. 1995); glutamic acid 229 is located quite a distance from the active site of thrombin, and from a priori considerations, one would not have guessed that replacement of glutamic acid at this location would have such a profound effect. Saturation mutagenesis at position E229 was performed and replacement of glutamic acid with lysine (E229K) gave the most desirable property of minimal clotting function (<1%) while retaining

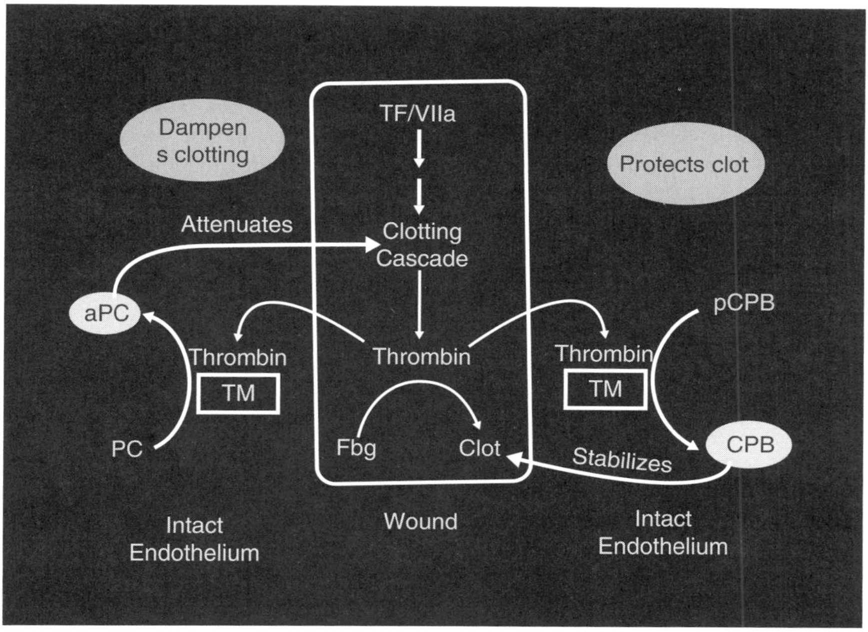

Fig. 1 Complementary roles for aPC and CPB at sites of vascular injury

significant protein C activation in the presence of thrombomodulin (~50% WT activity) (Tsiang et al. 1996). When infused into a monkey, E229K thrombin prolonged the PTT clotting time, and the anticoagulant activity correlated with the activation of protein C. It did not cause any consumption of platelets or fibrinogen. E229K thrombin also functions similarly in mice. Thus E229K is an engineered anticoagulant thrombin.

4 Thrombin-Activatable Procarboxypeptidase B as a Physiological Substrate for the Thrombin/Thrombomodulin Complex

Given the biological importance of the thrombin-thrombomodulin protein C activation pathway, does the thrombin-thrombomodulin complex activate any other physiological substrates? Thrombin-activatable procarboxypeptidase (proCPB, also termed thrombin-activatable fibrinolysis inhibitor or TAFI), is a plasma procarboxypeptidase and is now recognized as a second physiological substrate for the thrombin-thrombomodulin complex (its activation is enhanced by ~1,200-fold as compared to thrombin alone) (Redlitz et al. 1995; Bajzar et al. 1996; Broze and Higuchi 1996). When activated, CPB is specific for cleavage of carboxyl terminal arginine and lysine, and functions as a fibrinolysis inhibitor. During clot lysis, tissue type plasminogen activator (tPA) and plasminogen bind to the fibrin clot, plasmin is generated and digests the clot. In the process, new carboxyl terminal lysines are exposed in the partially digested clot, which serve as additional binding sites for plasminogen and tPA, and leads to enhanced clot lysis. The activated CPB removes the newly exposed carboxyl terminal lysines in the clot, diminishes additional tPA and plasminogen binding, thus protecting the clot and functioning as a fibrinolysis inhibitor.

From the hemostasis standpoint, activated protein C and CPB can be perceived as playing complementary roles (Fig. 1). At the site of vascular injury, thrombin is generated and forms the fibrin clot to stop bleeding. Thrombin binding to thrombomodulin on the neighboring intact endothelium will activate protein C and the activated protein C dampens the clotting cascade, preventing excessive fibrin clot formation. At the same time, the activation of CPB will stabilize and protect the clot already formed from premature lysis.

5 Thrombin-Activatable Carboxypeptidase B as an Anti-inflammatory Molecule

Given the fact that thrombin has well documented pro-inflammatory properties (via its activation of cellular PAR receptors), and activated protein C also has well established anti-inflammatory properties (via its protective effect on endothelial cells), CPB's activity may not be limited to fibrin protection, and it may have intrinsic anti-inflammatory functions. In contrast to carboxypeptidase N (CPN), the major plasma anaphylatoxin inhibitor that is constitutively active and stable, CPB is thermolabile (plasma t1/2 ~ 15 min) and requires activation by thrombin-thrombomodulin. It may thus function as a stimulus-responsive anti-inflammatory molecule that becomes activated locally at sites of tissue injury. In support of this, we and others have shown that CPB can efficiently inhibit bradykinin (BK), activated complement C3a, C5a, and thrombin-cleaved osteopontin (OPN) in vitro (Shinohara et al. 1994; Campbell et al. 2002; Myles et al. 2003).

OPN is a pro-inflammatory cytokine and immunoregulatory protein found in plasma and the extracellular matrix. It is involved in bone remodeling and wound healing. OPN interacts with many different cell types, including T cells, macrophages and neutrophils, and is upregulated at sites of injury and inflammation. Its cellular interactions are largely mediated by a RGD site that allows it to bind to many different integrins including $\alpha v\beta 3$ (vitronectin receptor). However there is also a thrombin-cleavage site within OPN, in close proximity to the RGD sequence, and thrombin cleavage leads to the exposure of a new integrin-binding site (SVVYGLR) at the carboxyl terminus, which is specific for $\alpha 4\beta 1$ and $\alpha 9\beta 1$, a subset of $\beta 1$ integrins that has a much more restricted cellular expression, so that the thrombin-cleaved OPN (OPN-R) becomes much more chemotactic for neutrophils and certain T cell subsets. We have shown that recombinant OPN-R is ~fivefold more potent than full length OPN in supporting Jurkat cell binding in vitro.

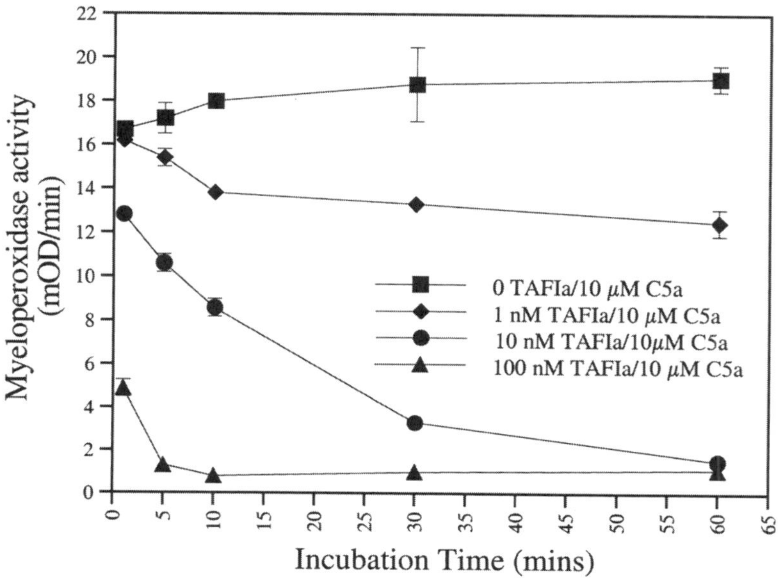

Fig. 2 Inhibitionof C5a by CPB (TAFIa) in vitro

Removal of the carboxyl terminus arginine from OPN-R, converting it to OPN-L (or OPN-des-Arg), renders it unable to bind to α4β1 and reduces its cell adhesion. It supports the notion that sequential thrombin and CPB cleavages of OPN will in turn up-regulate and down-modulate its cell binding activity.

CPB also inactivates C5a. Using an in vitro neutrophil myeloperoxidase release assay, CPB, at a molar ratio of 1:100 to C5a, effectively inhibited C5a, with complete inhibition of its neutrophil activation activity within 5 min of incubation (Fig. 2).

A formal comparison of the catalytic efficiencies (as measured by Kcat/km) of CPN and CPB, using a series of synthetic peptides to represent the various substrates, showed that CPB is about 10- to 25-fold more efficient than CPN in cleaving BK, thrombin-cleaved OPN, and C5a, while they are about the same in cleaving fibrin-based peptides (Myles et al. 2003). Thus the in vitro data show that CPB has broad substrate reactivity and may function as an anti-inflammatory molecule, in addition to its role as a fibrinolysis inhibitor.

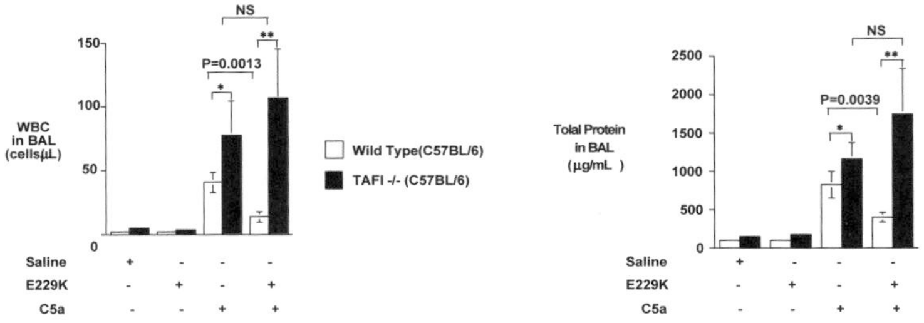

Fig. 3 E229K thrombin attenuated C5a-induced alveolitis in WT but not in proCPB (TAFI)-deficient mice in vivo

6 Carboxypeptidase B Reduces Bradykinin-Induced Hypotension In Vivo

To study the in vivo role of CPB, we used E229K thrombin to activate proCPB in vivo. E229K activates protein C and proCPB when infused intravenously in mice, without concomitant fibrinogen or platelet consumption. Pre-infusion of E229K thrombin into wild type (WT) mice reduced BK-induced hypotension but it had no effect in proCPB-deficient mice, indicating that the beneficial effect of E229K thrombin is mediated through its activation of proCPB and not protein C. It suggests that CPB can supplement CPN in inactivating BK.

7 Carboxypeptidase B Ameliorates C5a-Induced Alveolitis In Vivo

To test its role in C5a inhibition in vivo, we utilized a C5a-induced alveolitis mouse model (Nishimura et al. 2007). C5a was instilled intra-tracheally into the mouse; after 6 h, bronchoalveolar lavage was performed and the magnitude of pulmonary inflammation measured by cell counts and protein levels in the lavage fluid. In WT mice, C5a induced pulmonary alveolitis in a dose-dependent manner. Inflammation was significantly increased (by twofold) in the proCPB-deficient mice. E229K thrombin pre-infusion significantly reduced the pulmonary inflammation in WT but not proCPB-deficient mice (Fig. 3). Histological examination of the pulmonary tissues confirmed the finding of increased inflammation in the proCPB mice as compared to WT mice and its amelioration by E229K thrombin infusion in WT but not proCPB-deficient mice.

Delayed administration of E229K by 2 h negated its beneficial effect, suggesting that it was mediated by CPB inhibition of C5a, rather than CPB's effect(s) on some other downstream inflammatory mediators.

8 ProCPB-Deficient Mice are Predisposed to Abdominal Aneurysm Formation and Arthritis In Vivo

We have tested the proCPB-deficient mice in two additional models of inflammation – an abdominal aortic aneurysm (AAA) model and an inflammatory arthritis model.

The rationale for the AAA model is that OPN is known to be present in the human AAA lesions and increased plasma OPN is a risk factor for developing AAA. OPN-deficient mice are protected from AAA. We hypothesize that thrombin-cleaved OPN, and not just full length OPN, plays an important role in aneurysm formation and predicts that the proCPB-deficient mice will show enhanced AAA.

Preliminary studies, based on a porcine elastase-induced AAA model, showed that the proCPB-deficient mice developed much larger aortic aneurysms (Tedesco et al. Unpublished observations 2007). Histology revealed disruption of the internal elastic lamina, a hallmark of aneurysm formation. There was markedly enhanced pan-mural infiltration of macrophages in the proCPB-deficient mouse aortic wall. However, contrary to our prediction, OPN-deficient mice were not protected from porcine elastase-induced AAA formation, suggesting that OPN does not play a central role in this experimental AAA model. The in vivo target substrate for CPB remains to be defined.

The second model is an anti-collagen antibody-induced arthritis model. Both OPN, and specifically thrombin-cleaved OPN, as well as C5a have been reported to play an important role in this process. ProCPB-deficient mice developed enhanced arthritis in this model, with significant swelling in the paws and histology confirmed significant inflammatory cell infiltration in the synovial tissues (Song et al. 2007). C5a plays a key role in this model since C5-deficient mice were protected from arthritis development. Whether thrombin cleaved-OPN also plays a role remains to be determined.

9 Regulation of Thrombin's Inflammatory Properties by Thrombin-Activatable Procarboxypeptidase B

In summary, we have obtained in vitro and in vivo data supporting the thesis that thrombin-activatable CPB has broad anti-inflammatory properties. By specific cleavage of the carboxyl terminal lysines from C3a, C5a, BK and thrombin-cleaved OPN, it inactivates these active inflammatory mediators. Along with the activation

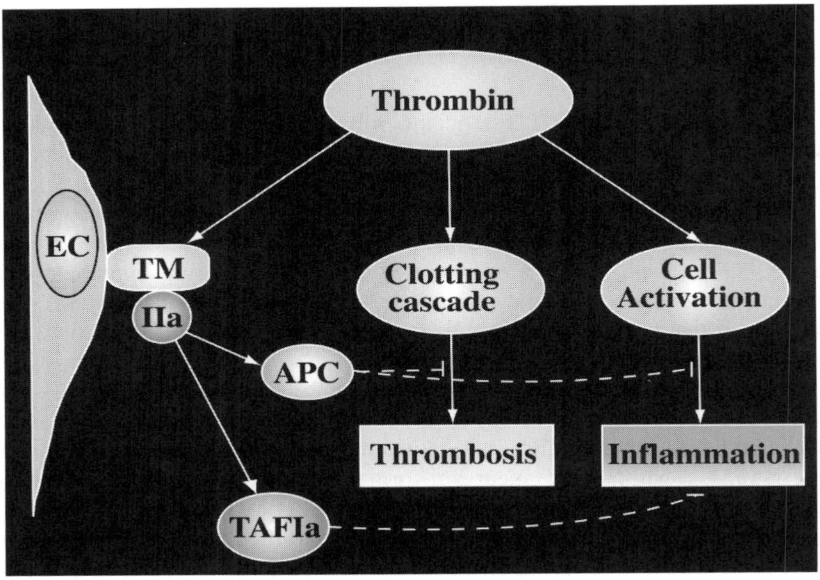

Fig. 4 CPB (TAFIa) as an anti-inflammatory molecule

of protein C, the activation of proCPB by the endothelial thrombin-thrombomodulin complex represents a homeostatic feedback mechanism in regulating thrombin's proinflammatory functions in vivo (Fig. 4).

Acknowledgement

This work was supported by NIH grant RO1 HL57530.

References

Bajzar, L., Morser, J., Nesheim, M. (1996). TAFI, or plasma procarboxypeptidase B, couples the coagulation and fibrinolytic cascades through the thrombin-thrombomodulin complex. *J Biol Chem* 271, 16603–16608

Bernard, G.R., Vincent, J.L., Laterre, P.F., LaRosa, S.P., et al. (2001). Efficacy and safety of recombinant human activated protein C for severe sepsis. *N Engl J Med* 344, 699–709

Broze, G., Higuchi, D.A. (1996). Coagulation-dependent inhibition of fibrinolysis: role of carboxypeptidase-U and the premature lysis of clots from hemophilic plasma. *Blood* 88, 3815–3823

Campbell, W.D., Lazoura, E., Okada, N., Okada, H. (2002). Inactivation of C3a and C5a octapeptides by carboxypeptidase R and carboxypeptidase N. *Microbiol Immunol* 46, 131–134

Coughlin, S.R. (2005). Protease-activated receptors in hemostasis, thrombosis and vascular biology. *J Thromb Haemost* 3, 1800–1814

Esmon, C.T. (2003). The protein C pathway. *Chest* 124, 26S–32S

Esmon, C.T. (2005). The interactions between inflammation and coagulation. *Br J Haematol* 131, 417–430

Gibbs, C.S., Coutre, S.E., Tsiang, M., Li, W.X., Jain, A.K., Dunn, K.E., Law, V.S., Mao, C.T., Matsumura, S.Y., Mejsa, S.J., Paborsky, L.R., Leung, L.L.K. (1995). Conversion of thrombin into an anticoagulant by protein engineering. *Nature* 378, 413–416

Leung, L.L.K., Gibbs, C.S. (1997). Modulation of thrombin's procoagulant and anticoagulant properties. *Thromb Haemost* 78, 577–580

Myles, T., Nishimura, T., Yun, T.H., Nagashima, M., Morser, J., Patterson, A.J., Pearl, R.G., Leung, L.L.K. (2003). Thrombin activatable fibrinolysis inhibitor, a potential regulator of vascular inflammation. *J Biol Chem* 278, 51059–51067

Nishimura, T., Myles, T., Piloponsky, P., Kao, P., Berry, G.J., Leung, L.L.K. (2007). Thrombin-activatable procarboxypeptidase B regulates activated complement C5a in vivo. *Blood* 109, 1992–1997

Redlitz, A., Tan, A.K., Eaton, D.L., Plow, E.F. (1995). Plasma carboxypeptidases as regulators of the plasminogen system. *J Clin Invest* 96, 2534–2538

Shinohara, T., Sakurada, C., Suzuki, T., Takeuchi, O., Campbell, W., Ikeda, S., Okada, N., Okasda, H. (1994). Pro-carboxypeptidase R cleaves bradykinin following activation. *Int Arch Allergy Immunol* 103, 400–404

Song, J., Nishimura, T., Garcia, M., Myles, T., Ho, P., Du, X.Y., Sharif, S., Leung, L., Robinson, W. (2007). Thrombin-activatable carboxypeptidase B prevents anti-collagen antibody induced arthritis. *Clinical Immunology* 123 Suppl, S98

Tsiang, M., Paborsky, L.R., Li, W.X., Jain, A.K., Mao, C.T., Dunn, K.E., Lee, D.W., Matsumura, S.Y., Matteucci, M.D., Coutre, S.E., Leung, L.L.K., and Gibbs, C.S. (1996). Protein engineering thrombin for optimal specificity and potency of anticoagulant activity in vivo. *Biochemistry* 35, 16449–16457

6. Interaction Between the Coagulation and Complement System

Umme Amara[1], Daniel Rittirsch[1], Michael Flierl[1], Uwe Bruckner[2], Andreas Klos[3], Florian Gebhard[1], John D. Lambris[4], and Markus Huber-Lang[1,*]

[1]Department of Traumatology, Hand-, Plastic-, and Reconstructive Surgery, University Hospital of Ulm, Ulm, Germany, markus.huber-lang@uni-klinik-ulm.de
[2]Division of Experimental Surgery, University Hospital of Ulm, Ulm, Germany, uwe.bruckner@uni-ulm.de
[3]Department of Medical Microbiology, Medical School Hannover, Hannover, Germany, klos.andreas@mh-hannover.de
[4]Department of Pathology, University of Pennsylvania, 401 Stellar Chance, Philadelphia, PA 19104, USA, lambris@mail.med.upenn.edu.

Abstract. The complement system as a main column of innate immunity and the coagulation system as a main column in hemostasis undergo massive activation early after injury. Interactions between the two cascades have often been proposed but the precise molecular pathways of this interplay are still in the dark. To elucidate the mechanisms involved, the effects of various coagulation factors on complement activation and generation of anaphylatoxins were investigated and summarized in the light of the latest literature. Own in vitro findings suggest, that the coagulation factors FXa, FXIa and plasmin may cleave both C5 and C3, and robustly generate C5a and C3a (as detected by immunoblotting and ELISA). The produced anaphylatoxins were found to be biologically active as shown by a dose-dependent chemotactic response of neutrophils and HMC-1 cells, respectively. Thrombin did not only cleave C5 (Huber-Lang et al. 2006) but also in vitro-generated C3a when incubated with native C3. The plasmin-induced cleavage activity could be dose-dependently blocked by the serine protease inhibitor aprotinin and leupeptine. These findings suggest that various serine proteases belonging to the coagulation system are able to activate the complement cascade independently of the established pathways. Moreover, functional C5a and C3a are generated, both of which are known to be crucially involved in the inflammatory response.

1 Introduction

The complement system as a key sentinel of innate immunity and the coagulation system as main actor in hemostasis belong both to the “first line of defense” against injurious stimuli and invaders (Choi et al. 2006). Being descended from a common ancestor, interactions between both cascades have often been proposed, but the precise molecular pathways of this cross-talk have remained elusive. Immediately after severe trauma, massive activation of a series of cascading enzymatic reactions results in fibrin deposition as well as synchronic fibrinolysis (Lampl et al. 1994),

J.D. Lambris (ed.), *Current Topics in Complement II.*
doi: 10:1007/978-0-387-78952-1_6, © Springer Science+Business Media, LLC 2008

which often causes an uncontrolled, systemic inflammatory response (SIRS) (Levi et al. 2004). Furthermore, both cascades contain series of serine-proteases with evidence of some shared activators and inhibitors, such as factor (F) XIIa, which is able to activate C1q, and thereby the classical pathway of complement. Similarly, the C1 esterase inhibitor acts not only as inhibitor of all three established complement pathways (classical: C1q/r/s, lectin: MBL, and alternative: C3b) but also of the endogenous coagulation activation path (kallikrein, FXIIa) (Davis 2004). Previously it has been shown that thrombin, generated at inflammatory sites in response to complement activation, is a physiological agonist for the PKC-dependent pathway of decay accelerating factor (DAF) regulation in terms of a negative feedback loop preventing thrombosis during inflammation (Lidington et al. 2000). The complement activation product C5a has also been reported to induce tissue factor (TF) activity in human endothelial cells (Ikeda et al. 1997) and may activate the exogenous (TF-dependent) coagulation pathway. In a recent study, a novel C5a receptor (C5aR)-TF cross-talk in neutrophils has been demonstrated (Ritis et al. 2006). Systemic inflammation is often triggered by severe trauma with subsequent extensive activation and depletion of the coagulation cascade. Findings from our laboratory suggest in accordance with other reports (Hecke et al. 1997; Ganter et al. 2007) that trauma leads not only to an early coagulopathy (DIC, disseminated intravascular coagulopathy) but also to an early hyper-activation of complement with generation of powerful anaphylatoxins, such as C3a and C5a, which may contribute to the disturbance of the coagulation system, and vice versa.

Therefore, it is tempting to speculate that these significant interactions between the coagulation and complement system may play an important role after trauma and for subsequent inflammatory reactions and complications.

2 Serine Protease Systems

2.1 Coagulation System

It is well known that any traumatic input rapidly activates the coagulation cascade to stop bleeding and to prevent invasion of microorganisms and the subsequent inflammatory response. Therefore, the clotting system has been considered as a crucial part of innate immunity. Most clotting factors (F) belong to the class of serine proteases (Fig. 1) with the final aim to induce fibrin polymerization in order to seal off leaking and injured vessels and also to wall off injured tissue and invading bacteria.

Thrombin (FIIa) represents the central serine protease cleaving fibrinogen to fibrin, the building block of a haemostatic plug. Generation of thrombin is initiated by tissue injury with corresponding vessel wall exposure to tissue factor (TF), which forms a complex with the serine protease FVIIa. The TF/FVIIa complex as part of the exogenous path activates the down-stream serine protease FX, where

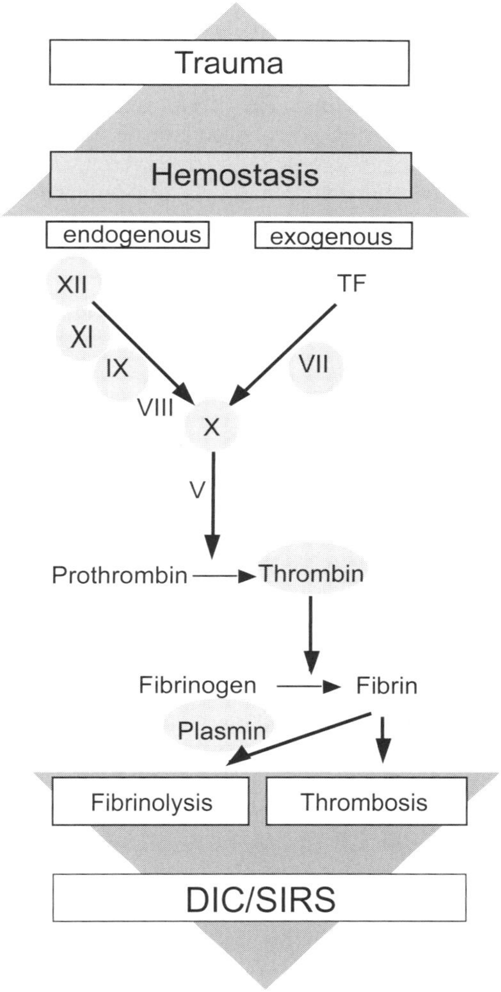

Fig. 1 Activation of the coagulation cascade by trauma with subsequent responses (*DIC* disseminated intravascular coagulopathy; *SIRS* systemic inflammatory response). Grey circles = serine protease

the exogenous and endogenous pathways of the coagulation system converge. The endogenous path involves other serine proteases, such as FXII, FXI, and FIX which are sequentially activated upon contact to negatively charged surfaces (e.g. collagen). Factor IXa and FXa assemble with their non-enzymatic protein cofactors (FVIIIa, FVa) resulting in thrombin generation and finally in conversion of fibrinogen to fibrin.

2.2 Fibrinolytic System

Fibrinolysis is a physiological regulatory process with breakdown of fibrin to limit clotting and to resolve blood clots. The fibrinolytic system is initiated when plasminogen is converted into the potent serine-protease plasmin, which leads to degradation of fibrin, fibrinogen, FV, and FVIII. Severe trauma and hemorrhage is associated with decreased plasminogen levels and enhanced plasma levels of plasmin and plasmin-anti-plasmin-complexes, indicating early fibrinolytic events (Lampl et al. 1994). It has been described that severe trauma induces fibrinolysis almost synchronically with the massive activation and subsequent depletion of the coagulation system. Hyperfibrinolysis due to excessive generation of plasmin seems to be the major cause of trauma-induced disseminated intravascular coagulopathy (DIC) as bleeding is most severe in trauma victims with low antiplasmin activity (Lampl et al. 1994). Fibrinolysis is also promoted by activated protein C (APC) which acts as a strong serine protease by interfering with important inhibitors of plasmin generation (plasmin activator inhibitor-1 [PAI-1], thrombin-activatable fibrinolysis inhibitor [TAFI]).

2.3 Complement System

There is increasing evidence that the rapid activation of the coagulation cascade after trauma is accompanied by a very early onset of an uncontrolled, progressive inflammatory response with often lethal consequences (Hierholzer and Billiar 2001). Obviously, acute blood loss and tissue trauma activate the complement cascade in humans (Hecke et al. 1997). Especially generation of the powerful anaphylatoxin C3a and C5a, and consumption of complement may play a detrimental role (Younger et al. 2001). Generally, the complement enzymes contain a single serine protease with an extremely restricted substrate specificity. The classical pathway proceeds through the sequential cleavage of C4 and C2 by active C1s and formation of the C3 convertase (C4b2a-complex). Additional binding of C3b leads to generation of the C5 convertase (C4b3b2a-complex). Both complexes develop their proteolytic activity via the serine-protease domain of C2a. Similarly, the active center of the C3- and C5-convertase of the alternative pathway (C3bBb- and $(C3b)_2Bb(P)$-complex, respectively) resides in the serine protease domain of factor B. Activation of the lectin pathway results in subsequent activation of mannose associated serine proteases (MASP), which in turn activate C4 and C2 to assemble C4b2a. Activated MASP-1 also reveals serine protease specificity for a direct C3 cleavage (Lambris et al. 1998).

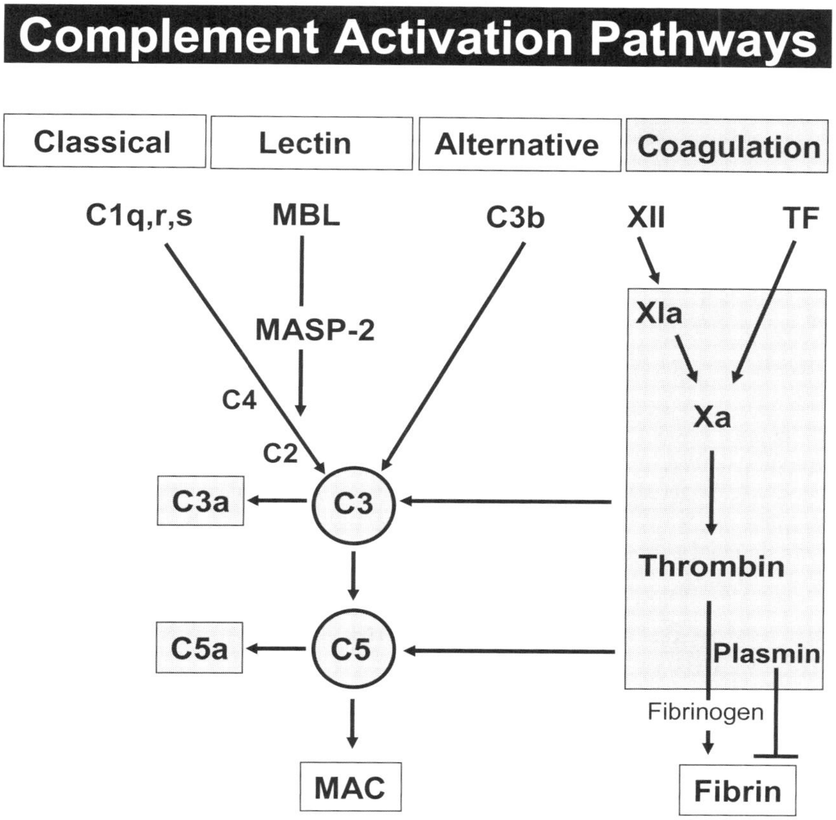

Fig. 2 Complement activation pathways (*MAC* membrane attack complex; *MBL* mannose binding lectin; *MASP-2* mannose associated serine protease-2; *TF* tissue factor.

3 C3a and C5a Generation by Coagulation Factors

Recently, we have shown that authentic C5a was generated in the absence of C3 with thrombin acting as a potent C5 convertase (Huber-Lang et al. 2006). In the presence of C3, thrombin did also generate C3a dose- and time-dependently, as assessed by immunoblotting and ELISA. The produced C3a dose-dependently increased the chemotactic response of the human mast cell line-1 (HMC-1) more than threefold, indicating biological activity of the thrombin-induced C3 cleavage product.

Investigations on other central molecules of the coagulation cascade revealed that FVIII and tissue factor failed to interact with C3 and C5. This was in striking contrast to the coagulation factors FXa and FXIa. Both serine proteases as representatives of the endogenous path cleaved C3 and C5 with generation of C3a and C5a, as detected by ELISA and Western blots. Plasmin as the strongest serine protease of the fibrinolytic system was capable of cleaving both C5 and C3, respectively. The produced anaphylatoxins were biologically active as shown by a dose-dependent chemotactic response of human neutrophils to C5a and HMC-1 cells to C3a. Furthermore, the plasmin-induced cleavage activity could be blocked by the serine protease inhibitor aprotinin and leupeptine.

4 Complement: New Activation Paths

It has been shown by our lab that phagocytic cells are capable of generating biologically active C5a from C5 independently of the plasma complement system (Huber-Lang et al. 2002). The actual findings indicate the existence of additional complement activation paths in the plasma (Huber-Lang et al. 2006) besides the three established pathways of complement activation (Fig. 2).

5 Interaction Between the Coagulation and Complement Cascade After Trauma

Many experimental and clinical studies have focused on the extreme coagulation challenge after severe trauma and during hemorrhagic shock (Kaplan et al. 1981). When reflecting a more philosophical consideration that "the end is in the beginning", the very early and rapid activation of both, the coagulation system and the complement system seem to be crucially involved in the initiation and progression of the systemic inflammatory response and also in the deathly escalation of posttraumatic organ failure (Hierholzer et al. 2001). During experimental hemorrhagic shock, pretreatment of rats with carboxypeptidase N inhibitor (blocker of C5a clearance) turned out lethally (Younger et al. 2001). Furthermore, pre-shock depletion of complement significantly improved post-resuscitation blood pressure. One in vitro study claimed that C5a induces TF activity on endothelial cells (Ikeda et al. 1997) and may thereby activate the exogenous coagulation pathway. Recently, a C5a-induced "switch" in mast cells from a pro-fibrinolytic (t-PA release) to a pro-thrombotic phenotype (PAI-1 release) has been reported (Wojta et al. 2004) thus modifying the balance between pro- and anti-coagulation. To ameliorate the inflammatory response, various immune modulators have been examined, some of which revealed regulatory effects on both the complement and the coagulation system (Weiler and Linhardt 1991; Campbell et al. 2001). The C1-esterase-inhibitor, which curbs not only complement components of all three pathways (C1r/s, mannose-associated serine protease-2 [MASP-2], C3b) but also the endogenous coagulation cascade (kallikrein, XIIa), evinced a cross-talk between both systems (Jansen et al. 1998; Davis 2004). Similarly, the application of the soluble complement receptor 1 (sCR1) revealed some protective effects during hemorrhagic shock by blocking the endothelial dysfunction and the post-shock vasoconstriction (Fruchterman et al. 1998). In addition, sCR1 has been shown to decrease rolling, adherence, and influx of neutrophils in the mesenteric circulation and gut after hemorrhagic shock (Spain et al. 1999). A nexus between complement and coagulation may also be suggested by the successful use of recombinant activated protein C (APC) during sepsis (Bernard et al. 2001; Rezende et al. 2004). The beneficial effects of APC during systemic inflammation seem not only to be based on the interaction with the fibrinolytic system but also on anti-inflammatory effects.

6 Conclusion

The findings suggest that various serine proteases belonging to the coagulation system are able to activate the complement cascade independently of the so far established pathways. Moreover, functional active C5a and C3a are generated, both of which are known to be crucially involved in the inflammatory response. Consequently, a misdirected interaction of the coagulation cascade and complement system may play a crucial role in hemorrhagic shock and the subsequent systemic inflammatory response.

Acknowledgement

We thank Sonja Albers and Barbara Acker for outstanding laboratory assistance. This work was supported by grants from the Deutsche Forschungsgemeinschaft (DFG HU 823/2-2, HU 823/2-3, European Shock Society/Novo Nordisk Grant 2005) and NIH AI068730.

References

Bernard, G.R., Vincent, J.L., Laterre, P.F., LaRosa, S.P., Dhainaut, J.F., Lopez-Rodriguez, A., Steingrub, J.S., Garber, G.E., Helterbrand, J.D., Ely, E.W., Fisher, C.J. (2001) Efficacy and safety of recombinant human activated protein C for severe sepsis. *N Engl J Med* 344, 699–709

Campbell, W., Okada, N., Okada, H. (2001) Caboxypeptidase R is an inactivator of complement-derived inflammatory peptides and an inhibitor of fibrinolysis. *Immunol Rev* 180, 162–167

Choi, G., Schultz, M.J., Levi, M., van der Poll, T. (2006) The relationship between inflammation and the coagulation system. *Swiss Med Wkly* 136, 139–144

Davis, A.E. III. (2004) Biological effects of C1 inhibitor. *Drug News Pespect* 17, 439–446

Fruchterman, T.M., Spain, D.A., Wilson, M.A., Harris, P.D., Garrison, R.N. (1998) Complement inhibition prevents gut ischemia and endothelial cell dysfunction after hemorrhage/resuscitation. *Surgery* 124, 782–791

Ganter, M.T., Brohi, K., Cohen, M.J., Shaffer, L.A., Walsh, M.C., Stahl, G.L., Pittet, J.F. (2007) Role of the alternative pathway in the early complement activation following major trauma. *Shock* 28, 29–34

Hierholzer, C., Billiar, T.R. (2001) Molecular mechanisms in the early phase of hemorrhagic shock. *Langenbecks Arch Surg*. 386, 302–308

Hecke, F., Schmidt, U., Kola, A., Bautsch, W., Klos, A., Kohl, J. (1997) Circulating complement proteins in multiple trauma patients – correlation with injury severity, development of sepsis, and outcome. *Crit Care Me*d 25, 2015–2024

Huber-Lang, M., Younkin, EM, Sarma, V.J., Riedemann, N., McGuire, S.R., Lu, K.T., Kunkel, R., Younger, J.G., Zetoune, F.S., Ward, P.A. (2002) Generation of C5a by phagocytic cells. *Am J Pathol* 161, 1849–1859

Huber-Lang, M., Sarma, J.V., Zetoune, F.S., Rittirsch, D., Neff, T.A., McGuire, S.R., Lambris, J.D., Warner, R.L., Flierl, M.A., Hoesel, L.M., Gebhard, F., Younger, J.G., Drouin, S.M., Wetsel, R.A., Ward, P.A. (2006) Generation of C5a in the absence of C3: a new complement activation pathway. *Nat Med* 12, 682–687

Ikeda, K., Nagasawa, K., Horiuchi, T., Tsuru, T., Nishizaka, H., Niho, Y. (1997) C5a induces tissue factor activity on endothelial cells. *Thromb Haemost* 77, 394–398

Jansen, P.M., Eisele, B., de Jong, I.W., Chang, A., Delvos, U., Taylor, F.B. Jr, Hack, C.E. (1998) Effect of C1 inhibitor on inflammatory and physiologic response patterns in primates suffering from lethal septic shock. *J Immunol* 160, 475–484

Kaplan, A.P., Ghebrehiwet, B., Silverberg, M., Sealey, J.E. (1981) The intrinsic coagulationn-kinin pathway, complement cascades, plasma renin-angiotensin system, and their interrelationships. *Crit Rev Immunol* 3, 75–93

Lambris, J.D., Sahu, A., Wetsel, R. (1998) The chemistry and biology of C3, C4, and C5. In: *The Human Complement System in Health and Disease*, Volanakis, J.E., Frank, M. (eds.). Marcel Dekker, New York, pp. 83–118

Lampl, L., Helm, M., Specht, A., Bock, K.H., Hartel, W., Seifried, E. (1994) Blood coagulation parameters as prognostic factors in multiple trauma: can clinical values be an early diagnostic aid? *Zentralblatt für Chirurgie* 119, 683–689

Levi, M., van der Poll, T., Büller, H.B. (2004) Relation between inflammation and coagulation. *Circulation* 109, 2698–704

Lidington, E.A., Haskard, D.O., Mason, J.C. (2000) Induction of decay-accelerating factor by thrombin through a protease-activated receptor 1 and protein kinase C-dependent pathway protects vascular endothelial cells from complement-mediated injury. *Blood* 96, 2784–2792

Rezende, S.M., Simmonds, R.E., Lane, D.A. (2004) Coagulation, inflammation, and apoptosis: different roles for protein S and the protein S-C4b binding protein complex. *Blood* 103, 1192–1201

Ritis, K., Doumas, M., Mastellos, D., Micheli, A., Giaglis, S., Magotti, P., Rafail, S., Kartalis, G., Sideras, P., Lambris, J.D. (2006) A novel C5a receptor-tissue factor cross-talk in neutrophils links innate immunity to coagulation pathways. *J Immunol* 177, 4794–4802

Spain, D.A., Fruchtermann, T.M., Matheson, P.J., Wilson, M.A., Martin, A.W., Garrison, R.N. (1999) Complement activation mediates intestinal injury after resuscitation from hemorrhagic shock. *J Trauma* 46, 224–233

Weiler, J.M., Linhardt, R.J. (1991) Antithrombin III regulates complement activity in vitro. *J Immunol* 146, 3889–3894

Wojta, J., Huber, K., Valent, P. (2004) New aspects in thrombotic research: complement induced switch in mast cells from a profibrinolytic to a prothrombotic phenotype. *Pathophysiol Haemost Thromb* 33, 438–441

Younger, J.G., Sasaki, N., Waite, M.D., Murray, H.N., Saleh, E.F., Ravage, Z.A., Hirschl, R.B., Ward, P.A., Till, G.O. (2001) Detrimental effects of complement activation in hemorrhagic shock. *J Appl Physiol* 90, 441–446

7. Platelet Mediated Complement Activation

Ellinor I.B. Peerschke[1], Wei Yin[2], and Berhane Ghebrehiwet[3]

[1]Department of Pathology, The Mount Sinai School of Medicine, New York, NY, USA
[2]Department of Mechanical and Aerospace Engineering, Oklahoma State University, Stillwater, OK, USA
[3]Departments of Medicine and Pathology, Stony Brook University, Stony Brook, NY, USA

Abstract. The complement system comprises a series of proteases and inhibitors that are activated in cascade-like fashion during host defense (Makrides 1998). A growing body of evidence supports the hypothesis that immune mechanisms, including complement activation, are involved in inflammatory conditions associated with vascular injury (Acostan et al. 2004; Giannakopoulos et al. 2007), and disseminated intravascular coagulation associated with massive trauma (Huber-Lang, this volume). We propose that platelets and platelet derived microparticles focus complement to sites of vascular injury where regulated complement activation participates in clearing terminally activated platelets and microparticles from the circulation, and dysregulated complement activation contributes to inflammation and thrombosis. Given the central role of platelets in hemostasis and thrombosis, it is not surprising that activated complement components have been demonstrated in many types of atherosclerotic and thrombotic vascular lesions (Torzewsjki et al. 2007; Niculescu et al. 2004).

1 Complement Activation on Platelets

Evidence for direct activation of both classical and alternative pathways of complement on platelets is emerging (Del Conde et al. 2005; Peerschke et al. 2006). For example, P-selectin, a platelet alpha granule membrane protein that contains a short consensus repeat domain common to many complement binding proteins (Kansas 1996), has been identified as an activator of the alternative complement cascade on platelets (Del Conde et al. 2005). In addition, platelet alpha granules contain Factor D, the serine protease that cleaves Factor B of the alternative pathway to its active form (Davis and Kenney 1979).

Platelets also express binding sites for classical complement components, most notably for C1q (Peerschke and Ghebrehiwet 1997; Peerschke et al. 1994). C1q interactions with platelets trigger a variety of cellular and biochemical responses that may contribute to inflammation and thrombosis. For example, the C-terminal

J.D. Lambris (ed.), *Current Topics in Complement II.*
doi: 10:1007/978-0-387-78952-1_7,

collagen like domain of C1q binds to the 60kDa platelet calreticulin (CR) homologue, cC1qR (Peerschke et al. 1997). This interaction is associated with induction of GPIIb-IIIa, platelet aggregation, P-selectin expression, and generation of platelet procoagulant activity (Peerschke et al. 1993).

In contrast, the amino terminal, globular domain of C1q binds the 33kDa gC1qR/p33 (gC1qR) (Ghebrehiwet et al. 1994). gC1qR is a ubiquitously expressed cellular protein (Ghebrehiwet and Peerschke 1998; Ghebrehiwet et al. 2001). Although its 73 amino acid presequence contains a mitochondrial targeting motif (Dedio et al. 1998), the expression of gC1qR in other cellular compartments, including the cell surface of platelets and endothelial cells, has been confirmed (Ghebrehiwet et al. 1994; Mahdi et al. 2002; Peerschke et al. 2003).

Mature gC1qR exhibits a noncovalent trimeric structure (Ghebrehiwet et al. 1998; Jiang et al. 1999). Multimerization is an essential process that increases the affinity of gC1qR for multivalent ligands such as C1q (Ghebrehiwet et al. 1998). The geometry and topography of the three dimensional structure of gC1qR indicates that gC1qR could engage C1q via at least two, if not three, of its globular heads. This could induce the subtle conformational change that is necessary to trigger C1 activation (Ghebrehiwet et al. 2006). Indeed, recombinant gC1qR has been shown recently to activate C1 (Peerschke et al. 2006).

The expression of C1q binding sites on human blood platelets (Peerschke et al. 1994, 1997), combined with the ability of gC1qR to engage the globular domain of C1q (Ghebrehiwet et al. 1994), suggests that platelets may possess an intrinsic capacity to initiate the classical complement pathway. Indeed, platelet mediated activation of C4 has been demonstrated recently using both solid phase and flow cytometric approaches (Peerschke et al. 2006).

Deposition of C3b and C5b-9 on activated platelets was detected also. Platelet mediated complement activation did not require thrombin generation (Peerschke et al. 2006). Thrombin has the potential to directly activate C5 (Huber-Lang et al. 2006) and other complement components (Huber-Lang, this volume), including C3. C3 activation on platelets was reduced in the absence of C1 or Factor B, suggesting a role, instead, for classical and alternative pathway C3 convertases (Peerschke et al. 2006).

Data from our laboratory (Peerschke et al. 2006) show that platelet mediated classical pathway C4 activation occurs in the presence of purified C1 and C4, and in diluted plasma or serum. Dilution is necessary, presumably to alter the balance between complement components and their circulating inhibitors. This is of interest, as it may have implications for clinical situations, such as massive trauma or cardiopulmonary bypass, in which significant hemodilution occurs.

C4 activation on platelets further requires divalent cations and platelet stimulation. C4 activation appears particularly enhanced following platelet exposure to shear stress (1,800 s^{-1}, 60 min), and correlates with gC1qR expression (Peerschke et al. 2006). Antibodies directed against gC1qR and exposure of

Table 1 Platelet mediated complement activation

Positive regulators	Negative regulators
Platelet activation	EDTA
Platelet granule secretion	C1q deficiency
gC1qR/p33	C4 deficiency
Platelet associated IgG or immune complexes	Plasmin
P-Selectin	Factor B depletion

platelets to plasmin, a protease previously shown to degrade gC1qR (Peerschke et al. 2004), inhibit C4 activation. Thus, plasmin generation at sites of inflammation or vascular injury may regulate platelet mediated classical complement activation. Further regulation of classical complement pathway activation may be achieved by secretion of platelet alpha granule C1 inhibitor (C1 INH) (Schmaier et al. 1993).

Requirements for platelet mediated complement activation are summarized in Table 1.

2 Complement Activation on Platelet Microparticles (PMP)

PMP are a storage pool for disseminating blood borne bioactive effectors (Morel, Toti, Hugel, and Freyssinet 2004) such as tissue factor, procoagulant phospholipids, and inflammatory mediators. Compared to activated platelets, PMP can express 50- to 100-fold higher procoagulant activity (Sinauridze et al. 2007). They retain selected platelet membrane constituents, including glycoproteins Ib, IIb-IIIa, and P-selectin (Gawaz et al. 1996; George et al. 1986) which support vascular inflammation by participating in heterotypic communication with leukocytes and vascular endothelial cells (Martinez et al. 2005; Barry et al. 1998).

Increasing evidence is emerging to support cross-talk between coagulation and complement systems (Markiewski and Lambris 2007; Markiewski et al. 2007). In this regard, PMP present a surface for the assembly and interaction of complement and coagulation cascades. PMP were recently shown to express gC1qR and to activate C4 in the absence of immune complexes (Yin et al. 2008). PMP further supported C3 activation and deposition of C5b-9. Amplification of complement activation on PMP may be achieved via exposure of P-selectin and engagement of the alternative pathway (Del Conde et al. 2005). Although the extent of complement activation on PMP was low compared to platelets, normalization of complement activation for PMP size suggests that PMP may be several orders of

Table 2 Pathophysiologic consequences of platelet and PMP mediated complement activation

Complement component	Pathophysiologic effect
C3a, C5a, C4a	Inflammatory cell recruitment
C3a	Enhanced platelet response to agonists (e.g. ADP, thrombin)
C5a	Endothelial cell activation, including tissue factor expression
C3b	Recognition and phagocytosis of apoptotic platelets and microparticles
C5b-9	Platelet activation and expression of procoagulant activity; endothelial cell activation

magnitude more active than platelets, and likely present concentrated activated complement components to vascular targets.

In addition, negatively charged phospholipids such as cardiolipin and phosphatidyl serine (PS) are expressed on PMP and have been shown to activate C1 (Kovacsovics et al. 1985). Since PS is present on early apoptotic cells (Vermes et al. 1995), we propose that complement activation on microparticles may be involved in physiologic clearance mechanisms. Indeed, complement components C1q, Factor B, and C3 have all been shown to participate in the phagocytosis of apoptotic cells (Mevorach et al. 1998).

3 Pathophysiology of Platelet Mediated Complement Activation

Potential pathophysiologic consequences of platelet mediated complement activation are summarized in Table 2. Activation of the complement system is associated with generation of potent proinflammatory peptides, C3a and C5a (Makrides 1998). C3a has been reported to enhance platelet stimulation by subthreshold concentrations of traditional agonists such as ADP or thrombin (Polley and Nachman 1983). In addition, both C3a and C5a bind receptors on endothelial cells and stimulate upregulation of interleukins 8 and 1β, in addition to RANTES, and strongly activate the MAP kinase signaling pathway (Monsinjon et al. 2003). Thus, the association of PMP with endothelial cells (Mause et al. 2005) may carry activated complement components along the vasculature and accelerate endothelial responses to vascular injury. In addition, C5a is known to induce tissue factor expression by endothelial cells and leukocytes in vitro (Ikeda et al. 1997; Muhlfelder et al. 1979).

When complement activation goes to completion, the terminal complement complex (TCC) (C5b-9) is formed. This complex is responsible for lysis of host pathogens and may contribute to tissue damage. Moreover, sublytic quantities of TCC have been reported to activate platelets and endothelial cells, leading to expression of procoagulant activity (Wiedmer et al. 1986). Thus, inhibition of potential injurious/inflammatory effects of complement activation by infusion of specific C5a antagonists (Allegretti et al. 2005) is under intense investigation.

In addition, the complement cascade is linked to other protease activated cascades of the blood including the coagulation (Polley and Nachman 1978), kinin (Kaplan et al. 1986), and fibrinolytic systems (Schaiff and Eisenberg 1997). Generation of a novel thrombin dependent C5 convertase (Polly et al. 1978; Polley and Nachman 1979; Zimmerman and Kolb 1976) has been described during blood coagulation, and direct activation of C5 and other complement components by thrombin has been reported (Huber-Lang 2006 and this volume).

Deposition of complement components, C1q, C3, and C4, and generation of the terminal complement complex C5b-9 has been shown in human atherosclerotic lesions (Niculescu and Rus 1999). The extent of C5b-9 deposition has been correlated with severity of the vascular lesion (Vlaicuet al. 1985). Moreover, iC3b deposition appears to be highest in vulnerable and ruptured plaques (Laine et al. 2002). Moreover, elevations in circulating C5a have been associated with increased cardiovascular risk in patients with advanced atherosclerosis (Seidl et al. 2005).

However, the role of complement in atherogenesis in animal models remains controversial. Deficiency of C5 fails to protect Apo E–/– mice from atherosclerosis (Patel et al. 2001). Although these results suggest that C5 activation and by extension C5b-9 generation are not required for atherogenesis, they do not rule out important contributions by complement components upstream of C5, including C3a, C3b, iC3b and C1q.

A role for C1q in the pathogenesis of inflammatory tissue injury in a murine stroke model is supported by accumulation of C1q in brain lesions within 3–6 h post ischemia (Mack et al. 2006). Further studies demonstrate C1q mediated amplification of ischemic cerebral injury in immature mice. Adult C1q deficient (–/–) mice, however, are not protected from stroke. In contrast, mature C3 deficient (–/–) mice demonstrated a marked resistance to cerebral injury. These observations are consistent with neuroprotective effects observed in mice treated with a C3a receptor antagonist.

Complement activation and deposition on platelets may also contribute to ongoing thrombosis. C4 fragments have been detected on platelets from a subset of patients with systemic lupus erythematosus (SLE), and this correlated with a history of neurologic events and the presence of antiphospholipid antibodies

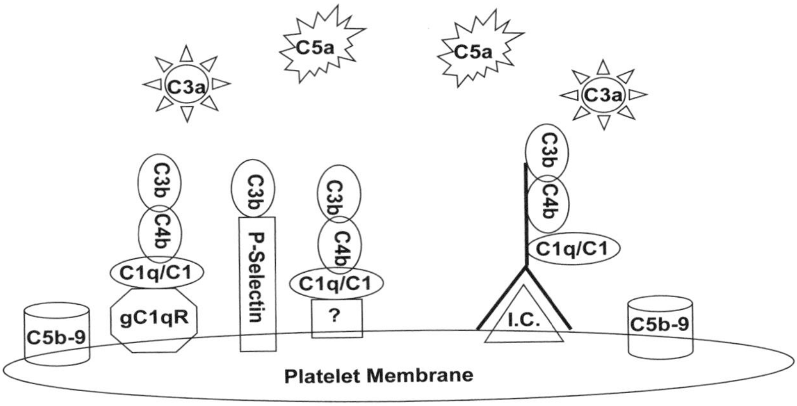

Fig. 1 Proposed mechanisms for complement activation on and by platelets and PMP. Complement activation may occur directly via expression of gC1qR, P-selectin, and as yet unidentified constituents expressed by or secreted from activated platelets. In autoimmune diseases, complement activation may occur via assembly of immune complexes (I.C.) on platelets. Complement activation leads to generation of inflammatory C3a and C5a peptides, as well as assembly of the C5b-9 membrane attack complex

(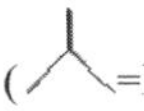 **=IgG)**

(Navratil et al. 2006).These observations are consistent with preliminary data from our laboratory which suggest a statistically significant correlation between classical complement pathway C1q and C4d deposition on platelets and arterial thrombosis in SLE patients with anti phospholipid antibody syndrome (APS) (Peerschke et al. 2007). Given the association of APS with circulating autoantibodies, the thrombogenic potential associated with APS may involve activation of the classical complement pathway on platelets and PMP.

4 Regulation of Complement Activity on Platelets

Platelets are armed with a variety of mechanisms to prevent in situ complement activation. Platelets contain C1 INH in their alpha granules and express C1 INH on their surface following granule secretion (Schmaier et al. 1993). C1 inhibitor is an alpha globulin that blocks the esterolytic activity of the first component of the classical complement cascade. The mechanism whereby C1 INH is expressed on the activated platelet surface is not yet known, but recent work with endothelial cells demonstrates C1 INH binding to P-selectin (Shenghe and Davie 2003). This interaction does not interfere with CI INH activity, but prevents leukocyte rolling on endothelial cells (Shenghe et al. 2003; Buerke et al. 1998).

Platelet alpha granules also contain releasable Factor H (Devine and Rosse 1987). Factor H binds cell associated C3b and acts as a cofactor for the cleavage of C3b by Factor I. Factor H can bind directly to platelets via GPIIb-IIIa and indirectly via thrombospondin-1 (Vaziri-Sani et al. 2004).

In addition to soluble inhibitors, platelets express a number of surface membrane anchored complement regulatory proteins. Under physiologic conditions, platelet damage or lysis by C5b-9 assembly is regulated by surface membrane CD55, CD59, and clusterin. Apparently clusterin is more highly expressed on platelets than on other mammalian cell types (Gnatenko et al. 2003). If stimulated platelets are indeed actively engaged in the initiation and propagation of both classical and alternative pathways, the arming of platelets with potent complement regulatory mechanisms would be necessary to prevent uncontrolled cytolysis and production of inflammatory mediators.

5 Summary and Conclusion

Potential mechanisms contributing to platelet mediated complement activation are summarized in Fig. 1. Evidence for both immune complex independent and dependent classical complement pathway activation on platelets and PMP has been reported, as well as amplification of complement activation via engagement of the alternative pathway. These observations provide strong support for the involvement of platelets in the generation of complement derived inflammatory mediators and complement induced platelet and vascular endothelial cell activation. We hypothesize that platelets focus complement activation to sites of vascular injury, and that interactions between platelets and the complement system contribute to acute and chronic inflammation and thrombosis. Thus, understanding mechanisms of complement activation on platelets will have a significant impact on identifying novel therapeutic targets for treating patients with thrombotic complications associated with atherosclerosis, ischemia/reperfusion injury, and APS.

Acknowledgements

This work was supported in part by grants HL67211 (EIBP) and AI060866 (BG) from the National Institutes of Health, and an American Heart Association Heritage Affiliate postdoctoral award # 0625900T (WY).

References

Acostan, J., Qin, X., and Halperin, J. (2004) Complement and complement regulatory proteins as potential molecular targets for vascular diseases. *Curr. Pharm. Dis.* 10, 203–211

Allegretti, M., Moriconi A., Beccari, A.R., Di Bitondo, R., Bizzarri, C., Bertini, R., and Colotta, F. (2005) Targeting C5a: recent advances in drug discover. *Curr. Med. Chem.* 12, 217–236

Arumugam, T.V., Shields, I.A., Woodruff, T.M., Granger, D.N., and Taylor, S.M. (2004) The role of the complement system in ischemia-reperfusion injury. *Shock* 21, 401–409

Barry, O.P., Pratico, D., Savani, R.C., and FitzGerald, G.A. (1998) Modulation of monocyte-endothelial cell interactions by platelet microparticles. *J. Clin. Invest.*102, 136–144

Buerke, M., Prufer, D., Dahm, M., Oelert, H., Meyer, J., and Darius, H. (1998) Blocking of classical complement pathway inhibits endothelial adhesion molecules expression and preserves ischemic myocardium from reperfusion injury. *J. Pharmacol. Exp. Ther.* 286, 429–438

Davis III, A.D. and Kenney, D.M. (1979) Properdin factor D; effects on thrombin-induced platelet aggregation. *J. Clin. Invest.* 64, 721–728

Dedio, J., Jahnen-Dechent, W., Bachmann, M., and Muller-Esterl, W. (1998) The multiligand binding protein gC1qR, putative C1q receptor, is a mitochondrial protein. *J. Immunol.* 160, 3534–3542

Del Conde, I., Cruz, M.A., Zhang, H., Lopez, J.A., and Afshar-Kharghan, V. (2005) Platelet activation leads to activation and propagation of the complement system. *J. Exp. Med.* 201, 871–879

Devine, D.V. and Rosse, W.F. (1987) Regulation of the activity of platelet-bound C3 convertase of the alternative pathway of complement by platelet factor H. *Proc. Natl. Acad. Sci. USA* 84, 5873–5877

Gawaz, M., Ott, I., Reininger, A.J., Heinzmann, U., and Neumann, F.J. (1996) Agglutination of isolated platelet membranes. *Arterioscler. Thromb. Vasc. Biol.* 16, 621–627

George, J.N., Pickett, E.B., Saucerman, S., McEver, R.P., Kunicki, T.J., Kieffer, N., and Newman, P.J. (1986) Platelet surface glycoproteins. Studies on resting and activated platelets and platelet membrane microparticles in normal subjects, and observations in patients during adult respiratory distress syndrome and cardiac surgery. *J. Clin. Invest.* 78, 340–348

Ghebrehiwet, B. and Peerschke, E.I.B. (1998) Structure and function of gC1qR a multiligand binding membrane protein. *Immunobiology* 199, 225–238

Ghebrehiwet, B., Lim, B.L., Peerschke, E.I.B., Willis, A.C., and Reid, K.B.M. (1994) Isolation of cDNA cloning, and overexpression of a 33-kDa cell surface glycoprotein that binds to the globular heads of C1q. *J. Exp. Med.* 179, 1809–1821

Ghebrehiwet, B., Lim, B.L., Kumar, R., Feng, X., and Peerschke, E.I.B. (2001) gC1qR/p33 a member of a new class of multifunctional and multicompartmental cellular proteins, is involved in inflammation and infection. *Immunol. Rev.* 180, 65–77

Ghebrehiwet, B., Cebada Mora, C., Tantral, L., Jesty, J., and Peerschke, E.I.B. (2006) gC1qR/p33 serves as a molecular bridge between the complement and contact activation systems and is an important catalyst in inflammation. *Adv. Exp. Med. Biol.* 586, 95–105

Giannakopoulos, B., Passam, F., Rahgozar, S., and Krillis, S.A. (2007) Current concepts on the pathogenesis of the antiphospholipid syndrome. *Blood* 109, 422–430

Gnatenko, D.V., Dunn, J.J., McCorkle, S.R., Weissmann, D., Perrotta, P.L., and Bahou, W.E. (2003) Transcript profiling of human platelets using microarray and serial analysis of gene expression. *Blood* 101, 2285–2293

Huber-Lang, M., Sarma, J.V., Zetoune, F.S., Rittirsch, D., Neff, T.A., McGuire, S.R., Lambris, J.D., Warner, R.L., Flierl, M.A., Hoesel, L.M., Gebhard, F., Younger, J.G., Drouin, S.M., Wetsel, R.A., and Ward, P.A. (2006) Generation of C5a in the absence of C3: a new complement activation pathway. *Nat. Med.* 12, 682–687

Ikeda, K., Nagasawa, K., Horiuchi, T., Nishizaka, H., and Niho, Y. (1997) C5a induced tissue factor activity on endothelial cells. *Thromb. Haemost.* 77, 394–398

Jiang, J., Zhang, Y., Krainer, A., and Xu, R.M. (1999) Crystal structure of p32, a doughnut-shaped acidic mitochondrial matrix protein. *Proc. Natl. Acad. Sci. U S A* 96, 3572–3577

Kansas, G.S. (1996) Selectins and their ligands: current concepts and controversies. *Blood* 88, 3259–3287

Kaplan, A.P., Silverberg, M., and Ghebrehiwet, B. (1986) The intrinsic coagulation/kinin pathway – the classical complement pathway and their interactions. *Adv. Exp. Med. Biol.* 198, 1311–1325

Kovacsovics, T., Tschopp, J., Kress, A., and Isliker, H. (1985) Antibody independent activation of C1, the first component of complement by cardiolipin. *J. Immunol.* 135, 2695–2700

Laine, P., Pentikainen, M.O., Wurzner, R., Penttila, A., Paavonen, T., Meri, S., and Kovanen, P.T. (2002) Evidence for complement activation in ruptured coronary plaques in acute myocardial infarction. *Am. J. Cardiol.* 90, 404–408

Mack, W.J., Sughrue, M.W., Ducruet, A.F., Mocco, J., Sosunov, S.A., Hassid, B.G., Silverberg, J.Z., Ten, V.S., Pinsky, D.J., and Connolly Jr., E.S. (2006). Temporal pattern of C1q deposition after transient focal cerebral ischemia. *Wiley InterScience* (www.interscience.wiley.com) doi:10, 1002/jnr.20775

Mahdi, F., Mader, Z.S., Figueroa, C.D., and Schmaier, A.H. (2002) Factor XII interacts with multiprotein assembly of urokinase plasminogen activator receptor, gC1qR and cytokeratin 1 on endothelial cell membranes. *Blood* 15, 3585–3596

Makrides, S.C. (1998) Therapeutic inhibition of the complement system. *Pharmacol. Rev.* 150, 59–87

Markiewski, M.M. and Lambris, J.D. (2007) The role of complement in inflammatory diseases from behind the scenes into the spotlight. *Am. J. Pathol.* 171, 715–727

Markiewski, M.M., Nilsson, B., Ekdahl, K.N., Mollnes, T.E., and Lambris, J.D. (2007) Complement and coagulation: strangers or partners in crime? *Trends Immunol.* 28, 184–192

Martinez, M.C., Tesse, A., Zobairi, F., and Andriantsitohaina, R. (2005) Shed membrane microparticles from circulating and vascular cells in regulating vascular function. *Am. J. Physiol. Heart Circ. Physiol.* 288, H1004–H1009

Mause, S.F., von Hundelshausen, P., Zernecke, A., Koenen, R.R., and Weber, C. (2005) Platelet microparticles: a transcellular delivery system for RANTES promoting monocyte recruitment on endothelium. *Arteriscler. Thromb. Vasc. Biol.* 25, 1512–1518

Mevorach, D., Mascarenhas, J.O., Gershoev, D., and Eldon, K.B. (1998) Complement dependent clearance of apoptotic cells by human macrophages. *J. Exp. Med.* 188, 2313–2320

Monsinjon, T., Gasque, P., Chan, P., Ischenko, A., Brady, J.J., and Fontaine, M.C. (2003) Regulation by complement C3a and C5a anaphylatoxins of cytokine production in human umbilical vein endothelial cells. *FASEB J.* 17, 1003–1014

Morel, O., Toti, F., Hugel, B., and Freyssinet, J.M. (2004) Cellular microparticles: a disseminated storage pool of bioactive vascular effectors. *Curr. Opin. Hematol.* 11, 156–164

Muhlfelder, T.W., Miemetz, J., Kreutzer, D., Beebe, D., Ward, P.A., and Rosenfeld, S.I. (1979) C5 chemotactic fragment induces leukocyte production of tissue factor activity: a link between complement and coagulation. *J. Clin. Invest.* 63, 147–150

Navratil, J.S., Manzi, S., Kao, A.H., Krishnaswami, S., Liu, C.C., Ruffing, M.J., Shaw, P.S., Nilson, A.C., Dryden, E.R., Johnson, J.J., and Ahearn, J.M. (2006). Platelet C4d is highly specific for systemic lupus erythematosus. *Arthritis and Rheum.* 54, 670–674

Niculescu, F., Niculescu, T., and Rus, H. (2004) C5b-9 terminal complement complex assembly on apoptotic cells in human arterial wall with atherosclerosis. *Mol. Immunol.* 36, 949–955

Niculescu, F., and Rus, H. (1999) Complement activation and atherosclerosis. *Mol. Immunol.* 36, 949–955

Patel, S., Thelander, E.M., Hernandez, M., Montenegro, J., Hassing, H., Burton, C., Mundt, S., Hermanowski-Vosatka, A., Wright, S.D., Chao, Y.S., and Detmers, P.A. (2001) ApoE−/− mice develop atherosclerosis in the absence of complement component C5. Biochem. Biophys. *Res. Commun.* 286, 164–170

Peerschke, E.I., and Ghebrehiwet, B. (1997) C1q augments platelet activation in response to aggregated Ig. *J. Immunol.* 159, 5594–5598

Peerschke, E.I.B., Reid, K.B.M., and Ghebrehiwet, B. (1993) Platelet activation by C1q results in the induction of alpha IIb/beta 3 integrins (GPIIb/IIIa) and the expression of P-selectin and procoagulant activity. *J. Exp. Med.* 178, 579–587

Peerschke, E.I.B., Reid, K.B.M., and Ghebrehiwet, B. (1994) Identification of a novel 33-kDa C1q-binding site on human blood platelets. *J. Immunol.* 152, 5896–5901

Peerschke, E.I.B., Murphy, T.K., and Ghebrehiwet, B. (2003) Activation-dependent surface expression of gC1qR/p33 on human blood platelets. *Thromb. Haemost.* 90, 331–339

Peerschke, E.I.B., Petrovan, R.J., Ghebrehiwet, B., and Ruf, W. (2004) Tissue factor pathway inhibitor-2 (TFPI-2) recognizes the complement and kininogen binding protein gC1qR/p33 (gC1qR): implications for vascular inflammation. *Thromb. Haemost.* 92, 811–819

Peerschke, E.I.B., Yin, W., Grigg, S.E., and Ghebrehiwet, B. (2006) Blood platelets activate the classical pathway of human complement. *J Thromb Haemost* 4, 2035–2042

Peerschke, E.I.B., Yin, W., Alpert, D.R., Salmon, J.E., Roubey, R.A.S., and Ghebrehiwet, B. (2007). Enhanced serum complement activation on platelets is associated with arterial thrombosis in patients with systemic lupus erythematosus (SLE) and antiphospholipid antibodies (aPL) (abstract). *Blood* 110, 488a

Perez-Pujol, S., Marker, P.H., and Key, N.S. (2007) Platelet microparticles are heterogeneous and highly dependent on the activation mechanism: studies using a new digital flow cytometer. *Cytometry* 71A, 38–45

Polley, M.J., and Nachman, R.L. (1978) The human complement system in thrombin-mediated platelet function. *J. Exp. Med.* 147, 1713–1726

Polley, M.J., and Nachman, R.L. (1979) Human complement in thrombin-mediated platelet function: uptake of the C5b-9 complex. *J. Exp. Med.* 150, 633–645

Polley, M.J., and Nachman, R.L. (1983) Human platelet activation by C3a and C3a des-arg. *J. Exp. Med.* 158, 603–615

Schaiff, W.T., and Eisenberg, P.R. (1997) Direct induction of complement activation by pharmacologic activation of plasminogen. *Coron. Artery Dis.* 8, 9–18

Schmaier, A.H., Amenta, S., Xiong, T., Heda, G.D., and Gewirtz, A.M. (1993) Expression of C1 Inhibitor. *Blood* 82, 465–474

Seidl, W.S., Exner, M., Amighi, J., Kastl, S.P., Zorn, G., Maurer, G., Wagner, O., Huber, K., Minar, E., Wojta, J., and Schillinger, M. (2005) Complement component C5a predicts future cardiovascular events in patients with advanced atherosclerosis. *Europ. Heart.* J. 26, 2294–2299

Sinauridze, E.I., Kireev, D.A., Popenko, N.Y., Pichugin, A.V., Panteleev, M.A., Krymskaya, O.V., and Ataullakhanov, F.I. (2007) Platelet microparticle membrane have 50 – 100 fold higher specific procoagulant activity than activated platelets. *Thromb Haemost*; 97, 425–434

Shenghe, C., and Davie III, A.E. (2003) Complement regulatory protein C1 inhibitor binds to selectins and interferes with endothelial-leukocyte adhesion. *J. Immunol.* 171, 4786–4791

Torzewsjki, J., Bowher, D.E., Wlatenberger, J., and Fitzsimmons, C. (2007) Processes in atherogenesis: complement activation. *Atherosclerosis* 132,131–138

Vaziri-Sani, F., Hellwage, J., Zipfel, P.F., Sjoholm, A.G., Iancu, R., and Karpman, D. (2004) Factor H binds to washed human platelets. *J. Thromb. Haemost.* 3, 154–162

Vermes, I., Haanen, C., Steffen-Nakken, H., and Reutelingsperger, C. (1995) A novel assay of apoptosis : flow cytometric detection of phosphatidylserine expression on early apoptotic cells using fluorescein-labeled Annexin V. *J. Immunol. Methods* 184, 39–51

Vlaicu, R., Niculescu, F., Rus, H.G., and Cristea A.(1985) Immunohistochemical localization of the terminal C5b-9 complement complex in human aortic fibrous plaque. *Atherosclerosis* 57, 163–177

Wiedmer, T., Esmon, C.T., and Sims, P.J. (1986) Complement proteins C5b-9 stimulate procoagulant activity through platelet prothrombinase. *Blood* 68, 875–880

Yin, W., Ghebrehiwet, B., and Peerschke, E.I.B. (2008) Expression of complement components and inhibitors on platelet microparticles. *Platelets* 19, 225–233

Zimmerman, T.S., and Kolb, N.P. (1976) Human platelet-initiated formation and uptake of the C5-9 complex of human complement. *J. Clin. Invest.* 57, 203–211

8. Adrenergic Regulation of Complement-Induced Acute Lung Injury

Michael A. Flierl[1], Daniel Rittirsch[1], J. Vidya Sarma[1], Markus Huber-Lang[2], and Peter A. Ward[1,*]

[1]Department of Pathology, University of Michigan Medical School, Ann Arbor, MI, USA, pward@umich.edu
[2]Department of Traumatology, Hand-, Plastic-, and Reconstructive Surgery University Hospital of Ulm, Ulm, Germany

Abstract. It is well established that catecholamines regulate immune and inflammatory responses. Until recently, they have been thought to derive from the adrenal medulla and from presynaptic neurons, when studies revealed that T cells, macrophages and neutrophils can also de novo synthesize and release endogenous catecholamines, which can then regulate immune cell functions in an autocrine/paracrine manner via engagement of adrenergic receptors. Accordingly, it appears that phagocytic cells and lymphocytes may represent a major, newly recognized source of catecholamines that regulate inflammatory responses.

1 Introduction

The brain and the immune systems are two major adaptive systems of the body (Sternberg 2006). During an immune response the brain and the immune system usually inter-communicate. The major pathway systems involved in this cross-talk are the hypothalamic-pituitary-adrenal (HPA) axis and the autonomic nervous system, including the sympathetic nervous system (SNS), the vagus-mediated parasympathetic nervous system and the enteric nervous system. Lately, especially the vagus-mediated parasympathetic nervous system has been shown to be an important player in the regulation of inflammation via cholinergic receptors on immune cells (Borovikova et al. 2000; Tracey 2002; Wang et al. 2003). Activation of these nicotinic acetylcholine receptors has been found to dampen inflammation and has therefore been termed the "cholinergic antiinflammatory pathway" (Gallowitsch-Puerta and Tracey 2005) or an "inflammatory reflex" (Tracey 2002). In contrast, the adrenergic nervous system's functional interplay with the immune system, aiming to counterbalance and antagonize the parasympathetic nervous system's actions, is less well understood. Being potent agonists of the adrenergic nervous system, catecholamines are released from the presynaptic sympathetic nerve terminals or the adrenal medulla, targeting postsynaptic nerve terminals and local immune cells that express adrenoreceptors. Locally released or circulating catecholamines stimulate these receptors, affecting lymphocyte trafficking (Kradin

J.D. Lambris (ed.), *Current Topics in Complement II.*
doi: 10:1007/978-0-387-78952-1_8, © Springer Science+Business Media, LLC 2008

et al. 2001), vascular perfusion, cell proliferation (Ackerman et al. 1991), cytokine production and, therefore, the wide spectrum of functional activity of immune cells. Systemically, catecholamines may cause selective suppression of Th_1 responses together with a shift toward a Th_2 dominance of humoral immunity (Elenkov et al. 2000). Such changes may reverse the detrimental effects of proinflammatory cytokines and other products of activated alveolar macrophages. In addition, classical noradrenergic actions on blood vessels include vasoconstriction, reducing the intensity of the inflammatory insult due to reduced vascular perfusion. On the other hand, catecholamines may boost local immune responses through α_2-adrenergic receptor-mediated enhancement of intracellular signaling pathways that lead to increased production of interleukin (IL)-1β and tumor necrosis factor (TNF-)α (Spengler et al. 1990, 1994). Therefore, activation of the adrenergic system during an immune response might be directed at enhancing the local inflammatory response, resulting in neutrophil accumulation (Morken et al. 2002) and stimulation of humoral immune responses, while attempting to prevent systemic infection.

2 Phagocytes: A New Adrenergic Organ

2.1 Evidence for De novo-Synthesis, Release and Inactivation of Catecholamines by Phagocytes

The synthesis of catecholamines relies on two key enzymes: The tyrosine-hydroxylase (TH) is known to be the rate-limiting step in the catecholamine synthesis, while the dopamine-β-hydroxylase (DBH) converts dopamine to norepinephrine (Elenkov et al. 2000) (Fig. 1). Thus, presence and expressional changes of these hydroxylase in phagocytic cells strongly suggest the ability to de novo synthesize catecholamines. Recently, phagocytes were found to contain mRNA for both, TH and DBH, which were clearly inducible by lipopolysaccharide (LPS) (Brown et al. 2003; Marino et al. 1999; Cosentino et al. 1999; Flierl et al. 2007). Inhibition of TH in vitro (with α-methyl-p-tyrosine or pargyline) affected intracellular dopamine, norepinephrine and their metabolites, suggesting catecholamine synthesis and degradation by these cells (Cosentino et al. 1999). Thus, it is now becoming clear that phagocytes possess the ability to de novo-produce catecholamines.

Catecholamines are found throughout adrenergic neurons but the highest concentrations of these biogenic amines are found in the peripheral nerve terminals where they are stored in membrane-bound granules to be temporarily inaccessible and protected from enzymatic destruction (Molinoff and Axelrod 1971; Shore 1972). Upon depolarizing stimulation of these neurons, rapid release of stored catecholamines is facilitated. One of the main modes to inactivate catecholamines is the cellular reuptake (see below). In neurons, this process is known to be inhibited by reserpine. In parallel, incubation of various immune cells with reserpine markedly reduced intracellular accumulation of catecholamines, while

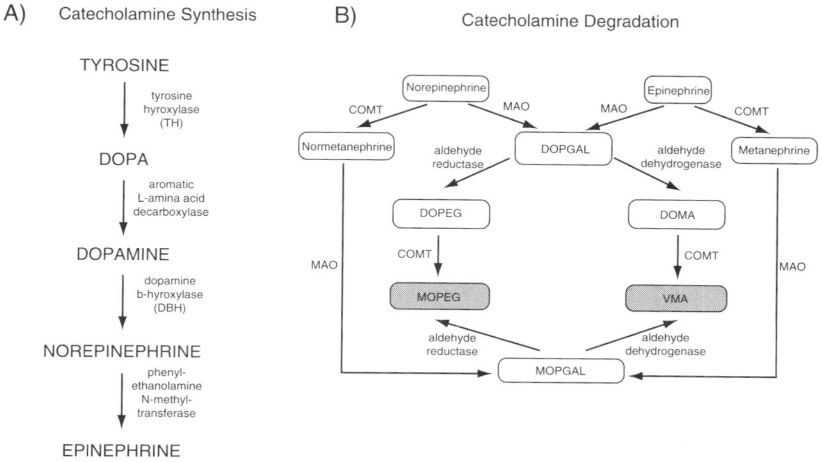

Fig. 1 Pathways of synthesis and inactivation of catecholamines. *DOPGAL* 3,4-dihydroxyphenyl-glycoaldehyde; *DOMA* 3,4-dihydroxy-mandelic acid; *DOPEG* 3,4-dihydroxyphenyl-ethylene glycol; *MOPEG* 3-methoxy-4-hydroxy-phenylethylene glycol; *VMA* 3-methoxy-4-hydroxy-mandelic acid; *COMT* catechol-*O*-methyltransferase; *MAO* monoamine oxidase

catecholamine levels in culture supernatants significantly increased (Cosentino 2000), suggesting that immune cells employ a mechanism similar to neurons. The functional significance of lymphocyte-generated catecholamines in inflammation is poorly understood. However, we are just beginning to understand the molecular mechanisms involved in the catecholamine release by immune cells. Achieving a more complete understanding of the precise involvement of various ion channels, neurotransmitters and other mediators triggering catecholamine-release need to be addressed in future studies. The actions of epinephrine and norepinephrine are terminated by (1) reuptake into nerve terminals, (2) dilution by diffusion into extracellular fluids and uptake at extraneuronal sites, and (3) metabolic transformation (Axelsson 1971). Two enzymes are essential in the initial steps of metabolic inactivation: monoamine oxidase (MAO) and catechol-O-methyl transferase (COMT) (Zhu 2002; Balsa et al. 1989) (Fig. 1). Lately, both, mRNA and protein of MAO and COMT were found in macrophages and neutrophils and were shown to be inducible by LPS (Flierl et al. 2007). Thus, phagocytes not only possess the full cellular machinery for de novo synthesis and release of catecholamines, but are also capable of intracellular inactivation of catecholamines by MAO and COMT, utilizing the same classical metabolic pathway described in nervous and endocrine systems.

2.2 Modulation of Phagocyte Functions by Catecholamines

Endogenous Catecholamines as Modulators of Phagocytic Cells
There is a large body of evidence indicating that, in addition to being crucial neurotransmitters and hormones, catecholamines are important immunomodulators in health and disease (Elenkov et al. 2000; Madden et al. 1995; Sanders and Kohm 2002; Sanders and Straub 2002; Ottaway and Husband 1994). Interestingly, the first report of the immunomodulating functions of catecholamines were published in 1904, describing a robust leukocytosis following subcutaneous administration of epinephrine (Loeper and Crouzon 1904). However, it was only in the mid 1990s when it was reported that immunocyte-derived catecholamines modulate the functions of lymphocytes, suggesting that immune cells can fine-tune their functions (Bergquist et al. 1994). Since the expression of TH and DBH mRNA has been found to be inducible in PMNs and macrophages, it seems likely that the catecholamine synthesis of phagocytes affects their functional state (Flierl et al. 2007). These auto-regulatory interactions between endogenous catecholamines and phagocytes exquisitely regulate cell functions via adrenoceptors expressed on phagocytes (Spengler et al. 1990, 1994; Flierl et al. 2007). Adrenoceptors are G-protein coupled, seven-transmembrane spanning receptors Pierce et al. 2002). Upon activation, they lead to activation and/or inhibition of intracellular second messengers such as cyclic AMP, calcium ions, diacylglycerol, and inositol 1,4, 5-triphosphate. Therefore, secreted endogenous catecholamines are able to activate the cell sources and their surrounding environment through autocrine/paracrine mechanisms stimulating cellular adrenergic receptors, activating intracellular second messengers and ultimately regulating cell functions (Flierl et al. 2007; Spengler et al. 1994; Engler et al. 2005). Phagocytes have been demonstrated to regulate their release of TNFα via these interactions of endogenous catecholamines with α_2-adrenergic receptors expressed on their cell surfaces (Spengler et al. 1990, 1994; Flierl et al. 2007). Blockade of α- and β-adrenoceptors on macrophages and neutrophils variably inhibited the cytokine/chemokine production of these cells (Spengler et al. 1990, 1994; Flierl et al. 2007; Starkie et al. 2001). These findings make it tempting to speculate that the inflammatory cytokine/chemokine network might be one of the important mediator systems tightly controlled by catecholamines via adrenergic receptors. It remains to be determined if catecholamine secretion by phagocytic cells is a ubiquitous phenomenon or if it is rather an ultimate weapon of choice for immune cells facing overpowering pathogenic insults.

3 Phagocyte-Derived Catecholamines Regulate Complement-Dependent Acute Lung Injury

Immune complex-induced lung injury is a useful model of acute lung injury, in which immunologic alveolitis is induced in rodents by intrapulmonary deposition of IgG immune complexes. The complement system as a part of innate immunity

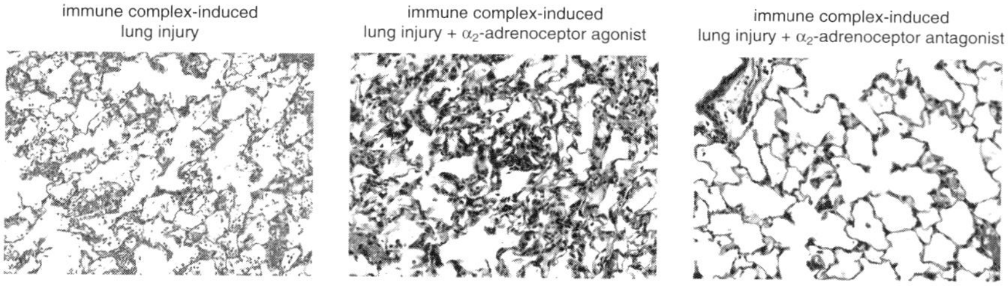

Fig. 2 Adrenergic regulation of complement-dependent, immune complex-induced lung injury. UK 14304 and RX 821002 were used as α_2-adrenergic agonist and antagonist, respectively

plays a central role in this model of acute lung injury. Immune complexes are potent activators of the classical pathway of the complement system. Activation of complement with its powerful downstream product, C5a, is required for the full development of injury in the IgG-IC model (Mulligan et al. 1996; Ward 1996). Most complement proteins can be produced in the lung by either type II pneumocytes, alveolar macrophages or lung fibroblasts (Hetland et al. 1986; Strunk et al. 1988; Rothman et al. 1989). Following IgG-IC deposition, complement activation, via C5a, enhances production of early response cytokines (TNFα and IL-1β). Therefore, C5a and the membrane attack complex appear to act synergistically with a co-stimulus (IgG-IC) to intensify the inflammatory response and tissue injury (Czermak et al. 1999a). Blockade of C5a by anti-C5a IgG greatly diminished tissue injury represented by a reduced capillary leak and neutrophil influx. These protective effects of anti-C5a were associated with reductions of various proinflammatory mediators together with decreased endothelial ICAM-1 upregulation, indicating a key role for C5a in the initiation of the inflammatory network (Mulligan et al. 1996; Czermak et al. 1999b). In a recent study, it has been demonstrated that this C5a-dependent model of acute lung injury is clearly modulated by α_2-adrenergic mechanisms (Flierl et al. 2007). While α_2-adrenergic agonists (RX 821002 or yohimbine) virtually abolished lung injury, co-presence of an α_2-adrenergic antagonist (UK 14304) greatly increased the intensity of inflammatory injury (Flierl et al. 2007) (Fig. 2). In addition, pharmacologic inhibition of both catecholamine-generating enzymes (TH or DBH) greatly attenuated levels of injury whereas blockade of catecholamine-degrading enzymes (COMT or MAO) resulted in exacerbation of lung injury (Flierl et al. 2007). Since C5a has been shown to be a key player in this experimental model of acute lung injury, we evaluated whether C5a itself would be able to induce norepinephrine production by human neutrophils (PMNs) in vitro. As shown in Fig. 3, there was a clear increase in norepinephrine release into supernatant fluids from human blood PMNs 15 min after exposure to 10 nM recombinant human C5a, followed by a fall and a second peak 4 h after exposure. It is tempting to speculate that the early norepinephrine peak at 15 min is due to the fact that cells store catecholamines in

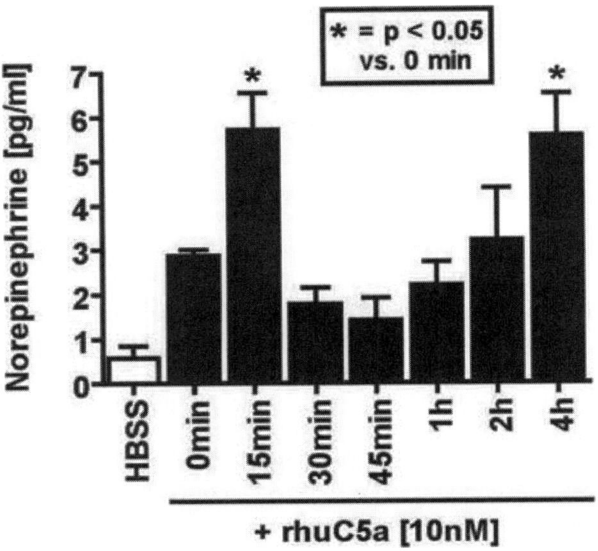

Fig. 3 Norepinephrine levels in supernatants of isolated human blood neutrophils following exposure to recombinant human C5a as a function of time. $n \geq 5$ per bar. Data are expressed as mean values ± s.e.m

intracellular secretory granules in a manner perhaps similar to neurons, chromaffin cells and lymphocytes and that this material is rapidly released upon C5a-exposure. The late peak of catecholamines at 4 h may be due to induction of mRNA for the catecholamine-generating enzymes, leading to an increased de novo production and release of catecholamines, as has been shown in rodents (Flierl et al. 2007).

4 Outlook

We are just beginning to understand the impact of phagocyte-derived catecholamines and their potential to fine-tune inflammation. A recent report has described how blockade of diverse adrenoceptors on phagocytes variably inhibits expression of different cytokines and chemokines released from LPS-stimulated macrophages (Flierl et al. 2007). TNFα production was completely inhibited by blockade of the α_2-adrenoceptor or high-dose blockade of the β_2-adrenoceptor, while remaining completely unaffected by α_1- or β_1-adrenoceptor blockade. In sharp contrast, IL-1β levels were greatly suppressed by pharmacological blockade of either α_1-, α_2-, β_1- or β_2-adrenoreceptors. IL-6 and CINC-1 production by

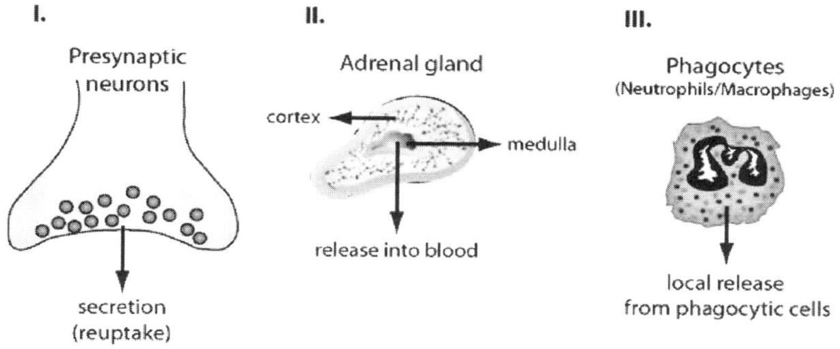

Fig. 4 Updated sources of endogenous catecholamines

phagocytes was regulated by either α_2-, β_1- or β_2-adrenoreceptors, respectively. Since norepinephrine has higher affinity for the α-adrenergic receptors while epinephrine exhibits superior affinity for β-adrenoceptors, it might well be that these two catecholamines display dissimilar effects on phagocytes which may define a distinct inflammatory mediator path that is most suitable for a particular inflammatory setting. The fact that phagocytic production of norepinephrine is inducible by C5a might in fact even suggest a co-involvement or regulation of endogenous catecholamines in various classic complement-dependent diseases such as rheumatoid arthritis, allergies and asthma or even sepsis. A large body of studies emphasizes the involvement of the sympathetic nervous system and catecholamines in rheumatoid arthritis (RA) (Imrich 2002; Wahle et al. 2002; Wilder 2002; Bijlsma et al. 1999). Accordingly, it was speculated that patients with RA presented with a dysfunctional autonomous nervous system (Baerwald et al. 2000). Lymphocytes from patients with chronic rheumatoid arthritis presented with decreased numbers of β-adrenergic receptors on lymphocytes, with the number of β-adrenoceptors inversely correlating with clinical activity of the disease (Baerwald et al. 1997; Straub et al. 2002). Accordingly, administration of the non-selective β-adrenergic blocker, propranolol, reduced the symptoms of rheumatoid arthritis (Lorton et al. 2003). Experimental animal studies confirmed the benefit of blocking the adrenergic system in RA. Systemic depletion of NA reduced inflammatory activity and joint destruction in an animal model of arthritis (Lorton et al. 2003). It has also been noted that patients with RA seem to lose sympathetic nerve fibers in their synovial tissue, which leads to compensatory increase of norepinephrine release from synovial macrophages (Miller et al. 2000). This loss of the influence of the sympathetic nervous system on inflammation, accompanied by an upregulation of the sensory input into the joint might contribute to the continuance of the disease. Yet, it remains to be seen if the complement anaphylatoxin C5a induces release of catecholamines from phagocytic cells during disease and whether disruption of this communication might result in a favorable clinical outcome.

5 Conclusion

It appears that, besides chromaffin cells of the adrenal medulla and neurons, phagocytes represent a third, diffusely expressed catecholamine-producing organ, which is able to de novo-synthesize and release endogenous catecholamines (Fig. 4). These new findings challenge traditional paradigms regarding the catecholamine system. The ample distribution of immune/inflammatory cells throughout the body, combined with their ability to migrate to the site of inflammation within minutes of tissue insult, adds unimagined immunomodulatory possibilities. Targeting immune cell-derived catecholamines or reducing catecholamine production by phagocytic cells, or blocking appropriate adrenoceptors poses challenging issues for researchers. Extending our understanding for extra- and intracellular adrenergic regulation of phagocytic cells might deepen our understanding of the pathogenesis of diverse diseases and might ultimately allow for clinical application.

Acknowledgement

This work was supported by NIH grants GM 29507, GM 61656 and HL-31963 (to P.A.W.) and Deutsche Forschungsgemeinschaft grants DFG HU 823/2-2 and HU 823/2-3 (to M.H.-L).

References

Ackerman, K. D., Madden, K. S., Livnat, S., Felten, S. Y., and Felten, D. L. (1991) Neonatal sympathetic denervation alters the development of in vitro spleen cell proliferation and differentiation. *Brain Behav Immun* 5, 235–261

Axelsson, J. (1971) Catecholamine functions. *Annu Rev Physiol* 33, 1–30

Baerwald, C. G., Laufenberg, M., Specht, T., von Wichert, P., Burmester, G. R., and Krause, A. (1997) Impaired sympathetic influence on the immune response in patients with rheumatoid arthritis due to lymphocyte subset-specific modulation of beta 2-adrenergic receptors. *Br J Rheumatol* 36, 1262–1269

Baerwald, C. G., Burmester, G. R., and Krause, A. (2000) Interactions of autonomic nervous, neuroendocrine, and immune systems in rheumatoid arthritis. *Rheum Dis Clin North Am* 26, 841–857

Balsa, M. D., Gomez, N., and Unzeta, M. (1989) Characterization of monoamine oxidase activity present in human granulocytes and lymphocytes. *Biochim Biophys Acta* 992, 140–144

Bergquist, J., Tarkowski, A., Ekman, R., and Ewing, A. (1994) Discovery of endogenous catecholamines in lymphocytes and evidence for catecholamine regulation of lymphocyte function via an autocrine loop. *Proc Natl Acad Sci U S A* 91, 12912–12916

Bijlsma, J. W., Cutolo, M., Masi, A. T., and Chikanza, I. C. (1999) The neuroendocrine immune basis of rheumatic diseases. *Immunol Today* 20, 298–301

Borovikova, L. V., Ivanova, S., Zhang, M., Yang, H., Botchkina, G. I., Watkins, L. R., Wang, H., Abumrad, N., Eaton, J. W., and Tracey, K. J. (2000) Vagus nerve stimulation attenuates the systemic inflammatory response to endotoxin. *Nature* 405, 458–462

Brown, S. W., Meyers, R. T., Brennan, K. M., Rumble, J. M., Narasimhachari, N., Perozzi, E. F., Ryan, J. J., Stewart, J. K., and Fischer-Stenger, K. (2003) Catecholamines in a macrophage cell line. *J Neuroimmunol* 135, 47–55

Cosentino, M., Marino, F., Bombelli, R., Ferrari, M., Lecchini, S., and Frigo, G. (1999) Endogenous catecholamine synthesis, metabolism, storage and uptake in human neutrophils. *Life Sci* 64, 975–981

Cosentino, M., Bombelli, R., Ferrari, M., Marino, F., Rasini, E., Maestroni, G. J., Conti, A., Boveri, M., Lecchini, S., and Frigo, G. (2000) HPLC-ED measurement of endogenous catecholamines in human immune cells and hematopoietic cell lines. *Life Sci* 68, 283–295

Czermak, B. J., Lentsch, A. B., Bless, N. M., Schmal, H., Friedl, H. P., and Ward, P. A. (1999a) Synergistic enhancement of chemokine generation and lung injury by C5a or the membrane attack complex of complement. *Am J Pathol* 154, 1513–1524

Czermak, B. J., Sarma, V., Bless, N. M., Schmal, H., Friedl, H. P., and Ward, P. A. (1999b) In vitro and in vivo dependency of chemokine generation on C5a and TNF-alpha. *J Immunol* 162, 2321–2325

Elenkov, I. J., Wilder, R. L., Chrousos, G. P., and Vizi, E. S. (2000) The sympathetic nerve – an integrative interface between two supersystems: the brain and the immune system. *Pharmacol Rev* 52, 595–638

Engler, K. L., Rudd, M. L., Ryan, J. J., Stewart, J. K., and Fischer-Stenger, K. (2005) Autocrine actions of macrophage-derived catecholamines on interleukin-1 beta. *J Neuroimmunol* 160, 87–91

Flierl, M. A., Rittirsch, D., Nadeau, B. A., Chen, A. J., Sarma, J. V., Zetoune, F. S., McGuire, S. R., List, R. P., Day, D. E., Hoesel, L. M., Gao, H., Van Rooijen, N., Huber-Lang, M. S., Neubig, R. R., and Ward, P. A. (2007) Phagocyte-derived catecholamines enhance acute inflammatory injury. *Nature* 449, 721–725

Gallowitsch-Puerta, M. and Tracey, K. J. (2005) Immunologic role of the cholinergic anti-inflammatory pathway and the nicotinic acetylcholine alpha 7 receptor. *Ann N Y Acad Sci* 1062, 209–219

Hetland, G., Johnson, E., and Aasebo, U. (1986) Human alveolar macrophages synthesize the functional alternative pathway of complement and active C5 and C9 in vitro. *Scand J Immunol* 24, 603–608

Imrich, R. (2002) The role of neuroendocrine system in the pathogenesis of rheumatic diseases (minireview). *Endocr Regul* 36, 95–106

Kradin, R., Rodberg, G., Zhao, L. H., and Leary, C. (2001) Epinephrine yields translocation of lymphocytes to the lung. *Exp Mol Pathol* 70, 1–6

Loeper, M. and Crouzon, O. (1904) L'action de l'adrenaline sur le sang. *Arch Med Exp Anat Pathol* 16, 83–108

Lorton, D., Lubahn, C., and Bellinger, D. L. (2003) Potential use of drugs that target neural-immune pathways in the treatment of rheumatoid arthritis and other autoimmune diseases. *Curr Drug Targets Inflamm Allergy* 2, 1–30

Madden, K. S., Sanders, V. M., and Felten, D. L. (1995) Catecholamine influences and sympathetic neural modulation of immune responsiveness. *Annu Rev Pharmacol Toxicol* 35, 417–448

Marino, F., Cosentino, M., Bombelli, R., Ferrari, M., Lecchini, S., and Frigo, G. (1999) Endogenous catecholamine synthesis, metabolism storage, and uptake in human peripheral blood mononuclear cells. *Exp Hematol* 27, 489–495

Miller, L. E., Justen, H. P., Scholmerich, J., and Straub, R. H. (2000) The loss of sympathetic nerve fibers in the synovial tissue of patients with rheumatoid arthritis is accompanied by increased norepinephrine release from synovial macrophages. *FASEB J* 14, 2097–2107

Molinoff, P. B. and Axelrod, J. (1971) Biochemistry of catecholamines. *Annu Rev Biochem* 40, 465–500

Morken, J. J., Warren, K. U., Xie, Y., Rodriguez, J. L., and Lyte, M. (2002) Epinephrine as a mediator of pulmonary neutrophil sequestration. *Shock* 18, 46–50

Mulligan, M. S., Schmid, E., Beck-Schimmer, B., Till, G. O., Friedl, H. P., Brauer, R. B., Hugli, T. E., Miyasaka, M., Warner, R. L., Johnson, K. J., and Ward, P. A. (1996) Requirement and role of C5a in acute lung inflammatory injury in rats. *J Clin Invest* 98, 503–512

Ottaway, C. A. and Husband, A. J. (1994) The influence of neuroendocrine pathways on lymphocyte migration. *Immunol Today* 15, 511–517

Pierce, K. L., Premont, R. T., and Lefkowitz, R. J. (2002) Seven-transmembrane receptors. *Nat Rev Mol Cell Biol* 3, 639–650

Rothman, B. L., Merrow, M., Despins, A., Kennedy, T., and Kreutzer, D. L. (1989) Effect of lipopolysaccharide on C3 and C5 production by human lung cells. *J Immunol* 143, 196–202

Sanders, V. M. and Kohm, A. P. (2002) Sympathetic nervous system interaction with the immune system. *Int Rev Neurobiol* 52, 17–41

Sanders, V. M. and Straub, R. H. (2002) Norepinephrine, the beta-adrenergic receptor, and immunity. *Brain Behav Immun* 16, 290–332

Shore, P. A. (1972) Transport and storage of biogenic amines. *Annu Rev Pharmacol* 12, 209–226

Spengler, R. N., Allen, R. M., Remick, D. G., Strieter, R. M., and Kunkel, S. L. (1990) Stimulation of alpha-adrenergic receptor augments the production of macrophage-derived tumor necrosis factor. *J Immunol* 145, 1430–1434

Spengler, R. N., Chensue, S. W., Giacherio, D. A., Blenk, N., and Kunkel, S. L. (1994) Endogenous norepinephrine regulates tumor necrosis factor-alpha production from macrophages in vitro. *J Immunol* 152, 3024–3031

Starkie, R. L., Rolland, J., and Febbraio, M. A. (2001) Effect of adrenergic blockade on lymphocyte cytokine production at rest and during exercise. *Am J Physiol Cell Physiol* 281, C1233–C1240

Sternberg, E. M. (2006) Neural regulation of innate immunity: a coordinated nonspecific host response to pathogens. *Nat Rev Immunol* 6, 318–328

Straub, R. H., Schaible, H. G., Wahle, M., Schedlowski, M., Neeck, G., and Buttgereit, F. (2002) [Neuroendocrine-immunologic mechanisms in rheumatic diseases – a congress report]. *Z Rheumatol* 61, 195–200

Strunk, R. C., Eidlen, D. M., and Mason, R. J. (1988) Pulmonary alveolar type II epithelial cells synthesize and secrete proteins of the classical and alternative complement pathways. *J Clin Invest* 81, 1419–1426

Tracey, K. J. (2002) The inflammatory reflex. *Nature* 420, 853–859

Wahle, M., Krause, A., Pierer, M., Hantzschel, H., and Baerwald, C. G. (2002) Immunopathogenesis of rheumatic diseases in the context of neuroendocrine interactions. *Ann N Y Acad Sci* 966, 355–364

Wang, H., Yu, M., Ochani, M., Amella, C. A., Tanovic, M., Susarla, S., Li, J. H., Wang, H., Yang, H., Ulloa, L., Al-Abed, Y., Czura, C. J., and Tracey, K. J. (2003) Nicotinic acetylcholine receptor alpha7 subunit is an essential regulator of inflammation. *Nature* 421, 384–388

Ward, P. A. (1996) Rous-Whipple Award Lecture. Role of complement in lung inflammatory injury. *Am J Pathol* 149, 1081–1086

Wilder, R. L. (2002) Neuroimmunoendocrinology of the rheumatic diseases: past, present, and future. *Ann N Y Acad Sci* 966, 13–19

Zhu, B. T. (2002) Catechol-*O*-Methyltransferase (COMT)-mediated methylation metabolism of endogenous bioactive catechols and modulation by endobiotics and xenobiotics importance in pathophysiology and pathogenesis. *Curr Drug Metab* 3, 321–349

9. Ficolins: Structure, Function and Associated Diseases

Xiao-Lian Zhang* and Mohammed A.M. Ali

*The State Key Laboratory of Virology, Department of Immunology, Hubei Province Key Laboratory of Allergy and Immune-related Diseases, Wuhan University School of Medicine, Wuhan, 430071, P.R. China, ZhangXL65@whu.edu.cn

Abstract. Innate immunity relies upon the ability of a few pattern recognition molecules to sense molecular markers. Ficolins are humoral molecules of the innate immune systems which recognize carbohydrate molecules on pathogens, apoptotic and necrotic cells. Three ficolins have been identified in humans: L-ficolin, H-ficolin and M-ficolin (also referred to as ficolin-2, -3 and -1, respectively). They are soluble oligomeric defence proteins with lectin-like activity and they are structurally similar to the human collectins, mannan-binding lectin (MBL) and surfactant protein A and D. Upon recognition of the infectious agent, the ficolins act through two distinct routes: initiate the lectin pathway of complement activation through attached serine proteases (MASPs), and a primitive opsonophagocytosis thus limiting the infection and concurrently orchestrating the subsequent adaptive clonal immune response. Recently a lot of reports showed that dysfunction or abnormal expressions of ficolins may play crucial roles in the pathogenesis of human diseases including: (1) infectious and inflammatory diseases, e.g., recurrent respiratory infections; (2) apoptosis, and autoimmune disease; (3) systemic lupus erythematosus; (4) IgA nephropathy; (5) clinical syndrome of preeclampsia; (6) other diseases associated factor e.g. C-reactive protein. Precise identification of ficolins functions will provide novel insight in the pathogenesis of these diseases and may provide novel innate immune therapeutic options to treat disease progression. This review discusses the structures, functions, and clinical implications of ficolins and summarizes the reports on the roles of ficolins in human diseases.

1 Introduction

The complement system functions as a cascade of binding interactions and enzymatic events, which help in the processing and elimination of microorganisms or non-self materials. The classical complement pathway was discovered 100 years ago as a system supplementing antibodies. Fifty years later it was proposed that complement could be activated by bacterial surfaces through an antibody-independent pathway, the alternative complement pathway. The recently described third pathway, lectin pathway, is activated by MBL and by ficolins (Holmskov et al. 2003). While the classical pathway functions as a component of both innate and adaptive parts of the immune system, and the lectin and alternative pathways of

J.D. Lambris (ed.), *Current Topics in Complement II.*
doi: 10:1007/978-0-387-78952-1_9, © Springer Science+Business Media, LLC 2008

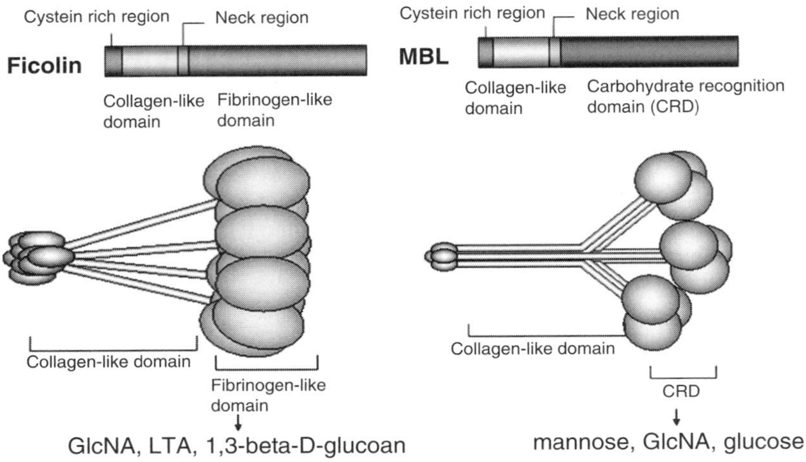

Fig. 1 Structures of ficolin and MBL

complement are essential components of the innate immune system. These three pathways merge at the proteolytic activation step of C3 and activate the complement late components, C5-C9, leading to cytolytic complex (membrane attack complex) formation.

Lectin pathway is initiated either by MBL or ficolins which are typical pattern recognition molecules and able to recognize conserved pathogen-associated molecules patterns (PAMPs) shared by large group of microorganisms, thereby successfully defending hosts against infection (Fujita, T. 2002). The binding of the lectin pathway recognition molecules (MBL or ficolins) to microbial carbohydrates can activate complement by a mechanism similar to the classical pathway, but using MBL-associated serine proteases (MASP-1, MASP-2, MASP3 as well as a truncated form of MASP-2 called small MBL-associated protein (sMAP) or Map 19) instead of C1r and C1s.

The importance of the MBL is underlined by a number of clinical studies linking MBL deficiency with increased susceptibility to a variety of infectious or inflammatory diseases (Eisen and Minchinton 2003; Takahashi et al. 2006) particularly, with greater risk in young children and immunocompromised hosts (Takahashi and Ezekowitz 2005). Deficiency and/or dysfunction of MBL have been also implicated in the pathophysiology of Systemic lupus erythematosus (SLE) (Roos et al. 2004).

In this review, we focus on the recently discovered important member of lectin family, the ficolins, and try to elucidate their contribution in different clinically important diseases.

2 Structures of Ficolins

The human ficolins are structurally similar to the C1q, mannan-binding lectin (MBL) and lung surfactant protein A and D (SP-A and SP-D), and are oligomeric proteins composed of carbohydrate-recognition domains (CRDs) attached to collagenous regions (Holmskov et al. 2003) (Fig. 1). The polypeptide chains of the ficolins comprised of different structural regions, an N-terminal region, a collagen-like region, and a globular fibrinogen-like domain (Matsushita et al. 1996). However, ficolins, unlike MBLs, they have a fibrinogen-like domain that is responsible for carbohydrate binding (Fig. 1). Similar to MBL, ficolins are capable of assembling as trimeric subunits at their collagen domains (Fig. 1), which multimerize to varying degrees forming active oligomers comprised of up to four and six trimers joined together by the N-terminal domain with the polypeptide chains radiating out in a sertiform structure (Lu and Le 1998; Yae et al. 1991). Mammalian ficolins associate with the MASPs through their collagen-like region, their target-binding activity are carried out by the fibrinogen-like domains which in contrast to the CRD of MBL, has a Ca^{2+}-independent lectin activity (Le et al. 1997).

The ficolins were originally identified as TGF-1-binding proteins on porcine uterus membranes (Ichijo et al. 1991, 1993). Several members of the ficolin family have been identified so far, including human L-ficolin/P35 (FCN2 or ficolin 2) (Matsushita et al. 1996), M-ficolin (FCN1 or ficolin 1) (Endo et al. 1996; Lu et al. 1996), and H-ficolin/Hakata antigen (FCN3 or ficolin 3) (Yae et al. 1991; Sugimoto et al. 1998; Aikawa et al. 1999), mouse ficolins A and B (Fujimori et al. 1998; Ohashi and Erickson 1998), pig ficolins α and β, and tachylectins 5A/5B (TL 5A/5B) from the horseshoe crab Tachypleus tridentatus (Gokudan et al. 1999). They all function as recognition molecules and trigger either the lectin pathway of complement or other effector mechanisms following binding to several ligands, such as GlcNAc (Krarup et al. 2004; Le et al. 1998), lipopolysaccharide (Tsujimura et al. 2001), lipoteichoic acid (Lynch et al. 2004) and several pathogenic microorganisms including bacteria (Krarup et al. 2005) and DNA (Jensen et al. 2007) and participate in the clearance of bacteria and dying host cells.

Human L-ficolin is an oligomeric protein assembled from 35-kDa subunits (Fig. 1). L-ficolin has a lectin-like activity for 1, 3-β-D glucan, a molecular marker of yeast and fungal cell walls (Ma et al. 2004), GlcNAc, and various acetylated compounds (Krarup et al. 2004). However, its binding specificity differs from that of MBL, in that L-ficolin binds to GlcNAc and does not bind to mannose. L-ficolin binds to GlcNAc residue next to galactose at the nonreducing terminal of the complex-type oligosaccharide (Table 1). Whereas, H-ficolin, which protein is the smallest among three human ficolins, binds to GlcNAc and GalNAc but not mannose or lactose. The M-ficolin on the other hand, is not considered being serum lectin, and its mRNA is expressed in monocytes, lung, and spleen (Table 1). Comparative analysis of the structures of the external S1 binding site in TL5A and

Table 1 Expressions of ficolins in human and mouse

Ficolins	Tissue expression	Protein indentified	Sugar specificity	Function	Gene localization
Human					
L-ficolin (FCN2, Hucolin)	liver	serum/ plasma	GlcNAc, LTA, LPS, 1,3-β-D glucan, Acetylated compounds	Opsonin Complement activation	9q34
H-ficolin (FCN3, Hakata antigen)	live lung	serum/ plasma	GlcNAc GalNAc fucose	Complement activation	1p35.3
M-ficolin (FCN1)	lung spleen leukocyte	secreted protein	GlcNAc GalNAc Salic acid	Complement activation	9q34
Mouse					
ficolin A	lung spleen	serum/ plasma	GlcNAc	Complement activation	2A3
ficolin B	bone marrow spleen	ND	ND	ND	2A3

ficolins L and H has been revealed some common features but also significant variations in the nature and orientation of the side chains (Garlatti et al. 2007) implying that these binding sites have different microorganism specificity.

Recently, it was found that M-ficolin is a secretory protein, and can recognize some common carbohydrate residues found in microbes, like its family members and functions as a recognition molecule of the lectin complement pathway and plays an important role in innate immunity (Liu et al. 2005b).

3 Ficolin Genetics

The human ficolin gene was found to locate on chromosome 9 (Lu et al. 1996) and L-ficolin is primarily expressed in the liver and subsequently released into the blood stream (Table 1). H-ficolin mRNA is found in liver, lung, and a glioma cell line (Yoshizawa 1997). H-ficolin is also produced in the lung by alveolar type II cells and unciliated bronchial epithelial cells and is secreted onto the epithelial surface (Akaiwa et al. 1999), indicating that H-ficolin plays a role on mucosal surfaces. M-ficolin is synthesized by monocytes (Hashimoto et al. 1999) and its expression is downregulated during monocyte differentiation and its mRNA is not detectable in macrophages (Lu et al. 1996) or monocyte-derived dendritic cells (Hashimoto et al. 1999). The median concentrations were found to be 0.8 μg per ml for MBL. 3.3 μg per ml for L-ficolin, and 18.4 μg per ml for H-ficolin.

The gene expression sites and protein localization of mouse ficolin A and ficolin B are different in distinct ontogenic patterns and cell types (Liu et al. 2005a). Ficolin A mRNA was detected only in liver and in spleen, and ficolin B mRNA was not detected in any of the adult tissues examined (Ng et al. 2007) but is expressed in more immature cells of the myeloid cell lineage than ficolin A and may function in hematopoiesis, or to be regulated in responses to the hematopoietic course. Ficolin B was found in lysosomes of activated macrophages (Runza et al. 2006). A ficolin is not expressed in mouse owing to alteration of the responsible gene to a pseudogene (Endo et al. 2004).

4 Ficolin and Infectious Diseases

Although the importance of MBL in antimicrobial host defense is well recognized, recently the role of the ficolins has been also emphasized. Human plasma L-ficolin (P35) is an important lectin that binds to the carbohydrate on bacterial surface and enhances opsonic activity of polymorphonuclear neutrophils (Aoyagi et al. 2005). Binding of L-ficolin/MASP complexes to type III group B *streptococci* (GBS) (presence of GlcNAc in the capsular polysaccharide of all serotype of GBS) leads to activation of the lectin pathway (Aoyagi et al. 2005). These findings give L-ficolin critical role in preventing neonatal GBS infection by complement activation through L-ficolin-mediated lectin pathway. Another reports also showed that L-ficolin could bind to LTAs purified from gram-positive bacteria (*S. aureus, Streptococcus agalactiae, Bifidobacterium animalis, Streptococcus pyogenes, and Bacillus subtilis*) (Lynch et al. 2004; Aoyagi et al. 2005).

Meanwhile, L-ficolin has been found to bind exclusively to some strains of capsulated *S. pneumoniae* serotypes (11A, 11D, and 11F) but does not bind to noncapsulated *S. aureus* variant (Wood) strain, suggesting that L-ficolin recognizes structures specific to a few capsular polysaccharides. On the other side, H-ficolin does not bind to any of the *S. pneumoniae* and *S. aureus* serotypes. The MBL however, binds strongly to the noncapsulated strain (Wood) (Krarup et al. 2005). These data suggest that MBL, L-ficolin and H-ficolin differ in ligand specificity.

The opsonophagocytosis can be also activated by the binding of ficolin with the carbohydrates on the pathogens. The binding of L-ficolin/MASP complexes to the capsular polysaccharide (CPS) generates C3 convertase C4b2a, which deposits C3b on streptococci and lead to increased opsonophagocytic killing by further deposition of C3b on the group B *streptococci* (GBS) (Aoyagi et al. 2005).

It appears that H-ficolin is a much powerful direct opsonin than L-ficolin. These can be explained by the differences in putative ligand specificity on the target cell surface (Jensen et al. 2007).

Recently, it is also reported that porcine plasma ficolin binds and reduces infectivity of porcine reproductive and respiratory syndrome virus (PRRS) in vitro (Keirstead et al. 2007). These studies indicate that ficolins might have antiviral roles similar to MBL and these might lead to new antiviral interventions.

5 Single Nucleotide Polymorphisms in Ficolins

Several polymorphisms were found in the promoter or structural regions of L-, M- and H-ficolin, but only polymorphisms in L-ficolin result in amino acid exchanges. L-ficolin promoter polymorphisms are associated with adverse effect on the plasma concentration, whereas two polymorphisms found in the exon of L-ficolin were associated either with increased or decreased GlcNAc binding (Hummelshoj et al. 2005). However, no significant associations were observed between the L-ficolin polymorphisms and susceptibility to invasive pneumococcal disease (Chapman et al. 2007), or Behcet's disease (Chen et al. 2006). The possession of mutant alleles has been linked to the outcome of a variety of bacterial and recurrent respiratory infections.

6 Ficolins and Apoptosis

Several reports have clearly showed that L-ficolin, H-ficolin, and other collagen-like defence molecules, such as MBL and C1q, can bind to late apoptotic cells and mediates phagocytosis possibly through the calreticulin-CD91 receptor complex present on the surface of macrophages (Ogden et al. 2001). DNA on permeable dying cells is also a plausible ligand for L-ficolin (Jensen et al. 2007). It has been found that H-ficolin binding to late apoptotic cells resulted in a significant increase in adhesion/uptake by macrophages, a strong and uniform binding of H-ficolin to necrotic cells was also observed, and the binding properties differed from those of MBL and L-ficolin (Honoré et al. 2007). These molecules mediate the clearance of apoptotic cells and involved in the maintenance of tissue homeostasis. Dysfunction of molecules important for the removal of dying host cell might involve in the development of autoimmune diseases.

7 Ficolins and Systemic Lupus Erythematosus

Systemic lupus erythematosus (SLE) is a chronic inflammatory autoimmune disease with of unknown cause, but immune dysfunction is clearly one of its central features. H-ficolin was firstly identified as a serum antigen target for an autoantibody present in the sera of some patients with SLE. That might participate in the pathogenesis of SLE. Anti-H-ficolin antibody appears during the active hypocomplementemic phase in some SLE patients and its concentration was found to correlate positively with diseases activity in SLE (Yoshizawa et al. 1997).

8 Ficolins and IgA Nephropathy (IgAN)

IgA nephropathy (IgAN) is a common renal disease that is characterized primarily by mesangial deposition of IgA (Floege and Feehally 2000; Donadio and Grande 2002). Earlier studies showed that IgA activates the alternative pathway of complement, whereas more recent data also indicate activation of the lectin pathway. Activation of the lectin pathway of complement is associated with more severe renal damage (Roos et al. 2006). It was also demonstrated that both MBL and ficolin were found to contribute to the progression of the to IgA nephropathy (IGAN) (Roos et al. 2006). All Glomerular deposition of MBL was positively associated with co-deposition of L-ficolin. Renal biopsies of patients with IgAN showed mesangial deposition of IgA1 but not IgA2. Circulating IgA1 from patients with IgAN was reported to have aberrant glycosylation of *O*-linked glycans, which potentially are involved in recognition by lectins. Precise identification of the ligand for MBL and L-ficolin in the mesangium, which presumably is presented and/or accessible in only some of the patients with IgAN, will provide novel insight in the pathogenesis of IgAN and may provide novel therapeutic options to treat disease progression.

9 Ficolins and Preeclampsia

Recently it has also been found that ficolins were differentially expressed in plasma from preeclamptic pregnancies (Wang et al. 2007). Ficolins were present in low concentrations in plasma but at high concentrations in the placenta, particularly in syncytiotrophoblasts undergoing apoptosis. The binding of ficolins in apoptotic trophoblasts induced innate immunity through local and systemic cytokine activation and correlated with the clinical manifestation of preeclampsia (Wang et al. 2007).

10 Ficolin and C-Reactive Protein

Although human C-reactive protein (CRP) becomes upregulated during septicemia, its role remains unclear, since purified CRP showed no binding to many common pathogens. Recent study showed that human CRP interacts with ficolin and this interaction stabilizes CRP binding to bacteria and activates the lectin-mediated complement pathway and target a wide range of bacteria for destruction (Ng et al. 2007).

11 Concluding Remarks

All data suggest that ficolins function in clearance of non-self, based on their location sites and their molecular features. Ficolin is an important aspect in host innate defences, while defective or abnormal expression will cause the pathogenological activities. Our recent data also demonstrated that L-ficolin can contribute host defense against *Salmonella* bacteria (a major intracellular pathogen) infection by enhancing IFN-γ expression in vivo (Ma et al. 2007). This new information may stimulate the development of a new immunotherapy approach using L-ficolin for the stimulation of innate immunity against anti-intracellular bacteria. Assessment of L-ficolin is warranted to study its regulatory role in the innate immune system and it will not be surprising to see clinical data addressing the contribution of lectin ficolin to host innate immunity and immunopathology in the near future.

Acknowledgments

This work was supported by grants from the National Natural Science Foundation of China (30370310, 20532020, and 30670098), the 973 Program of China (2006CB504300), the Ministry of Education Scientific Research Foundation for the New Century Outstanding Scholars (NCET-04-0685), the Hubei Ministry of Public Health (JX1B074), and Hubei Province Science and Technology Department (2006ABD007, 2007ABC010, 2005AA304B04).

References

Akaiwa, M., Yae, Y., Sugimoto, R., Suzuki, S. O., Iwaki, T., Izuhara, K., and Hamasaki, N. (1999). Hakata antigen, a new member of the ficolin/opsonin P35 family, is a novel human lectin secreted into bronchus/alveolus and bile. *J Histochem Cytochem* 47, 777–785

Aoyagi, Y., Adderson, E. E., Min, J. G., Matsushita, M., Fujita, T., Takahashi, S., Okuwaki, Y., and Bohnsack, J. F. (2005). Role of L-ficolin/mannose-binding lectin-associated serine protease complexes in the opsonophagocytosis of type III group B Streptococci. *J Immunol* 174, 418–425

Chen, X., Katoh, Y., Nakamura, K., Oyama, N., Kaneko, F., Endo, Y., Fujita, T., Nishida, T., and Mizuki, N. (2006). Single nucleotide polymorphisms of ficolin 2 gene in Behçet's disease. *J Dermatol Sci* 43, 201–205

Chapman, S. J., Vannberg, F. O., Khor, C. C., Segal, S., Moore, C. E., Knox, K., Day, N. P., Davies, R. J., Crook, D. W., and Hill, A. V. (2007). Functional polymorphisms in the FCN2 gene are not associated with invasive pneumococcal disease. *Mol Immunol* 44, 3267–3270

Donadio, J. V. and Grande, J. P. (2002). IgA nephropathy. *N Engl J Med* 347, 738–748

Eisen, D. P., and Minchinton, R. M. (2003). Impact of mannose-binding lectin on susceptibility to infectious diseases. *Clin Infect Dis* 37, 1496–1505

Endo, Y., Sato, Y., Matsushita, M., and Fujita, T. (1996). Cloning and characterization of the human lectin P35 gene and its related gene. *Genomics* 36, 515–521

Endo, Y., Liu, Y., Kanno, K., Takahashi, M., Matsushita, M., and Fujita, T. (2004). Identification of the mouse H-ficolin gene as a pseudogene and orthology between mouse ficolins A/B and human L-/M-ficolins. *Genomics* 84, 737–744

Floege, J. and Feehally, J. (2000). IgA nephropathy: recent developments. *J Am Soc Nephrol* 11, 2395–2403.

Fujimori, Y., Harumiya, S., Fukumoto, Y., Miura, Y., Yagasaki, K., Tachikawa, H. and Fujimoto, D. (1998). Molecular cloning and characterization of mouse ficolin-A. *Biochem Biophys Res* Co 244, 796–800

Fujita, T. (2002). Evolution of the lectin-complement pathway and its role in innate immunity. *Nat Rev Immunol* 2, 346–353

Garlatti, V., Belloy, N., Martin, L., Lacroix, M., Matsushita, M., Endo, Y., Fujita, T., Fontecilla-Camps, J. C., Arlaud, G. J., Thielens, N. M., et al. (2007). Structural insights into the innate immune recognition specificities of L- and H-ficolins. *EMBO J* 26, 623–633

Gokudan, S., Muta, T., Tsuda, R., Koori, K., Kawahara, T., Seki, N., Mizunoe, Y., Wai, S. N., Iwanaga, S., and Kawabata, S. (1999). Horseshoe crab acetyl group-recognizing lectins involved in innate immunity are structurally related to fibrinogen. *Proc Natl Acad Sci USA* 96, 10086–10091

Hashimoto, S., Suzuki, T., Dong, H. Y., Nagai, S., Yamazaki, N., and Matsushima, K. (1999). Serial analysis of gene expression in human monocyte-derived dendritic cells. *Blood* 94, 845–852

Holmskov, U., Thiel, S., and Jensenius, J. C. (2003). Collections and ficolins: lectins of the innate immune defence. *Annu Rev Immunol* 21, 547–578

Honoré, C., Hummelshoj, T., Hansen, B. E., Madsen, H. O., Eggleton, P., and Garred, P. (2007). The innate immune component ficolin 3 (Hakata antigen) mediates the clearance of late apoptotic cells. *Arthritis Rheum* 5, 1598–1607

Hummelshoj, T., Munthe-Fog, L., Madsen, H. O., Fujita, T., and Matsushita, M. (2005). Polymorphisms in the FCN2 gene determine serum variation and function of ficolin-2. *Hum Mol Gen* 14, 1651–1658

Ichijo, H., Rönnstrand, L., Miyagawa, K., Ohashi, H., Heldin, C., and Miyazono, K. (1991). Purification of transforming growth factor-ß1 binding proteins from porcine uterus membranes. *J Biol Chem* 266, 22459–22464

Ichijo, H., Hellman, U., Wernstedt, C., and Gonez, L. J. (1993). Molecular cloning and characterization of ficolin, a multimeric protein with fibrinogen- and collagen-like domains. *J Biol Chem* 268, 14505–14513

Jensen, M. L., Honore, C., Hummelshoj, T., Hansen, B. E., Madsen, H. O., and Garred, P. (2007). Ficolin-2 recognizes DNA and participates in the clearance of dying host cells. *Mol Immunol* 44, 856–865

Keirstead, N. D., Lee, C., Yoo, D., Brooks, A. S., and Hayes, M. A. (2008). Porcine plasma ficolin binds and reduces infectivity of porcine reproductive and respiratory syndrome virus (PRRSV) in vitro. *Antiviral Res* 77, 28–38

Krarup, A., Thiel, S., Hansen, A., Fujita, T., and Jensenius, J. C. (2004). L-ficolin is a pattern recognition molecule specific for acetyl groups. *J Biol Chem* 279, 47513–47519

Krarup, A., Sørensen, U. B., Matsushita, M., Jensenius, J. C., and Thiel, S. (2005). Effect of capsulation of opportunistic pathogenic bacteria on binding of the pattern recognition molecules mannan-binding lectin, L-ficolin, and H-ficolin. *Infect Immun* 73, 1052–1060

Le, Y., Tan, S. M., Lee, S. H., Kon, O. L., and Lu, J. (1997). Purification and binding properties of a human ficolin-like protein. *J Immunol Methods* 204, 43–49

Le, Y., Lee, S. H. Kon, O. L., and Lu, J. (1998). Human 1-ficolin: plasma levels, sugar specificity, and assignment of its lectin activity to the fibrinogen-like (FBG) domain. *FEBS Lett* 425, 367–370

Liu, Y., Endo, Y., Homma, S., Kanno, K., Yaginuma, H., and Fujita, T. (2005a). Ficolin A and ficolin B are expressed in distinct ontogenic patterns and cell types in the mouse. *Mol Immunol* 42, 1265–1273

Liu, Y., Endo, Y., Iwaki, D., Nakata, M., Matsushita, M., Wada, I., Inoue, K., Munakata, M., and Fujita. T. (2005b). Human M-ficolin Is a Secretary Protein That Activates the Lectin Complement Pathway. *J Immunol* 175, 3150–3156

Lu, J. and Le, Y. (1998). Ficolins and the fibrinogen-like domain. *Immunobiology* 199, 190–199

Lu, J., Tay, P. N., Kon, O. L., and Reid, K. B. M. (1996). Human ficolin: cDNA cloning, demonstration of peripheral blood leucocytes as the major site of synthesis and assignment of the gene to chromosome 9. *Biochem J* 313, 473–478

Lynch, N. J., Roscher, S., Hartung, T., Morath, S., Matsushita, M., Maennel, D. N., Kuraya, M., Fujita, T., and Schwaeble, W. J. (2004). L-ficolin specifically binds to lipoteichoic acid, a cell wall constituent of gram-positive bacteria, and activates the lectin pathway of complement. *J Immunol* 172, 1198–1202

Ma, Y. G., Cho, M. H., Zhao, M., Park, J. W., Matsushita, M., Fujita, M., and Lee, B. (2004). Human mannose-binding lectin and L-ficolin function as specific pattern recognition proteins in the lectin pathway of complement. *J Biol Chem* 279, 25307–25312

Ma, Y. F., Luo, F. L., Xiang, T., Li, X. Y., Pan, Q., Chen, M., Fujita, T., and Zhang, X. L. (2007). Effects of L-ficolin on host resistance, gamma interferon production and phagocytosis against *Salmonella* infection. *Mol Immunol* 44, 211

Matsushita, M., Endo,Y., Taira, S., Sato, Y., Fujita, T., Ichikawa, N., Nakata, M., and Mizuochi,T. (1996). A novel human serum lectin with collagen- and fibrinogen-like domains that functions as an opsonin. *J Biol Chem* 271, 2448–2454

Ng, P. M., Le Saux, A., Lee, C. M., Tan, N. S., Lu, J., Thiel, S., Ho, B., and Ding, J. L. (2007). C-reactive protein collaborates with plasma lectins to boost immune response against bacteria. *EMBO J* 26, 3431–3440

Ogden, C. A., deCathehneau, A., Hoffmann, P. R., Bratton, D., Ghebrehiwet, B., Fadok, V. A., and Henson, P. M. (2001). C1q and mannose binding lectin engagement of cell surface calreticulin and CD91 initiates macropinocytosis and uptake of apoptotic cells. *J Exp Med* 194, 781–795

Ohashi, T. and Erickson, H. P. (1997). Two oligomeric forms of plasma ficolin have differential lectin activity. *J Biol Chem 22*, 14220–14226

Ohashi, T. and Erickson, H. P. (1998). Oligomeric structure and tissue distribution of ficolins from mouse, pig and human. *Arch Biochem Biophys* 360, 223–232

Roos, A., Xu, W., Castellano, G., Nauta, A. J., Garred, P., Daha, M. R., and Kooten, C. (2004). A pivotal role for innate immunity in the clearance of apoptotic cells. *Eur J Immunol 34*, 921–929

Roos, A., Rastaldi, M. P., Calvaresi, N., Oortwijn, B. D., Schlagwein, N., van Gijlswijk-Janssen, D. J., Stahl, G. L., Matsushita, M., Fujita, T., van Kooten, C., et al. (2006). Glomerular activation of the lectin pathway of complement in IgA nephropathy is associated with more severe renal disease. *J Am Soc Nephrol* 17, 1724–1734

Runza, V. L., Hehlgans, T., Echtenacher, B., Zähringer, U., Schwaeble, W. J., and Männel, D. N. (2006). Localization of the mouse defense lectin ficolin B in lysosomes of activated macrophages. *J Endotoxin Res* 12, 120–126

Sugimoto, R., Yae, Y., Aikawa, M., Kitajima, S., Shibata, Y., Sato, H., Hirata, J., Okochi, K., Izuhara, K., and Hamasaki, N. (1998). Cloning and characterization of the Hakata antigen, a member of the ficolin/opsonin p35 lectin family. *J Biol Chem* 273, 20721–20727

Takahashi, K., Ip, W. E., Michelow, I. C., and Ezekowitz, R. A. (2006). The mannose-binding lectin: a prototypic pattern recognition molecule. *Curr Opin Immunol* 18, 16–23

Takahashi, K. and Ezekowitz, R. A. (2005). The role of the mannose-binding lectin in innate immunity. *Clin Infect Dis* 41(Suppl 7), S440–S444

Tsujimura, M., Ishida, C., Sagara, Y., Miyazaki, T., Murakami, K., Shiraki, H., Okochi, K., and Maeda, Y. (2001). Detection of serum thermolabile beta-2 macroglycoprotein (Hakata antigen) by enzyme-linked immunosorbent assay using polysaccharide produced by *Aerococcus viridans*. *Clin Diagn Lab Immunol* 8, 454–459

Wang, C. C., Yim, K. W., Poon, T. C., Choy, K. W., Chu, C. Y., Lui, W. T., Lau, T. K., Rogers, M. S., and Leung, T. N. (2007). Innate immune response by ficolin binding in apoptotic placenta is associated with the clinical syndrome of preeclampsia. *Clin Chem* 53, 42–52

Yae, Y., Inaba, S., Sato, H., Okochi, K., Tokunaga, F., and Iwanaga, S. (1991). Isolation and characterization of a thermolabile β-2 macroglycoprotein ("thermolabile substrate" or "Hakata antigen") detected by precipitating (auto) antibody in sera of patients with systemic lupus erythematosus. *Biochim Biophys Acta* 1078, 369–376

Yoshizawa, S., Nagasawa, K., Yae, Y., Niho, Y., and Okochi, K. (1997). A thermolabile b2-macroglobulin (TGM) and the antibody against TMG in patients with system lupus erythematosus. *Clin Chim Acta* 264, 219–225

10. Complement Factor H: Using Atomic Resolution Structure to Illuminate Disease Mechanisms

Paul N. Barlow[1], Gregory S. Hageman[2], and Susan M. Lea[3]

[1]Schools of Chemistry and Biological Sciences, Joseph Black Chemistry Building, University of Edinburgh, Edinburgh EH9 2PB, UK, Paul.Barlow@ed.ac.uk
[2]Department of Ophthalmology and Visual Sciences, University of Iowa, Iowa City, IA, USA, gregory-hageman@uiowa.edu
[3]Sir William Dunn School of Pathology, University of Oxford, South Parks Road, Oxford, OX1 3RE, UK, susan.lea@path.ox.ac.uk

Abstract. Complement Factor H has recently come to the fore with variant forms implicated in a range of serious disease states. This review aims to bring together recent data concerning the structure and biological activity of this molecule to highlight the way in which a molecular understanding of function may open novel therapeutic possibilities. In particular we examine the evidence for and against the hypothesis that sequence variations in factor H may predispose to disease if they perturb its ability to recognise and respond appropriately to polyanionic carbohydrates on host surfaces that require protection from complement-mediated damage.

1 Introduction

The human complement factor H (CFH) protein is a key regulator of the alternative pathway of the complement system. The alternative pathway entails ubiquitous and continuous generation of relatively small quantities of C3b molecules. These bind indiscriminately and have the potential to self-amplify. Any surface exposed to complement, and not protected by regulatory molecules, will therefore quickly become coated in C3b and subject to multiple potentially destructive complement-mediated processes. CFH acts to specifically prevent C3b amplification on self-surfaces while permitting complement to proceed unchecked on foreign surfaces. It thereby selectively protects host tissues and it appears to be essential for the healthy functioning of at least some organs.

This approximately 155-kDa (Ripoche et al. 1988) plasma (~150–550 μg/ml) (Esparza-Gordillo et al. 2004) glycoprotein is encoded by the CFH gene within the regulators of complement activation (RCA) gene cluster on chromosome 1q32 (Rodriguez de Cordoba et al. 1985) where it is linked closely to genes CFHR1-CFHR5, which code for five smaller and less abundant CFH-related proteins (Zipfel et al. 1999). A seventh protein – FH-like or CFHT – is a splice-variant of CFH

J.D. Lambris (ed.), *Current Topics in Complement II.*
doi: 10:1007/978-0-387-78952-1_10,

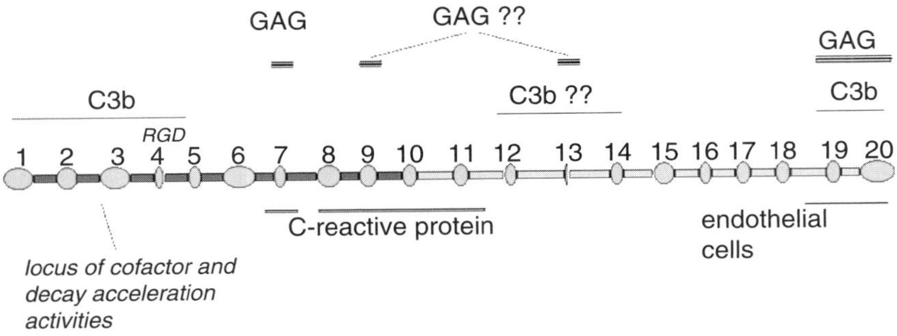

Fig. 1 Modular composition of CFH and putative binding sites. In this schematic, the CCPs (ovals) of CFH are drawn to emphasize their variation in size (51–63 residues) and the linkers (rectangles) are proportional to their lengths (3–8 residues). Functional/binding sites are indicated. The question marks are used where no direct proof is available

present at appreciable levels in the blood (Schwaeble et al. 1987). Some of the CFH-related proteins and CFH-like-1 also exhibit complement regulatory activity (Zipfel and Skerka 1994) although their physiological properties have not yet been clearly defined. Most CFH in the circulation has a hepatic origin but many cell-types are endogenous sources (Schlaf et al. 2001).

The CFH protein is composed from a total of 20 domains, each containing approximately 60 amino acid residues and termed "complement control protein modules" (CCPs) or short consensus repeats (Ripoche et al. 1986), that are joined by short linkers consisting of 3–8 residues (Fig. 1). Each module is encoded by a single exon out of the 23 in CFH, except CCP module 2 that is encoded by two exons (Male et al. 2000). Exon one encodes an 18 amino acid signal peptide that is cleaved during processing. Exon ten does not contribute to CFH but encodes the last of seven modules in the splice-variant, CFH-like-1, plus its unique C-terminal sequence Ser-Phe-Leu-Thr (Estaller et al. 1991).

By utilising its multiple protein and carbohydrate binding sites (Schmidt et al. 2008), CFH binds to C3b and its complexes C3bBb and C3b2Bb – the C3- and C5-convertases of the alternative pathway. It engages with these binding partners both in fluid phase and when they are immobilised. Critically, CFH has the remarkable property of acting selectively on C3b and C3bBb on self vs. non-self surfaces (Meri and Pangburn 1990). It competes for binding of complement factor B (CFB) to C3b (Farries et al. 1990), acts as a cofactor for factor I-catalysed proteolytic cleavage of C3b (Pangburn et al. 1977), and it accelerates the irreversible dissociation of C3bBb and C3b2Bb into their separate components (Pangburn and Muller-Eberhard 1983). Thus CFH not only inhibits formation of the convertases but it also shortens the lifespan of any convertase complex that forms.

The discovery of mutations and single nucleotide polymorphisms (SNPs) in CFH that can be linked to human diseases has led to a surge of interest in this protein; an update is presented in the next section. We will then summarise the current state of knowledge of structure-function relationships of CFH, with an emphasis on recent structural studies of a key disease-linked, glycosamioglycan (GAG)-binding region in CCPs 6, 7 and 8. Finally, both established and emerging biochemical attributes/properties of CFH will be discussed in the context of the consequences for human health of its genetically encoded sequence variations.

2 Involvement of Factor H in Human Disease

Mutations and polymorphisms that lead to amino-acid residue substitutions (or a deletion in one case) within CFH have been linked to several human diseases, including atypical haemolytic uremic syndrome (aHUS), age-related macular degeneration (AMD), and membranoproliferative glomulonephritis type II (MPGNII also known as dense deposit disease) (de Cordoba and de Jorge 2008). More tentatively and controversially the AMD-linked Y402H SNP (see below) has been linked to an altered risk of three other late-onset conditions – myocardial infarction (Kardys et al. 2006; Mooijaart et al. 2007; Nicaud et al. 2007; Pai et al. 2007; Stark et al. 2007), CAD/CHD (Meng et al. 2007; Pulido et al. 2007; Topol et al. 2006) and Alzheimer's Disease (Hamilton et al. 2007; Zetterberg et al. 2007).

Hemolytic uremic syndrome (HUS) is a leading cause of pediatric kidney failures and is generally a sequela of bacterial infections. It is characterized by thrombocytopenia, microangiopathic hemolytic anemia and acute renal failure. The rare atypical HUS (aHUS) is not linked to infection, but is sporadic or familial. Full recovery from typical HUS is the norm, but the long-term diagnosis for sufferers of aHUS – 5–10 % of cases of HUS – is unfavorable. Single and double amino-acid changes that occur predominantly in the C-terminal segment of CFH (CCPs 19 and 20) have been identified in 10–15% of patients with aHUS (Buddles et al. 2000; Heinen et al. 2006; Richards et al. 2001; Warwicker et al. 1998) (http://www.fh-hus.org/). In many cases the variant form of CFH is present in plasma at approximately normal levels (Kavanagh et al. 2007). Mutations in membrane cofactor protein and factor I have also been linked to aHUS. This topic has been the subject of a recent review (Kavanagh et al. 2007), hence our emphasis here is on AMD.

Age-related macular degeneration is characterized by a progressive loss of central vision attributable to degenerative and, in advanced cases, neovascular changes in the macula, a highly specialized region of the ocular retina comprising only 4% of total retinal surface area (Gehrs et al. 2006). This region contains the highest density of cone photoreceptor cells, is the only portion of the retina where 20/20 vision is attainable, and accounts for approximately 10% of the visual field. Thus, the pathological lesions that develop in this region have a major impact on visual function and productivity.

Age-related macular degeneration is characterized in its earliest clinical stages by the accumulation of drusen, hallmark extracellular deposits that accumulate between the retinal pigmented epithelium (RPE) and Bruch's membrane, a unique basement membrane complex separating the retina and choroid. In recent years, a variety of complement activators, complement components, and complement regulatory proteins have been identified as molecular constitutents of drusen (Anderson et al. 2001; Crabb et al. 2002; Hageman et al. 1999; Johnson et al. 2000, 2001; Mullins et al. 2000). This compositional profile of drusen formed the basis for the general conclusion that drusen are a byproduct of chronic, local inflammatory events at the level of the RPE-Bruch's membrane interface and that complement activation and immune responsiveness are important facets of AMD pathogenesis (Hageman et al. 2001; Penfold et al. 2001; Zarbin 2004). In this model, RPE atrophy and the subsequent deposition of RPE-derived debris in the sub-RPE space is construed as a local pro-inflammatory event, leading to activation and amplification of the complement alternative pathway which, in turn, induces substantial bystander damage to macular cells and tissues over time.

Strong support for this new paradigm of AMD pathobiology emerged from subsequent genetic discoveries reported in early 2005 by four independent research teams (Edwards et al. 2005; Hageman et al. 2005; Haines et al. 2005; Klein et al. 2005). These studies revealed a highly significant association between AMD and common variants in the CFH gene on human chromosome 1q31. Three of these studies focused on a single SNP in the coding region of CFH, Y402H, as the causal variant (Edwards et al. 2005; Haines et al. 2005; Klein et al. 2005). However, it was clear from the fourth study (Hageman et al. 2005) that multiple CFH variants (and their tagged haplotypes), rather than the single Y402H variant, confer either an elevated or reduced risk for one's propensity to develop AMD. Three significant haplotypes; a major risk haplotype, tagged by the Y402H coding SNP (cSNP) and two protective haplotypes, tagged by intronic variants and/or cSNPs, were identified (Hageman et al. 2005). Iwata and colleagues subsequently identified an additional risk haplotype in the Japanese population (Okamoto et al. 2006). More recent studies of Caucasian populations have provided evidence for an additional risk haplotype, tagged by rs2274700 [A473A] or an intronic rs1410996 SNP (Li et al. 2006; Seddon et al. 2006) and a strongly protective haplotype tagged by a complete deletion of the two CFH-related genes, CFHR1 and CFHR3 (Hageman et al. 2006; Hughes et al. 2006). Interestingly, the same AMD-associated risk haplotype has also been shown to be associated with MPGNII, a rare renal disease characterized by uncontrolled activation of the alternative complement pathway and the development of ocular drusen and neovascularization that are indistinguishable from those that occur in AMD (Hageman et al. 2005). These data provided additional strong support for the notion that specific CFH variants are associated with drusen deposition and AMD.

The AMD-CFH discovery was followed by the identification of an association of AMD with three additional complement regulators – Complement Factor B and component 2 (CFB/C2) on chromosome 6p21 (Gold et al. 2006) and C3 on chromosome 19p13 (Maller et al. 2007; Yates et al. 2007) – as well as a significant association of three tightly linked genes (*PLEKHA1*, *LOC387715* and *PRSS11/ HTRA1*) on human chromosome 10q26 (Jakobsdottir et al. 2005; Rivera et al. 2005;

Weeks et al. 2004). Genetic variation at all four loci defines a major proportion of AMD disease burden, making this disease one of the most well-defined complex traits.

All accumulated data at this time support the concept that dysregulation of the complement alternative pathway is an early step in AMD (Gehrs et al. 2006) regardless of ocular phenotype. This contrasts with genes within the 10q26 locus, which is associated mainly with the advanced stages of AMD. The specific variants, or their combinations, within the *CFH*, *CFB/C2* and *C3* genes, which result in functional consequences at the protein and functional levels, must be more precisely defined. Nonetheless, it seems clear from the available data that genetically pre-determined variation in genes associated with the complement pathway, when combined with an as yet undefined triggering event, underlie a major proportion of AMD in the human population.

Thus, although aHUS and MPGNII affect glomeruli while AMD is an ocular disease, it is MPGNII and AMD that share pathological characteristics. Both MPGNII and AMD involve the deposition of complement-containing material (Mullins et al. 2000) onto specialised regions of extracellular matrix (the glomerular basement membrane in the case of the kidney or Bruch's membrane in the macula) accessible to plasma via fenestrations within the endothelial lining of the microvasculature; and indeed MPGNII patients develop ocular drusen as mentioned earlier (Mullins et al. 2001). The phenotype of CFH–/– knockout mice includes both MPGNII (Pickering et al. 2002) and abnormalities in retinal ultrastructure (Coffey et al. 2007). On the other hand aHUS, which is characterised by endothelial cell activation, swelling and detachment from the basement membrane, does not arise in CFH–/– mice unless a gene encoding CFH- 1-15 (i.e. a CFHΔ16-20 mutant) is re-introduced (Pickering et al. 2007). As discussed below, such a mutant will retain fluid-phase C3b-binding activity but will be unable to recognise surfaces. This suggests that the fluid-phase convertase-regulating activity of CFH is required for development of aHUS while surface-specific convertase regulation by CFH inhibits the development of mouse model of aHUS. This is discussed further below in the light of recent structural and functional results.

3 CFH Binding Sites

Early work identified the N-terminal five or seven CCPs of CFH as the locus of its fluid-phase C3b-binding and cofactor activities (Alsenz et al. 1984, 1985; Misasi et al. 1989). Subsequent studies localised this functionality (Gordon et al. 1995; Kuhn et al. 1995), along with a proportion of the native decay accelerating activity (Kuhn and Zipfel 1996), to CCP modules 1–4. Additional C3b-binding sites were identified using module-deletion CFH mutants (Sharma and Pangburn 1996). Thus, while CFHΔ1-5 retained some binding affinity for cell-surface -bound C3b (csbC3b), deletions of CCPs 16–20 decimated its binding affinity for csbC3b. This observation implicated the C-terminal region of CFH as a second C3b-binding site, which was later pinpointed to CCPs 19 and 20 REF. A third C3b-binding site has been suggested on the basis that CFHΔ6-10 exhibited a decreased affinity for csbC3b, similar to that of CFHΔ1-5. Additional studies (Jokiranta et al. 2000) implied but did not confirm a third C3b binding site in CCPs 12-14. This information is summarised in Fig. 1.

Fundamental to the ability of CFH to regulate efficiently complement on surfaces is its affinity for polyanions (Meri et al. 1990). A photoaffinity-tagging analogue (Pangburn et al. 1991) implicated the very basic CCP 13 in heparin binding. But while deletion of CCP 13 (Sharma et al. 1996) resulted in only very slightly reduced ability to bind a heparin-agarose column, deletion of CCPs 6–10 showed significantly weaker heparin affinity (Sharma et al. 1996) implying that a stronger GAG/sialic acid-binding site exists in the CCPs 6–10 region; CCP 7 was subsequently demonstrated to play a prominent role (Blackmore et al. 1996). A CFHΔ7Δ13 deletion mutant still bound heparin but a CFHΔ7Δ20 construct did not (Blackmore et al. 1998) implicating CCP 20 as a further locus for polyanion interaction. Convincingly, non-heparin-binding CFH-1-5 was converted to heparin-binding CFH-1-5,20 by inclusion of CCP 20 in the construct (Blackmore et al. 1998). Moreover, highly purified, structurally characterised CFH-19,20 (Herbert et al. 2006) binds well to a heparin-agarose column. More recently, a set of short recombinant CFH constructs having in common the presence of CCP 9 were all found to bind heparin (Ormsby et al. 2006), but the additional presence of an N-terminal cloning artefact containing arginine residues might have contributed to this affinity. In summary (Fig. 1), while GAG-binding sites in module 7 (with contributions likely from CCPs 6 and 8) and module 20 (with possible contributions from CCP 19) are well-established, current evidence for involvement of either CCP 9 or CCP 13 is inconclusive. Finally, it should be mentioned that the N-linked glycans of CFH could theoretically have an elecrostatic influence on CFH-polyanion interactions although they are dispensable for complement regulation (Jouvin et al. 1984).

The importance of the C-terminal heparin-binding site for self vs. non-self discrimination was clearly demonstrated by Ferreira et al. (2006) who found that highly purified CFH-19,20 competitively inhibited the action of CFH on cell surfaces. This double-module construct was able to overcome the protective effects of full-length CFH and thereby promote aggressive complement-mediated lysis of sheep erythrocytes. This finding is reinforced by studies showing the ability of monoclonal CCP 20-specific antibodies (Oppermann et al. 2006) to block interactions of CFH with endothelial cells. Thus this C-terminal polyanion- and C3b-binding site is critical for the ability of CFH to recognise and protect host cells bearing sialic acids and GAGs.

4 The Structure of Factor H

The multiple, flexibly-linked modules of CFH render it a problematic target for high-resolution structural methods. Below we outline low-resolution structural studies of intact, full-length CFH molecule, which have provided us with valuable insights into its overall architecture. Our understanding of the atomic-resolution structure of CFH derives from studies of constructs consisting of only a few CCPs. Some of these studies will be described subsequently, with an emphasis on recent work on the AMD-related and GAG-binding CCP 7 and its neighbouring modules.

4.1 Low Resolution Structural Information on CFH

Determination of 3-D structure from extended, multi-domain, proteins is a significant challenge and the level of accuracy of such models is low compared to the atomic structures of fragments. However, major progress has been made by the group of Perkins using a combination of biophysical techniques (primarily neutron and X-ray solution-scattering combined with analytical ultracentrifugation, reviewed in (Perkins et al. 2002)) capable of giving low-resolution information about molecular shape combined with using homology models of the domains to aid interpretation of the shape information. Although these studies lead to sets of atomic coordinates for the proteins studied it is important to reflect on the fact that the data can only inform about **overall** shape and have no ability to validate the atomic detail of the homology models used, as discussed further below. When applied to study of full length CFH (Aslam and Perkins 2001) the data revealed that the maximum length of the CFH in solution was approximately 40 nm compared to a length of 73 nm theoretically possible for a fully extended structure. These data strongly suggest that, on the average, the 20 CCP modules within CFH are folded back upon themselves in solution. Such folding back has been supported by later studies (Okemefuna et al. 2008) using the same techniques, which suggest short truncated segments of CFH are likewise not entirely linear with regard to their module arrangements.

4.2 Atomic Structure of the CFH Carboxy-Terminus

Both NMR (Herbert et al. 2006) and X-ray crystallography (Jokiranta et al. 2006) have been employed to study the 3-D structure of the C-terminal pair of CCPs, CFH-19,20 (Fig. 2a). Module 19 is similar in structure to other CCPs of known structure such as CFH modules 5 (Barlow et al. 1992), 15 (Barlow et al. 1993) and 16 (Barlow et al. Campbell 1991). In all these CCPs, the polypeptide chain contributes to five extended regions that are joined by turns or loops. The extended regions, which form short beta-strands and small sheets to an extent that varies between CCP structures (Soares et al. 2005), are approximately aligned forming an oblate structure with N and C termini at opposite poles. Each 40 Å-long module has its own small hydrophobic core containing highly conserved non-polar side-chains and is further stabilised by two disulfides that form between four invariant Cys residues. CCP 19 is tightly coupled, in a head-to-tail arrangement, to CCP 20 via a three-residue linking sequence – the shortest linker in CFH. This C-terminal module also has many features typical of CCP modules but it is shorter and broader than any other CCP of known structure. When CFH-19,20 is titrated with the GAG-like

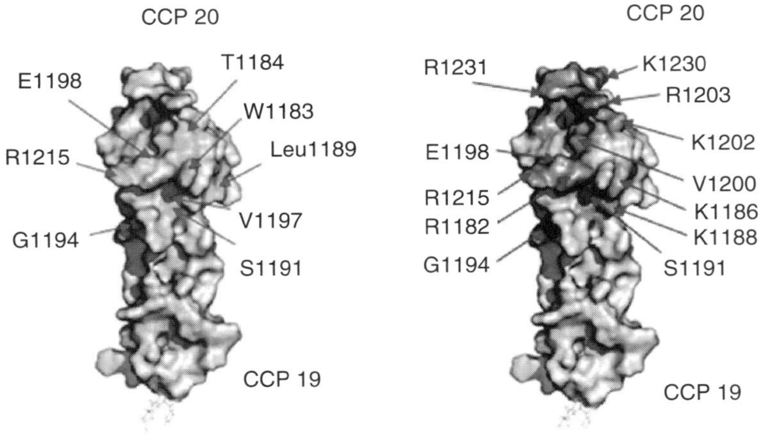

Fig. 2 Surface representations of the structure of FH-19,20. Highlighting: (**a**) The positions of mutations in CFH found in aHUS patients. (**b**) The residues that are involved in interacting with a model GAG compund, dp4

compound, dp4, perturbations are observed of CCP 20 NMR chemical shifts but not CCP 19 ones (Herbert et al. 2006), thus identifying module 20 as the dp4-binding site. Closer scrutiny of which residues within CCP 20 are most affected in terms of their chemical shifts upon addition of dp4 implicates the presence of a dp4-binding patch on one face of CCP 20 (Fig. 2b). There is a notable degree of overlap between this putative GAG-binding region, which is predominantly electropositive, and a set of residues considered important in aHUS (see above).

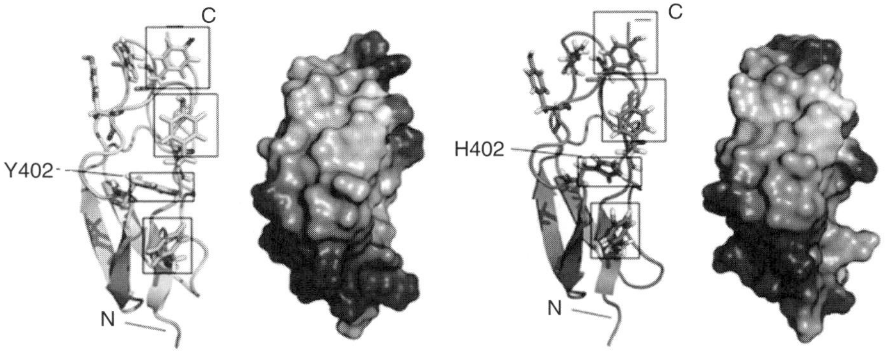

Fig. 3 Comparison of structures of FH-7. Secondary structure and electrostatic surface representations (red = negative, blue = positive) are shown for the Y402 (left-hand panels) and H402 (right-hand panels) allotypic variants. Residue 402 is labeled and boxed and three tyrosine side-chains on the same face of the module are also boxed

4.3 Atomic Structures for both Tyr and His Variants of CCP 7

The first detailed structural information on an AMD-associated CCP 7, emerged from NMR-derived structures of both the Tyr402 and His402 allotypic variants of the recombinantly expressed single module (Herbert et al. 2007) (Fig. 3) i.e. CFH-7. This study revealed that both variants of CFH-7 adopt the classical CCP-fold (Norman et al. 1991), and that the side-chain of residue 402 lies on the surface between two Tyr side-chains (Tyr390 and Tyr393). With the exception of the altered side-chain, the two variants are identical in structure; indeed even the orientations of the altered side-chain are the same. The authors further demonstrated that residue 402 lies on the edge of a patch of amino acid residues whose NMR chemical shifts are sensitive to addition of a fully sulphated heparin-derived tetrasaccharide, dp4, which is a model-GAG compound. Thus they concluded that the AMD-linked substitution could alter the interaction between CCP 7 and GAGs on the self-surfaces.

4.4 Low-Resolution Models for CFH-6,7,8

Subsequent insights into the structure of this AMD-associated region of CFH came from low-resolution techniques: small-angle X-ray scattering (SAXS) measurements in solution and analytical ultracentrifugation. These methods were employed (Fernando et al. 2007) to investigate the structure of a recombinantly expressed construct (Clark et al. 2006) consisting of the seventh CCP flanked by its immediate neighbours i.e. CFH-6,7,8. As pointed out earlier, neither of these techniques is able to yield atomic-level information and both are dependent in their interpretation upon construction of homology-based computer-generated models for the domains or modules under investigation. Their potential advantage is that they can provide information about the dimensions of the molecule and hence inferences may be made regarding orientation of the domains with respect to each other and the overall molecular architecture. Both His402 and Tyr402 variants of CFH-6,7,8 were shown to adopt similar, bent conformations (Fernando et al. 2007). The authors noted that the His402 variant had a slightly higher tendency to self-associate (Fernando et al. 2007; Nan et al. 2008), although there is no evidence that this occurs in vivo.

By exploring numerous possible relative arrangements of the three homology-modelled CCPs for best fit to the data, restrained models of CFH-6,7,8 were created. These allowed the tentative mapping of positively charge and therefore putative GAG-interacting residues, and led to the claim that many of these are exposed on the convex surfaces of the models. Additionally, study of CFH-6,7,8 in complex with a heparin decamer by SAXS (Fernando et al. 2007) suggested that subtle structural rearrangements resulting in a more linear arrangement of the three CCPs might accompany the binding of heparin-like ligands. Presumably these could arise from the involvement of more than one module in contacting a single heparin molecule.

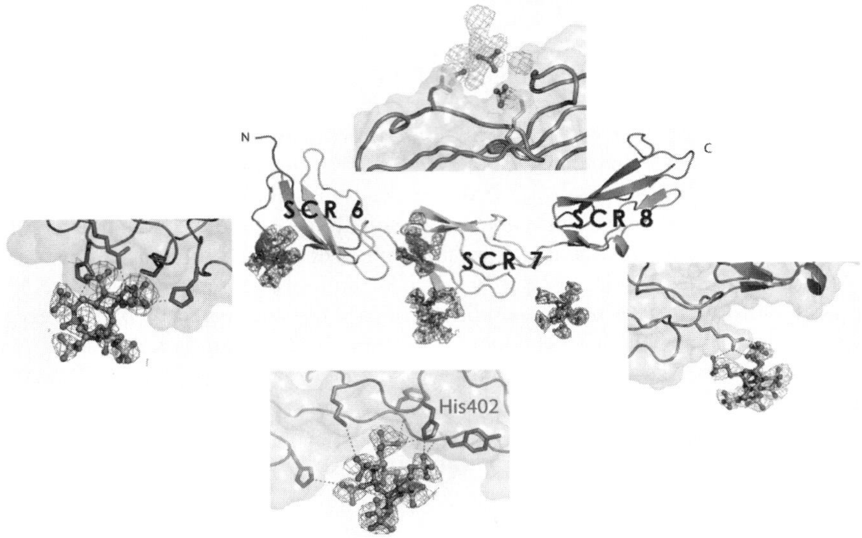

Fig. 4 Centre panel shows a cartoon representation of the structure of CFH-6,7,8_{402H} (Prosser et al. 2007b) with the sulphated sugar ligands displayed as electron density and modeled atoms (ball and stick representation). The panels arranged around the centre show close up views of the GAG binding sites

4.5 Atomic Structure for CFH-678

While low-resolution studies inform the overall shape of CFH-6,7,8 they are difficult or impossible to interpret reliably in terms of individual side-chains. To gain true insight into the implications of the Y402H substitution for this region, high-resolution data were required. In particular, there was a need for structural information on the interaction of CFH-6,7,8 with anionic carbohydrates. Such information has become available in the form of a near-atomic resolution crystal structure (Prosser et al. 2007b) of the His402 variant of CFH-6,7,8 in complex with sucrose octasulfate (SOS), a highly sulfated sugar analogue of GAGs (Fig. 4). Justification for the use in this study of SOS as a structurally amenable mimic of GAGs was provided by NMR work on CFH-7,8, which indicated similar binding modes for both the heparin-derived tetrasaccharide, dp4, and SOS.

As was predicted on the basis of low-resolution methods (Fernando et al. 2007) The CFH-6,7,8_{402H} structure (Prosser et al. 2007b) adopts an extended but slightly curved conformation. The crystal structure reveals that the overall conformation is maintained by a tight interface between CCPs 6 and 7, but there are weaker interactions between CCPs 7 and 8 that may allow some flexibility between the

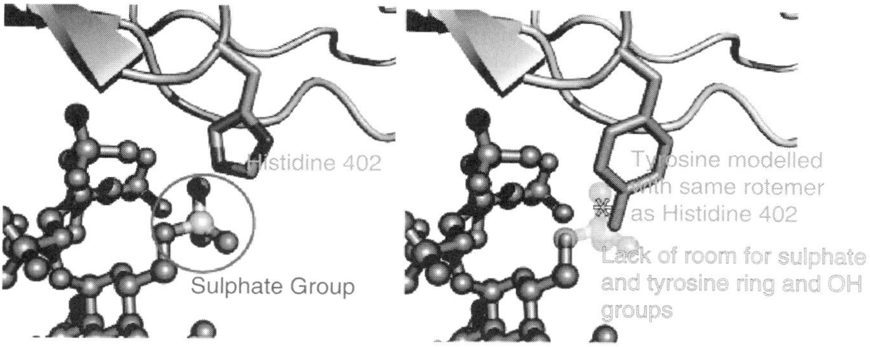

Fig. 5 Cartoon to illustrate incompatibility of sulphated sugar recognition by the Histidine at position 402 with a tyrosine at the same position

domains. However, the scattering and homology based model was found to be wrong with respect to inter-modular angles, the structures of individual modules, and the details of how side-chains are arranged at the modular surfaces. These discrepancies likely reflect the difficulties of making detailed interpretations of inherently low-resolution data and there is not necessarily a conflict between the crystal structure and the solution scattering and sedimentation data of Fernando et al. (Fernando et al. 2007).

Within the crystal structure of CFH-6,7,8_{402H} (Prosser et al. 2007b) the structure of CCP 7 is identical to the NMR-derived solution structure of CFH-7 (Herbert et al. 2007), which increases confidence in the reliability of both solution and crystal structures. Comparison of the NMR structure (solved in absence of a GAG-like ligand) and the crystal structure (solved in the presence of bound SOS) reveals that the His402 side-chain adopts the same conformation in the absence or presence of bound SOS. The Tyr402 side-chain in the NMR structure of CFH-7_{402Y} also adopts this conformation. These observations justified the generation of a reliable model for the CFH-6,7,8_{402Y} structure, which unfortunately could not be crystallized (Prosser et al. 2007a).

4.6 Sulfated Sugar Recognition at the Polymorphic Residue

The earlier NMR chemical-shift perturbation studies had implied a direct role for the polymorphic residue in recognition of GAGs (Herbert et al. 2007). This was confirmed by the crystallographic structure of the complex that revealed a direct interaction between the AMD-associated His402 residue in CCP 7 and the sulfated-sugar ligand; one of the histidine ring nitrogens is hydrogen-bonded to a sulfate group of the ligand (Figs. 4 and 5). Together with His360 from CCP 6 His402 forms a histidine-clamp around the SOS sulfate groups, docking the ligand into the

CCP 6–7 interface. This mode of SOS recognition by His402 is incompatible – both chemically and sterically – with the presence of a bulkier tyrosine at the same position. This is not to imply that the Tyr402 variant is incapable of interacting with other sulfated-sugar ligands that are subtly different at the molecular level; indeed experimental data suggests CFH-6,7,8_{402Y} can bind to some GAGs more tightly than CFH-6,7,8_{402Y} depending upon GAG-type and level of sulfation (Clark et al. 2006; Herbert et al. 2007; Skerka et al. 2007). Other protein-sugar contacts in this region consist mainly of hydrogen-bonds and Van der Waals contacts with the protein backbone; there is an additional contribution from the side chain of the more distant Lys405, which has been previously implicated in heparin binding by mutagenesis (Clark et al. 2006; Giannakis et al. 2003).

It is noteworthy that the SOS at this site is present in two, slightly different conformations in the native protein crystals, related by a rotation of the sulfated fructose ring (a major conformation representing 65% of the population and a minor conformation representing 35%). This conformational difference between two forms of SOS is accompanied by a reorientation of the side-chain of Tyr390 to accommodate a repositioned sulfate group. In the ligand-free solution structure of CFH-7, only the minor conformation of the tyrosine side-chain is observed (Herbert et al. 2007), implying that movement of this side-chain may contribute to recognition of specific ligands. Such a role for Tyr390 in binding heparin fits well with NMR chemical shift perturbation studies on CFH-7 (Herbert et al. 2007) that also implicated this residue in GAG binding.

4.7 Additional Binding Sites for Sulfated-Sugar Within CFH-6,7,8

In addition to a SOS-binding site centred on the polymorphic residue 402, the co-crystal structure of CFH-6,7,8 showed that other parts of the surface of this region of CFH could be involved in specific GAG recognition. Furthermore, earlier work (Clark et al. 2006; Giannakis et al. 2003; Herbert et al. 2007), along with mutagenesis experiments designed specifically to test these additional sites (Prosser et al. 2007b), support a novel picture of GAG binding that incorporates extensive interactions over this region.

A secondary SOS-binding site is observed on the opposite face of CCP 7 in which the sugar sulfates contact the side-chains of Arg404 and Lys410, previously implicated in heparin binding by mutagenesis (Clark et al. 2006; Giannakis et al. 2003), and consistent with NMR chemical-shift perturbation studies on CFH-7 (Herbert et al. 2007). In addition to the involvement of H360 from CCP 6 in the major CCP 7-centered SOS-binding site, the structure also revealed a novel SOS-binding site contained entirely within CCP 6. This third site is chemically similar to that seen in CCP 7, with a central positively charged residue (Arg341) flanked by two histidine sidechains (His337 and His371) that directly coordinate two ligand sulfate groups. Finally, a fourth SOS-binding site in the crystal was observed in the linker between CCPs 7 and 8, and involves a salt bridge between sulfate and Arg444.

Previously, GAG-binding sites have been localised primarily to single modules of CFH – see above (Blackmore et al. 1996, 1998; Ormsby et al. 2006; Pangburn et al. 1991). Calculation of the electrostatic potential of the crystal structure reveals that the binding sites for SOS observed in the complex occupy a positively charged groove extending over all three modules of CFH-6,7,8_{402H} with the AMD-related polymorphism lying within this channel. Many other residues (e.g. Arg387, Lys388, Lys405) previously implicated in polyanion recognition by mutagenesis (Clark et al. 2006; Giannakis et al. 2003) also line this groove. Taken together this suggests an extended mode of interaction between GAGs and this region of CFH with the polyanion lying along the positively charged groove to form a large interaction surface. This hypothesis was supported by an analytical ultracentrifugation analysis of CFH-6,7,8 in complex with heparin dp18 or dp24.

5 Structure: Insights into Disease Mechanism

As discussed above, three regions of the CFH molecule have been highlighted in respect of its role in disease. To summarise: single and double amino-acid changes occurring predominantly in the C-terminal segment of factor H (CCPs 19 and 20) have been identified in some aHUS patients. (Buddles et al. 2000; Heinen et al. 2006; Richards et al. 2001; Warwicker et al. 1998) (http://www.fh-hus.org/); an SNP at the CFH locus (rs1061170) corresponding to the presence of His at position 402 (in CCP 7 of CFH), is one component of a major risk haplotype for AMD (Edwards et al. 2005; Hageman et al. 2005; Haines et al. 2005; Klein et al. 2005) There is evidence for at least one aHUS-linked mutation (Kavanagh et al. 2007) and an AMD-protective SNP in the N-terminal portion of factor H (CCPs 1-2) (Hageman et al. 2005). The SNPs linked to AMD (Abrera-Abeleda et al. 2006) are also linked to DDD/MPGNII (Hageman et al. 2005) along with a deletion in CCP 4 (Licht et al. 2006).

The results of structural and functional work carried out on various regions of CFH (outlined in the previous sections) afford the opportunity to explore the molecular basis of some of these diseases. A "GAG recognition" hypothesis is emerging as one mechanism amongst several possibilities for linking molecular changes to disease (Schmidt et al. 2008). This is founded upon the putative capability of CFH to discriminate between subtly different GAG molecules through specific contacts with, for example, sugar rings or sulfate groups. For the host, the advantage is that a specific recognition capability of CFH deters subversion by a pathogen that adopts a molecular mimicry strategy (Lindahl et al. 2000). A potential disadvantage is that GAG-specific CFH would have a diminished capacity to bind to the diversity of GAGs that is necessitated by unrelated aspects of cell and tissue physiology or that may arise from loss of homeostasis in stressed or ageing individuals (Verdugo and Ray 1997). GAGs form an extremely diverse group of molecules that vary according to tissue, cell type and age (Turnbull et al. 2001). Thus according to the GAG recognition hypothesis, CFH has different affinities for different tissues and this will vary over the lifetime of an individual. It is difficult to prove or disprove the notion that CFH can discriminate between closely related GAGs due to the difficulty of preparing a panel of chemically pure model GAG compounds with which to test the idea. Nonetheless, the available structural and functional data can be interpreted in a way that is broadly supportive of this model as follows:

(i) Different allotypic variations (at 402) of CFH (in the context of the CFH-6,7,8 fragment) have different affinities for certain types of GAG (Clark et al. 2006).

(ii) Disease-linked amino-acid substitutions in modules 7 and 20 coincide with discrete patches on the surface of these modules that have been identified as GAG-interaction sub-sites (Herbert et al. 2006, 2007). Therefore they are likely to modulate specific aspects of factor H-GAG interactions.

(iii) The two modules, CCP 7 and CCP 20, each lie within different segments of CFH that are unique in their proven affinities for heparin (Schmidt et al. 2008), which is a widely accepted although crude model for GAGs. It seems unlikely that the correspondence of the two principal GAG interaction sites with the two principal sites of disease-linked sequence variations is a mere coincidence.

(iv) The structure of CFH-6,7,8 (Prosser et al. 2007b) shows an interaction between His402 and a sulfate group in the ligand, SOS – another GAG model compound – that is incompatible with the presence of a Tyr at this position. This is direct evidence for the ability of CFH to "read" a pattern of sulfates and for such a pattern recognition functionality to be modulated by a disease-linked amino-acid switch.

(v) The presence of several sub-sites able to contact SOS within CFH-6,7,8 (Prosser et al. 2007b) implies binding of a GAG molecule across multiple modules and a consequent alteration in the shape of this segment of the protein. Scattering studies support this idea to some extent (Fernando et al. 2007). This scenario creates the potential for CFH to respond in different ways to different GAGs – thus adding an extra layer of specificity or selectivity to this interaction.

(vi) The absence of proven binding functionality amongst the CCPs 8-18 (Schmidt et al. 2008) of CFH implies they could have an architectural rather than a binding role. The preponderance of long linkers and small modules in the CCPs 12-14 region is consistent with a role as a "hinge" promoting the bending back upon itself of CFH. This could bring the two GAG-interaction sites into proximity creating the potential for higher specificity via a multi-site interaction.

(vii) Scattering studies indicate that CFH does indeed have a bent back structure (Aslam et al. 2001). Other studies indicate that antibodies to the C-terminus can modulate complement regulatory activity at the N-terminus adding further support to functionally critical proximity of the two ends of CFH (Oppermann et al. 2006).

Thus an attractive model, supported by most data but requiring further testing, is that the likelihood of excessive complement activation and attendant inflammation is inversely correlated to the affinity of CFH for self-surfaces. A diminishment in affinity could arise either from an amino-acid substitution in CFH; or it could stem from inappropriate levels or patterns of sulfation amongst surface GAGs. A third factor to consider is the level of complement activation in the immediate vicinity generating short-lived C3b molecules that will stretch local complement-regulatory capacity including that contributed by CFH. Some combination of two or all of these factors may be required for pathogenic levels of complement activation to occur. The variation of GAGs between tissue-types might explain organ-to-organ variations in the consequences (or lack of consequences) of CFH sequence variations; and given that GAG composition also changes with age, this would additionally explain the late onset of some symptoms. Moreover, certain locations may be particularly susceptible due to higher levels of apoptotic or necrotic activity –the macular RPE-choroid interface is one such region since Bruch's membrane separates the RPE from the epithelial cells of the choriocapallaris, thereby creating a potential subRPE entrapment site that accumulates debris (Sivaprasad et al. 2005).

Not all observations, however, fully support a GAG-recognition model. Several aHUS-linked mutations – within the context of the double-module fragment CFH-19,20 – do not appear to affect binding to heparin columns or defined-length, sulphated, heparin fragments (Jokiranta et al. 2006). This may simply reflect the inadequacy of heparin as an emulator of the physiological GAG ligand. On the other hand, several of these mutants appear to exhibit lower affinity for C3b (Jokiranta et al. 2006). Given that CFH probably binds to a composite binding surface consisting both of C3b and the surrounding GAGs, it is not unreasonable to propose that disturbance of either its protein binding or carbohydrate binding components could have similar physiological outcomes. Thus it is not necessary to abandon the idea of specific GAG recognition to accommodate these results. A more radical suggestion originating from the observation of a crystallographic tetramer for CFH-19,20 (Jokiranta et al. 2006) is that some mutations could interfere with a putative ologomerisation of CFH via the C-terminal region. There is little support so far in the literature for oligomerisation of CFH but it is an enticing idea (Aslam et al. 2001; Perkins et al. 1991). CFH recognition of CRP (Jarva et al. 1999) (Aronen et al. 1990) is also mapped to this region and, in the case of Y402H in module 7, there are multiple reports of defective C-reactive protein (CRP) recognition (Laine, Jarva et al. 2007; Sjoberg et al. 2007; Skerka et al. 2007; Yu et al. 2007) but these have been challenged on the basis that some were performed on CRP in a physiologically questionable non-pentameric state (Hakobyan 2007). On the other hand a "CRP hypothesis" to explain the mechanism of the AMD pathogenesis is seductive in that the putative CRP-CFH interaction has been reported to form part of a strategy for the recruitment of CFH to apoptotic cells so as to protect against alternative pathway activation and inflammation (Gershov et al. 2000). The presence of higher levels of CRP in the choroids of AMD patients (Johnson et al. 2006) further fuels this theory.

6 Options for Therapy

Activation of complement is part of a normal homeostatic mechanism that includes opsonization and lysis of microorganisms, removal of foreign particles and dead cells, recruitment and activation of inflammatory cells, regulation of antibody production, and the elimination of immune complexes (Kinoshita 1991; Markiewski and Lambris 2007). Under normal conditions, local complement activation is beneficial by promoting the rapid clearance of cell debris and facilitating the removal of toxic protein aggregates. Uncontrolled or aberrant activation and/or regulation of this system, however, can lead to bystander damage of host cells and tissues, thereby contributing significantly to the pathogenesis of a number of diseases in addition to those discussed above including rheumatoid arthritis, IgA nephropathy, asthma, systemic lupus erythematosus, and ischemia reperfusion injury (Markiewski et al. 2007; Thurman 2007; Thurman and Holers 2006). Much of the tissue damage in these diseases results from complement-mediated lysis of bystander host cells by the membrane attack complex or downstream effects resulting from excess production of the anaphylatoxin C3a by C3 convertases.

Based upon a surge of complement-related research over the last two decades, it has become apparent that modulation of the complement system is a promising strategy for drug discovery (Liszewski and Atkinson 1998; Ricklin and Lambris 2007; Ryan 1995; Sahu and Lambris 2000). Additional evidence supporting the development of complement modulating strategies comes from studies showing that complement inhibitors can reduce inflammation and tissue destruction and ameliorate disease progression in a variety of animal models of human disease (Bao et al. 2003; Holers 2003; Holers et al. 2002; Thurman et al. 2006). Importantly, the first complement pathway-directed therapeutic, a humanized monoclonal antibody directed against complement component C5 (Eculizumab; Soliris) was approved in March 2007 for use in the treatment of paroxysmal nocturnal hemoglobinuria (PNH), a debilitating disease characterized by chronic complement-mediated intravascular hemolysis (Rother et al. 2007). This breakthrough validates the complement system as a realistic therapeutic target and has provided a strong rationale for the investigation of other indications for this and other drugs in complement-related diseases.

Many promising complement pathway-directed therapeutics have been developed; these include various recombinant human complement inhibitors, monoclonal antibodies, synthetic peptides, and peptidomimetics designed to block activation of specific complement components, neutralize complement activation fragments, or antagonize complement receptors (Ricklin et al. 2007). Additional therapeutic compounds are currently in various investigational and developmental stages (Ricklin et al. 2007). None of the agents developed to date have been designed to specifically target CFH in diseases such as AMD, MPGNII, and aHUS. The more comprehensive understanding of the functional attributes of the risk and protective forms of CFH that has emerged from studies conducted over the past 2 years will no doubt hasten the rapid conception and development of CFH-directed drug candidates. That being said, important forethought will have to be given regarding length of administration (chronic vs. acute), mode of delivery (systemic vs. local) and method of delivery. These parameters will certainly affect decisions related to the nature (i.e. small molecule, large protein) of candidate drugs.

In contrast to a strategy of inhibition, one interesting operational paradigm being developed is based upon the concept that replacement or augmentation of the complement modulating activity of dysfunctional “risk” CFH protein with functional "protective" protein might be an effective strategy for preventing or delaying the pathology associated with CFH-mediated disease. Indeed, replacement of defective CFH with normal CFH protein through plasma exchange has been shown to be effective in preventing further disease episodes and normalizing kidney function in MPGN II and atypical hemolytic uremic syndrome (aHUS) patients harboring loss of function CFH mutations (Gerber et al. 2003; Licht et al. 2005; Pickering and Cook 2008; Stratton and Warwicker 2002). Similar results have been obtained in Yorkshire pigs with an inherited deficiency of CFH that develop lethal glomerulonephritis (Hegasy et al. 2002; Jansen et al. 1998).

Collectively, the robust efforts being expended to understand the structures, functions and roles of CFH and other complement-associated molecules will likely expedite the transition from early stage discovery and identification of therapeutic targets, to the translation of this information into clinically effective diagnostics and pharmaceutical treatment modalities for a number of diseases associated with aberrant regulation of the complement system.

References

Abrera-Abeleda, M.A., Nishimura, C., Smith, J.L., Sethi, S., McRae, J.L., Murphy, B.F., Silvestri, G., Skerka, C., Jozsi, M., Zipfel, P.F., et al. (2006). Variations in the complement regulatory genes factor H (CFH) and factor H related 5 (CFHR5) are associated with membranoproliferative glomerulonephritis type II (dense deposit disease). *Journal of Medical Genetics 43*, 582–589

Alsenz, J., Lambris, J.D., Schulz, T.F., and Dierich, M.P. (1984). Localization of the complement-component-C3b-binding site and the cofactor activity for factor I in the 38kDa tryptic fragment of factor H. *The Biochemical Journal 224*, 389–398

Alsenz, J., Schulz, T.F., Lambris, J.D., Sim, R.B., and Dierich, M.P. (1985). Structural and functional analysis of the complement component factor H with the use of different enzymes and monoclonal antibodies to factor H. *The Biochemical Journal 232*, 841–850

Anderson, D.H., Ozaki, S., Nealon, M., Neitz, J., Mullins, R.F., Hageman, G.S., and Johnson, L.V. (2001). Local cellular sources of apolipoprotein E in the human retina and retinal pigmented epithelium: implications for the process of drusen formation. *American Journal of Ophthalmology 131*, 767–781

Aronen, M., Leijala, M., and Meri, S. (1990). Value of C-reactive protein in reflecting the magnitude of complement activation in children undergoing open heart surgery. *Intensive Care Medicine 16*, 128–132

Aslam, M. and Perkins, S.J. (2001). Folded-back solution structure of monomeric factor H of human complement by synchrotron X-ray and neutron scattering, analytical ultracentrifugation and constrained molecular modelling. *Journal of Molecular Biology 309*, 1117–1138

Bao, L., Haas, M., Kraus, D.M., Hack, B.K., Rakstang, J.K., Holers, V.M., and Quigg, R.J. (2003). Administration of a soluble recombinant complement C3 inhibitor protects against renal disease in MRL/lpr mice. *Journal of American Society of Nephrology 14*, 670–679

Barlow, P.N., Baron, M., Norman, D.G., Day, A.J., Willis, A.C., Sim, R.B., and Campbell, I.D. (1991). Secondary structure of a complement control protein module by two-dimensional 1H NMR. *Biochemistry 30*, 997–1004

Barlow, P.N., Norman, D.G., Steinkasserer, A., Horne, T.J., Pearce, J., Driscoll, P.C., Sim, R.B., and Campbell, I.D. (1992). Solution structure of the fifth repeat of factor H: a second example of the complement control protein module. *Biochemistry 31*, 3626–3634

Barlow, P.N., Steinkasserer, A., Norman, D.G., Kieffer, B., Wiles, A.P., Sim, R.B., and Campbell, I.D. (1993). Solution structure of a pair of complement modules by nuclear magnetic resonance. *Journal of Molecular Biology 232*, 268–284

Blackmore, T.K., Sadlon, T.A., Ward, H.M., Lublin, D.M., and Gordon, D.L. (1996). Identification of a heparin binding domain in the seventh short consensus repeat of complement factor H. *Journal of Immunology 157*, 5422–5427

Blackmore, T.K., Hellwage, J., Sadlon, T.A., Higgs, N., Zipfel, P.F., Ward, H.M., and Gordon, D.L. (1998). Identification of the second heparin-binding domain in human complement factor H. *Journal of Immunology 160*, 3342–3348

Buddles, M.R., Donne, R.L., Richards, A., Goodship, J., and Goodship, T.H. (2000). Complement factor H gene mutation associated with autosomal recessive atypical hemolytic uremic syndrome. *American Journal of Human Genetics 66*, 1721–1722

Clark, S.J., Higman, V.A., Mulloy, B., Perkins, S.J., Lea, S.M., Sim, R.B., and Day, A.J. (2006). His-384 allotypic variant of factor H associated with age-related macular degeneration has different heparin binding properties from the non-disease-associated form. *The Journal of Biological Chemistry 281*, 24713–24720

Coffey, P.J., Gias, C., McDermott, C.J., Lundh, P., Pickering, M.C., Sethi, C., Bird, A., Fitzke, F.W., Maass, A., Chen, L.L., et al. (2007). Complement factor H deficiency in aged mice causes retinal abnormalities and visual dysfunction. *Proceedings of the National Academy of Sciences of the United States of America 104*, 16651–16656

Crabb, J.W., Miyagi, M., Gu, X., Shadrach, K., West, K.A., Sakaguchi, H., Kamei, M., Hasan, A., Yan, L., Rayborn, M.E., et al. (2002). Drusen proteome analysis: an approach to the etiology of age-related macular degeneration. *Proceedings of the National Academy of Sciences of the United States of America 99*, 14682–14687

de Cordoba, S.R. and de Jorge, E.G. (2008). Translational mini-review series on complement factor H: genetics and disease associations of human complement factor H. *Clinical and Experimental Immunology 151*, 1–13

Edwards, A.O., Ritter, R., III, Abel, K.J., Manning, A., Panhuysen, C., and Farrer, L.A. (2005). Complement factor H polymorphism and age-related macular degeneration. *Science 308*, 421–424

Esparza-Gordillo, J., Soria, J.M., Buil, A., Almasy, L., Blangero, J., Fontcuberta, J., and Rodriguez de Cordoba, S. (2004). Genetic and environmental factors influencing the human factor H plasma levels. *Immunogenetics 56*, 77–82

Estaller, C., Koistinen, V., Schwaeble, W., Dierich, M.P., and Weiss, E.H. (1991). Cloning of the 1.4-kb mRNA species of human complement factor H reveals a novel member of the short consensus repeat family related to the carboxy terminal of the classical 150-kDa molecule. *Journal of Immunology 146*, 3190–3196

Farries, T.C., Seya, T., Harrison, R.A., and Atkinson, J.P. (1990). Competition for binding sites on C3b by CR1, CR2, MCP, factor B and Factor H. *Complement and Inflammation 7*, 30–41

Fernando, A.N., Furtado, P.B., Clark, S.J., Gilbert, H.E., Day, A.J., Sim, R.B., and Perkins, S.J. (2007). Associative and structural properties of the region of complement factor H encompassing the Tyr402His disease-related polymorphism and its interactions with heparin. *Journal of Molecular Biology 368*, 564–581

Ferreira, V.P., Herbert, A.P., Hocking, H.G., Barlow, P.N., and Pangburn, M.K. (2006). Critical role of the C-terminal domains of factor H in regulating complement activation at cell surfaces. *Journal of Immunology 177*, 6308–6316

Gehrs, K.M., Anderson, D.H., Johnson, L.V., and Hageman, G.S. (2006). Age-related macular degeneration – emerging pathogenetic and therapeutic concepts. *Annals of Medicine 38*, 450–471

Gerber, A., Kirchhoff-Moradpour, A.H., Obieglo, S., Brandis, M., Kirschfink, M., Zipfel, P.F., Goodship, J.A., and Zimmerhackl, L.B. (2003). Successful (?) therapy of hemolytic-uremic syndrome with factor H abnormality. *Pediatric Nephrology (Berlin, Germany) 18*, 952–955

Gershov, D., Kim, S., Brot, N., and Elkon, K.B. (2000). C-Reactive protein binds to apoptotic cells, protects the cells from assembly of the terminal complement components, and sustains an antiinflammatory innate immune response: implications for systemic autoimmunity. *The Journal of Experimental Medicine 192*, 1353–1364

Giannakis, E., Jokiranta, T.S., Male, D.A., Ranganathan, S., Ormsby, R.J., Fischetti, V.A., Mold, C., and Gordon, D.L. (2003). A common site within factor H SCR 7 responsible for binding heparin, C-reactive protein and streptococcal M protein. *European Journal of Immunology 33*, 962–969

Gold, B., Merriam, J.E., Zernant, J., Hancox, L.S., Taiber, A.J., Gehrs, K., Cramer, K., Neel, J., Bergeron, J., Barile, G.R., et al. (2006). Variation in factor B (BF) and complement component 2 (C2) genes is associated with age-related macular degeneration. *Nature Genetics 38*, 458–462

Gordon, D.L., Kaufman, R.M., Blackmore, T.K., Kwong, J., and Lublin, D.M. (1995). Identification of complement regulatory domains in human factor H. *Journal of Immunology 155*, 348–356

Hageman, G.S., Mullins, R.F., Russell, S.R., Johnson, L.V., and Anderson, D.H. (1999). Vitronectin is a constituent of ocular drusen and the vitronectin gene is expressed in human retinal pigmented epithelial cells. *FASEB J 13*, 477–484

Hageman, G.S., Luthert, P.J., Victor Chong, N.H., Johnson, L.V., Anderson, D.H., and Mullins, R.F. (2001). An integrated hypothesis that considers drusen as biomarkers of immune-mediated processes at the RPE-Bruch's membrane interface in aging and age-related macular degeneration. *Progress in Retinal and Eye Research 20*, 705–732

Hageman, G.S., Anderson, D.H., Johnson, L.V., Hancox, L.S., Taiber, A.J., Hardisty, L.I., Hageman, J.L., Stockman, H.A., Borchardt, J.D., Gehrs, K.M., et al. (2005). A common haplotype in the complement regulatory gene factor H (HF1/CFH) predisposes individuals to age-related macular degeneration. *Proceedings of the National Academy of Sciences of the United States of America 102*, 7227–7232

Hageman, G.S., Hancox, L.S., Taiber, A.J., Gehrs, K.M., Anderson, D.H., Johnson, L.V., Radeke, M.J., Kavanagh, D., Richards, A., Atkinson, J., et al. (2006). Extended haplotypes in the complement factor H (CFH) and CFH-related (CFHR) family of genes protect against age-related macular degeneration: characterization, ethnic distribution and evolutionary implications. *Annals of Medicine 38*, 592–604

Haines, J.L., Hauser, M.A., Schmidt, S., Scott, W.K., Olson, L.M., Gallins, P., Spencer, K.L., Kwan, S.Y., Noureddine, M., Gilbert, J.R., et al. (2005). Complement factor H variant increases the risk of age-related macular degeneration. *Science* (New York, NY) *308*, 419–421

Hakobyan, S., Harris, C.L., van den Berg, C., Pepys, M.B., Morgan, B.P. (2007). Binding of factor H to C-reactive protein occurs only when the latter has undergone non-physiologic denaturation. *Molecular Immunology 44*, 3983–3984

Hamilton, G., Proitsi, P., Williams, J., O'Donovan, M., Owen, M., Powell, J., and Lovestone, S. (2007). Complement factor H Y402H polymorphism is not associated with late-onset Alzheimer's disease. *Neuromolecular Medicine 9*, 331–334

Hegasy, G.A., Manuelian, T., Hogasen, K., Jansen, J.H., and Zipfel, P.F. (2002). The molecular basis for hereditary porcine membranoproliferative glomerulonephritis type II: point mutations in the factor H coding sequence block protein secretion. *The American Journal of Pathology 161*, 2027–2034

Heinen, S., Sanchez-Corral, P., Jackson, M.S., Strain, L., Goodship, J.A., Kemp, E.J., Skerka, C., Jokiranta, T.S., Meyers, K., Wagner, E., et al. (2006). De novo gene conversion in the RCA gene cluster (1q32) causes mutations in complement factor H associated with atypical hemolytic uremic syndrome. *Human Mutation 27*, 292–293

Herbert, A.P., Uhrin, D., Lyon, M., Pangburn, M.K., and Barlow, P.N. (2006). Disease-associated sequence variations congregate in a polyanion recognition patch on human factor H revealed in three-dimensional structure. *The Journal of Biological Chemistry 281*, 16512–16520

Herbert, A.P., Deakin, J.A., Schmidt, C.Q., Blaum, B.S., Egan, C., Ferreira, V.P., Pangburn, M.K., Lyon, M., Uhrin, D., and Barlow, P.N. (2007). Structure shows that a glycosaminoglycan and protein recognition site in factor H is perturbed by age-related macular degeneration-linked single nucleotide polymorphism. *The Journal of Biological Chemistry 282*, 18960–18968

Holers, V.M. (2003). The complement system as a therapeutic target in autoimmunity. *Clinical Immunology* (Orlando, Fla) *107*, 140–151

Holers, V.M., Girardi, G., Mo, L., Guthridge, J.M., Molina, H., Pierangeli, S.S., Espinola, R., Xiaowei, L.E., Mao, D., Vialpando, C.G., et al. (2002). Complement C3 activation is required for antiphospholipid antibody-induced fetal loss. *The Journal of Experimental Medicine 195*, 211–220

Hughes, A.E., Orr, N., Esfandiary, H., Diaz-Torres, M., Goodship, T., and Chakravarthy, U. (2006). A common CFH haplotype, with deletion of CFHR1 and CFHR3, is associated with lower risk of age-related macular degeneration. *Nature Genetics 38*, 1173–1177

Jakobsdottir, J., Conley, Y.P., Weeks, D.E., Mah, T.S., Ferrell, R.E., and Gorin, M.B. (2005). Susceptibility genes for age-related maculopathy on chromosome 10q26. *American Journal of Human Genetics 77*, 389–407

Jansen, J.H., Hogasen, K., Harboe, M., and Hovig, T. (1998). In situ complement activation in porcine membranoproliferative glomerulonephritis type II. *Kidney International 53*, 331–349

Jarva, H., Jokiranta, T.S., Hellwage, J., Zipfel, P.F., and Meri, S. (1999). Regulation of complement activation by C-reactive protein: targeting the complement inhibitory activity of factor H by an interaction with short consensus repeat domains 7 and 8–11. *Journal of Immunology 163*, 3957–3962

Johnson, L.V., Ozaki, S., Staples, M.K., Erickson, P.A., and Anderson, D.H. (2000). A potential role for immune complex pathogenesis in drusen formation. *Experimental Eye Research 70*, 441–449

Johnson, L.V., Leitner, W.P., Staples, M.K., and Anderson, D.H. (2001). Complement activation and inflammatory processes in Drusen formation and age related macular degeneration. *Experimental Eye Research 73*, 887–896

Johnson, P.T., Betts, K.E., Radeke, M.J., Hageman, G.S., Anderson, D.H., and Johnson, L.V. (2006). Individuals homozygous for the age-related macular degeneration risk-conferring variant of complement factor H have elevated levels of CRP in the choroid. *Proceedings of the National Academy of Sciences of the United States of America 103*, 17456–17461

Jokiranta, T.S., Hellwage, J., Koistinen, V., Zipfel, P.F., and Meri, S. (2000). Each of the three binding sites on complement factor H interacts with a distinct site on C3b. *The Journal of Biological Chemistry 275*, 27657–27662

Jokiranta, T.S., Jaakola, V.P., Lehtinen, M.J., Parepalo, M., Meri, S., and Goldman, A. (2006). Structure of complement factor H carboxyl-terminus reveals molecular basis of atypical haemolytic uremic syndrome. *The EMBO Journal 25*, 1784–1794

Jouvin, M.H., Kazatchkine, M.D., Cahour, A., and Bernard, N. (1984). Lysine residues, but not carbohydrates, are required for the regulatory function of H on the amplification C3 convertase of complement. *Journal of Immunology 133*, 3250–3254

Kardys, I., Klaver, C.C., Despriet, D.D., Bergen, A.A., Uitterlinden, A.G., Hofman, A., Oostra, B.A., Van Duijn, C.M., de Jong, P.T., and Witteman, J.C. (2006). A common polymorphism in the complement factor H gene is associated with increased risk of myocardial infarction: the Rotterdam Study. *Journal of the American College of Cardiology 47*, 1568–1575

Kavanagh, D., Richards, A., Fremeaux-Bacchi, V., Noris, M., Goodship, T., Remuzzi, G., and Atkinson, J.P. (2007). Screening for complement system abnormalities in patients with atypical hemolytic uremic syndrome. *Clinical Journal of American Society of Nephrology 2*, 591–596

Kinoshita, T. (1991). Biology of complement: the overture. Immunology today *12*, 291–295

Klein, R.J., Zeiss, C., Chew, E.Y., Tsai, J.Y., Sackler, R.S., Haynes, C., Henning, A.K., SanGiovanni, J.P., Mane, S.M., Mayne, S.T., et al. (2005). Complement factor H polymorphism in age-related macular degeneration. *Science 308*, 385–389

Kuhn, S. and Zipfel, P.F. (1996). Mapping of the domains required for decay acceleration activity of the human factor H-like protein 1 and factor H. *European Journal of Immunology 26*, 2383–2387

Kuhn, S., Skerka, C., and Zipfel, P.F. (1995). Mapping of the complement regulatory domains in the human factor H-like protein 1 and in factor H1. *Journal of Immunology 155*, 5663–5670

Laine, M., Jarva, H., Seitsonen, S., Haapasalo, K., Lehtinen, M.J., Lindeman, N., Anderson, D.H., Johnson, P.T., Jarvela, I., Jokiranta, T.S., et al. (2007). Y402H polymorphism of complement factor H affects binding affinity to C-reactive protein. *Journal of Immunology 178*, 3831–3836

Li, M., Atmaca-Sonmez, P., Othman, M., Branham, K.E., Khanna, R., Wade, M.S., Li, Y., Liang, L., Zareparsi, S., Swaroop, A., et al. (2006). CFH haplotypes without the Y402H coding variant show strong association with susceptibility to age-related macular degeneration. *Nature Genetics 38*, 1049–1054

Licht, C., Weyersberg, A., Heinen, S., Stapenhorst, L., Devenge, J., Beck, B., Waldherr, R., Kirschfink, M., Zipfel, P.F., and Hoppe, B. (2005). Successful plasma therapy for atypical hemolytic uremic syndrome caused by factor H deficiency owing to a novel mutation in the complement cofactor protein domain 15. *American Journal of Kidney Disease 45*, 415–421

Licht, C., Heinen, S., Jozsi, M., Loschmann, I., Saunders, R.E., Perkins, S.J., Waldherr, R., Skerka, C., Kirschfink, M., Hoppe, B., et al. (2006). Deletion of Lys224 in regulatory domain 4 of Factor H reveals a novel pathomechanism for dense deposit disease (MPGN II). *Kidney International 70*, 42–50

Lindahl, G., Sjobring, U., and Johnsson, E. (2000). Human complement regulators: a major target for pathogenic microorganisms. *Current Opinion in Immunology 12*, 44–51

Liszewski, M.K., and Atkinson, J.P. (1998). Novel complement inhibitors. *Expert Opinion on Investigational Drugs 7*, 323–331

Male, D.A., Ormsby, R.J., Ranganathan, S., Giannakis, E., and Gordon, D.L. (2000). Complement factor H: sequence analysis of 221 kb of human genomic DNA containing the entire fH, fHR-1 and fHR-3 genes. *Molecular Immunology 37*, 41–52

Maller, J.B., Fagerness, J.A., Reynolds, R.C., Neale, B.M., Daly, M.J., and Seddon, J.M. (2007). Variation in complement factor 3 is associated with risk of age-related macular degeneration. *Nature Genetics 39*, 1200–1201

Markiewski, M.M. and Lambris, J.D. (2007). The role of complement in inflammatory diseases from behind the scenes into the spotlight. *The American Journal of Pathology 171*, 715–727

Meng, W., Hughes, A., Patterson, C.C., Belton, C., Kamaruddin, M.S., Horan, P.G., Kee, F., and McKeown, P.P. (2007). Genetic variants of complement factor H gene are not associated with premature coronary heart disease: a family-based study in the Irish population. *BMC Medical Genetics 8*, 62

Meri, S. and Pangburn, M.K. (1990). Discrimination between activators and nonactivators of the alternative pathway of complement: regulation via a sialic acid/polyanion binding site on factor H. *Proceedings of the National Academy of Sciences of the United States of America 87*, 3982–3986

Misasi, R., Huemer, H.P., Schwaeble, W., Solder, E., Larcher, C., and Dierich, M.P. (1989). Human complement factor H: an additional gene product of 43 kDa isolated from human plasma shows cofactor activity for the cleavage of the third component of complement. *European Journal of Immunology 19*, 1765–1768

Mooijaart, S.P., Koeijvoets, K.M., Sijbrands, E.J., Daha, M.R., and Westendorp, R.G. (2007). Complement Factor H polymorphism Y402H associates with inflammation, visual acuity, and cardiovascular mortality in the elderly population at large. *Experimental Gerontology 42*, 1116–1122

Mullins, R.F., Russell, S.R., Anderson, D.H., and Hageman, G.S. (2000). Drusen associated with aging and age-related macular degeneration contain proteins common to extracellular deposits associated with atherosclerosis, elastosis, amyloidosis, and dense deposit disease. *FASEB J 14*, 835–846

Mullins, R.F., Aptsiauri, N., and Hageman, G.S. (2001). Structure and composition of drusen associated with glomerulonephritis: implications for the role of complement activation in drusen biogenesis. *Eye* (London, England) *15*, 390–395

Nan, R., Gor, J., and Perkins, S.J. (2008). Implications of the progressive self-association of wild-type human factor H for complement regulation and disease. *Journal of Molecular Biology 375*, 891–900

Nicaud, V., Francomme, C., Ruidavets, J.B., Luc, G., Arveiler, D., Kee, F., Evans, A., Morrison, C., Blankenberg, S., Cambien, F., et al. (2007). Lack of association between complement factor H polymorphisms and coronary artery disease or myocardial infarction. *Journal of Molecular Medicine 85*, 771–775

Norman, D.G., Barlow, P.N., Baron, M., Day, A.J., Sim, R.B., and Campbell, I.D. (1991). Three-dimensional structure of a complement control protein module in solution. *Journal of Molecular Biology 219*, 717–725

Okamoto, H., Umeda, S., Obazawa, M., Minami, M., Noda, T., Mizota, A., Honda, M., Tanaka, M., Koyama, R., Takagi, I., et al. (2006). Complement factor H polymorphisms in Japanese population with age-related macular degeneration. *Molecular Vision 12*, 156–158

Okemefuna, A.I., Gilbert, H.E., Griggs, K.M., Ormsby, R.J., Gordon, D.L., and Perkins, S.J. (2008). The regulatory SCR-1/5 and cell surface-binding SCR-16/20 fragments of factor H reveal partially folded-back solution structures and different self-associative properties. *Journal of Molecular Biology 375*, 80–101

Oppermann, M., Manuelian, T., Jozsi, M., Brandt, E., Jokiranta, T.S., Heinen, S., Meri, S., Skerka, C., Gotze, O., and Zipfel, P.F. (2006). The C-terminus of complement regulator Factor H mediates target recognition: evidence for a compact conformation of the native protein. *Clinical and Experimental Immunology 144*, 342–352

Ormsby, R.J., Jokiranta, T.S., Duthy, T.G., Griggs, K.M., Sadlon, T.A., Giannakis, E., and Gordon, D.L. (2006). Localization of the third heparin-binding site in the human complement regulator factor H1. *Molecular Immunology 43*, 1624–1632

Pai, J.K., Manson, J.E., Rexrode, K.M., Albert, C.M., Hunter, D.J., and Rimm, E.B. (2007). Complement factor H (Y402H) polymorphism and risk of coronary heart disease in US men and women. *European Heart Journal 28*, 1297–1303

Pangburn, M.K., and Muller-Eberhard, H.J. (1983). Kinetic and thermodynamic analysis of the control of C3b by the complement regulatory proteins factors H and I. *Biochemistry 22*, 178–185

Pangburn, M.K., Schreiber, R.D., and Muller-Eberhard, H.J. (1977). Human complement C3b inactivator: isolation, characterization, and demonstration of an absolute requirement for the serum protein beta1H for cleavage of C3b and C4b in solution. *The Journal of Experimental Medicine 146*, 257–270

Pangburn, M.K., Atkinson, M.A., and Meri, S. (1991). Localization of the heparin-binding site on complement factor H. *The Journal of Biological Chemistry 266*, 16847–16853

Penfold, P.L., Madigan, M.C., Gillies, M.C., and Provis, J.M. (2001). Immunological and aetiological aspects of macular degeneration. *Progress in Retinal and Eye Research 20*, 385–414

Perkins, S.J., Nealis, A.S., and Sim, R.B. (1991). Oligomeric domain structure of human complement factor H by X-ray and neutron solution scattering. *Biochemistry 30*, 2847–2857

Perkins, S.J., Gilbert, H.E., Aslam, M., Hannan, J., Holers, V.M., and Goodship, T.H. (2002). Solution structures of complement components by X-ray and neutron scattering and analytical ultracentrifugation. *Biochemical Society Transactions 30*, 996–1001

Pickering, M.C. and Cook, H.T. (2008). Translational mini-review series on complement factor H: renal diseases associated with complement factor H: novel insights from humans and animals. *Clinical and Experimental Immunology 151*, 210–230

Pickering, M.C., Cook, H.T., Warren, J., Bygrave, A.E., Moss, J., Walport, M.J., and Botto, M. (2002). Uncontrolled C3 activation causes membranoproliferative glomerulonephritis in mice deficient in complement factor H. *Nature Genetics 31*, 424–428

Pickering, M.C., de Jorge, E.G., Martinez-Barricarte, R., Recalde, S., Garcia-Layana, A., Rose, K.L., Moss, J., Walport, M.J., Cook, H.T., de Cordoba, S.R., et al. (2007). Spontaneous hemolytic uremic syndrome triggered by complement factor H lacking surface recognition domains. *The Journal of Experimental Medicine 204*, 1249–1256

Prosser, B.E., Johnson, S., Roversi, P., Clark, S.J., Tarelli, E., Sim, R.B., Day, A.J., and Lea, S.M. (2007a). Expression, purification, cocrystallization and preliminary crystallographic analysis of sucrose octasulfate/human complement regulator factor H SCRs 6-8. *Acta Crystallographica 63*, 480–483

Prosser, B.E., Johnson, S., Roversi, P., Herbert, A.P., Blaum, B.S., Tyrrell, J., Jowitt, T.A., Clark, S.J., Tarelli, E., Uhrin, D., et al. (2007b). Structural basis for complement factor H linked age-related macular degeneration. *The Journal of Experimental Medicine 204*, 2277–2283

Pulido, J.S., McConnell, J.P., Lennon, R.J., Bryant, S.C., Peterson, L.M., Berger, P.B., Somers, V., and Highsmith, W.E. (2007). Relationship between age-related macular degeneration-associated variants of complement factor H and LOC387715 with coronary artery disease. *Mayo Clinic Proceedings 82*, 301–307

Richards, A., Buddles, M.R., Donne, R.L., Kaplan, B.S., Kirk, E., Venning, M.C., Tielemans, C.L., Goodship, J.A., and Goodship, T.H. (2001). Factor H mutations in hemolytic uremic syndrome cluster in exons 18–20, a domain important for host cell recognition. *American Journal of Human Genetics 68*, 485–490

Ricklin, D. and Lambris, J.D. (2007). Complement-targeted therapeutics. *Nature Biotechnology 25*, 1265–1275

Ripoche, J., Day, A.J., Willis, A.C., Belt, K.T., Campbell, R.D., and Sim, R.B. (1986). Partial characterization of human complement factor H by protein and cDNA sequencing: homology with other complement and non-complement proteins. *Bioscience Reports 6*, 65–72

Ripoche, J., Erdei, A., Gilbert, D., Al Salihi, A., Sim, R.B., and Fontaine, M. (1988). Two populations of complement factor H differ in their ability to bind to cell surfaces. *The Biochemical Journal 253*, 475–480

Rivera, A., Fisher, S.A., Fritsche, L.G., Keilhauer, C.N., Lichtner, P., Meitinger, T., and Weber, B.H. (2005). Hypothetical LOC387715 is a second major susceptibility gene for age-related macular degeneration, contributing independently of complement factor H to disease risk. *Human Molecular Genetics 14*, 3227–3236

Rodriguez de Cordoba, S., Lublin, D.M., Rubinstein, P., and Atkinson, J.P. (1985). Human genes for three complement components that regulate the activation of C3 are tightly linked. *The Journal of Experimental Medicine 161*, 1189–1195

Rother, R.P., Rollins, S.A., Mojcik, C.F., Brodsky, R.A., and Bell, L. (2007). Discovery and development of the complement inhibitor eculizumab for the treatment of paroxysmal nocturnal hemoglobinuria. *Nature Biotechnology 25*, 1256–1264

Ryan, U.S. (1995). Complement inhibitory therapeutics and xenotransplantation. *Nature Medicine 1*, 967–968

Sahu, A., and Lambris, J.D. (2000). Complement inhibitors: a resurgent concept in anti-inflammatory therapeutics. *Immunopharmacology 49*, 133–148

Schlaf, G., Demberg, T., Beisel, N., Schieferdecker, H.L., and Gotze, O. (2001). Expression and regulation of complement factors H and I in rat and human cells: some critical notes. *Molecular Immunology 38*, 231–239

Schmidt, C.Q., Herbert, A.P., Hocking, H.G., Uhrin, D., and Barlow, P.N. (2008). Translational mini-review series on complement factor H: structural and functional correlations for factor H. *Clinical and Experimental Immunology 151*, 14–24

Schwaeble, W., Zwirner, J., Schulz, T.F., Linke, R.P., Dierich, M.P., and Weiss, E.H. (1987). Human complement factor H: expression of an additional truncated gene product of 43 kDa in human liver. *European Journal of Immunology 17*, 1485–1489

Seddon, J.M., George, S., Rosner, B., and Klein, M.L. (2006). CFH gene variant, Y402H, and smoking, body mass index, environmental associations with advanced age-related macular degeneration. *Human Heredity 61*, 157–165

Sharma, A.K. and Pangburn, M.K. (1996). Identification of three physically and functionally distinct binding sites for C3b in human complement factor H by deletion mutagenesis. *Proceedings of the National Academy of Sciences of the United States of America 93*, 10996–11001

Sivaprasad, S., Chong, N.V., and Bailey, T.A. (2005). Serum elastin-derived peptides in age-related macular degeneration. *Investigative Ophthalmology and Visual Science 46*, 3046–3051

Sjoberg, A.P., Trouw, L.A., Clark, S.J., Sjolander, J., Heinegard, D., Sim, R.B., Day, A.J., and Blom, A.M. (2007). The factor H variant associated with age-related macular degeneration (His-384) and the non-disease-associated form bind differentially to C-reactive protein, fibromodulin, DNA, and necrotic cells. *The Journal of Biological Chemistry 282*, 10894–10900

Skerka, C., Lauer, N., Weinberger, A.A., Keilhauer, C.N., Suhnel, J., Smith, R., Schlotzer-Schrehardt, U., Fritsche, L., Heinen, S., Hartmann, A., et al. (2007). Defective complement control of factor H (Y402H) and FHL-1 in age-related macular degeneration. *Molecular Immunology 44*, 3398–3406

Soares, D.C., Gerloff, D.L., Syme, N.R., Coulson, A.F., Parkinson, J., and Barlow, P.N. (2005). Large-scale modelling as a route to multiple surface comparisons of the CCP module family. *Protein Engineering Design and Selection 18*, 379–388

Stark, K., Neureuther, K., Sedlacek, K., Hengstenberg, W., Fischer, M., Baessler, A., Wiedmann, S., Jeron, A., Holmer, S., Erdmann, J., et al. (2007). The common Y402H variant in complement factor H gene is not associated with susceptibility to myocardial infarction and its related risk factors. *Clinical Science* (Lond) *113*, 213–218

Stratton, J.D., and Warwicker, P. (2002). Successful treatment of factor H-related haemolytic uraemic syndrome. *Nephrology Dialysis Transplantation 17*, 684–685

Thurman, J.M. (2007). Triggers of inflammation after renal ischemia/reperfusion. *Clinical Immunology* (Orlando, Fla) *123*, 7–13

Thurman, J.M., and Holers, V.M. (2006). The central role of the alternative complement pathway in human disease. *Journal of Immunology 176*, 1305–1310

Topol, E.J., Smith, J., Plow, E.F., and Wang, Q.K. (2006). Genetic susceptibility to myocardial infarction and coronary artery disease. *Human Molecular Genetics 15 Spec No 2*, R117–R123

Turnbull, J., Powell, A., and Guimond, S. (2001). Heparan sulfate: decoding a dynamic multifunctional cell regulator. *Trends in Cell Bbiology 11*, 75–82

Verdugo, M.E. and Ray, J. (1997). Age-related increase in activity of specific lysosomal enzymes in the human retinal pigment epithelium. *Experimental Eye Research 65*, 231–240

Warwicker, P., Goodship, T.H., Donne, R.L., Pirson, Y., Nicholls, A., Ward, R.M., Turnpenny, P., and Goodship, J.A. (1998). Genetic studies into inherited and sporadic hemolytic uremic syndrome. *Kidney International 53*, 836–844

Weeks, D.E., Conley, Y.P., Tsai, H.J., Mah, T.S., Schmidt, S., Postel, E.A., Agarwal, A., Haines, J.L., Pericak-Vance, M.A., Rosenfeld, P.J., et al. (2004). Age-related maculopathy: a genomewide scan with continued evidence of susceptibility loci within the 1q31, 10q26, and 17q25 regions. *American Journal of Human Genetics 75*, 174–189

Yates, J.R., Sepp, T., Matharu, B.K., Khan, J.C., Thurlby, D.A., Shahid, H., Clayton, D.G., Hayward, C., Morgan, J., Wright, A.F., et al. (2007). Complement C3 variant and the risk of age-related macular degeneration. *The New England Journal of Medicine 357*, 553–561

Yu, J., Wiita, P., Kawaguchi, R., Honda, J., Jorgensen, A., Zhang, K., Fischetti, V.A., and Sun, H. (2007). Biochemical analysis of a common human polymorphism associated with age-related macular degeneration. *Biochemistry 46*, 8451–8461

Zarbin, M.A. (2004). Current concepts in the pathogenesis of age-related macular degeneration. *Archives of Ophthalmology 122*, 598–614

Zetterberg, M., Tasa, G., Palmer, M.S., Juronen, E., Teesalu, P., Blennow, K., and Zetterberg, H. (2007). Apolipoprotein E polymorphisms in patients with primary open-angle glaucoma. *American Journal of Ophthalmology 143*, 1059–1060

Zipfel, P.F. and Skerka, C. (1994). Complement factor H and related proteins: an expanding family of complement-regulatory proteins? *Immunology Today 15*, 121–126

Zipfel, P.F., Jokiranta, T.S., Hellwage, J., Koistinen, V., and Meri, S. (1999). The factor H protein family. *Immunopharmacology 42*, 53–60

11. Role of Complement in Motor Neuron Disease: Animal Models and Therapeutic Potential of Complement Inhibitors

Trent M. Woodruff[1], Kerina J. Costantini[2], Steve M. Taylor[3], and Peter G. Noakes[4]

School of Biomedical Sciences, University of Queensland, St Lucia, QLD 4072, Australia
[1]t.woodruff@uq.edu.au
[2]k.costantini@uq.edu.au
[3]s.taylor@uq.edu.au
[4]p.noakes@uq.edu.au

Abstract. Amyotrophic lateral sclerosis (ALS) is one of the major forms of motor neuron disease (MND), a group of degenerative disorders causing progressive motor neuron death leading to eventual paralysis and death. The pathogenesis of MND is poorly understood and may include genetic and/or environmental factors, with a common end-stage outcome. The majority of cases are sporadic, with a small percentage of familial cases identified. Mutations in the copper/zinc superoxide dismutase (SOD1) enzyme are frequent in familial ALS, and have allowed for the development of transgenic SOD1 rodent models of ALS. There has been evidence for immune system involvement in the disease, and activated components of the classical complement pathway have been observed in the serum, cerebrospinal fluid and neuronal tissue of diseased individuals. Furthermore, motor neurons and spinal cord tissue from SOD1 transgenic mice show an upregulation in C1q mRNA transcript and protein, in some cases prior to disease onset. Our laboratory has preliminary data indicating a specific pathogenic role for the activation fragment of complement C5 (C5a) in this disease. Using selective C5a receptor antagonists, we dosed SOD1 transgenic rats and observed an extension in survival and reduced motor symptoms compared to untreated rats. Collectively, these clinical and experimental findings suggest that targeting complement using specific inhibitors may represent a novel therapeutic approach to treating MND. Further experimental and clinical studies are required to validate this hypothesis. This review will summarize the clinical and experimental evidence to date implicating complement in the pathogenesis of MND.

1 Background

Motor neuron disease (MND) refers to a group of progressive, neurodegenerative conditions characterized by the selective loss of upper and/or lower motor neurons in the central nervous system. Amyotrophic lateral sclerosis (ALS) is the most prevalent form of MND; other forms include progressive muscular atrophy and progressive bulbar palsy. Afflicted individuals experience severe weakness, spasticity, and atrophy of the musculature and, ultimately, denervation of the

J.D. Lambris (ed.), *Current Topics in Complement II.*
doi: 10:1007/978-0-387-78952-1_11, © Springer Science+Business Media, LLC 2008

muscles necessary for respiration results in death (Brown 1995). ALS has a prevalence of approximately 6 per 100 000, with males more susceptible than females (1.6:1). The majority of cases (~90%) are sporadic with unknown etiology and no robust environmental risk factors – although remarkably, professional Italian soccer players have been found to have a substantially raised risk of developing the disease (Chio et al. 2005). The remaining ~10% of cases can be attributed to heredity, with familial forms being clinically indistinguishable from sporadic cases. The precise pathogenic mechanisms underlying MND are unclear, however glutamatergic neurotoxicity (Spalloni et al. 2004), mitochondrial dysfunction (Dupuis et al. 2003), aberrant protein aggregation (Shaw and Valentine 2007), and inflammation (Moreau et al. 2005) are all likely to contribute and interact in ways that are not yet clear. The present review focuses on the role of complement – a key component of innate immunity – in ALS pathophysiology, and the potential for specific complement inhibitors to be used as a novel treatment for ALS patients.

2 Animal Models of Motor Neuron Disease

2.1 Early Models of Motor Neuron Disease

Animal models of MND have been an indispensable tool for understanding both MND pathophysiology and for testing novel therapeutics. One of the first animal models of MND was the *wobbler* mouse, which exhibits a selective and progressive loss of motor neurons, forelimb weakness, paralysis and eventual death at 1 year of age (Andrews et al. 1974). The *wobbler* mouse remains a reliable model of selective anterior horn cell disease, despite disparities in amino acid and receptor distribution compared to MND patients (Krieger et al. 1991; Tomiyama et al. 1994). The *mnd* mouse was described by Messer and colleagues in 1986 and was originally thought to be a model for MND, however, the accumulation of adenosine triphosphate synthase subunit c has demonstrated that this mouse is instead a model for cerebral ceroid lipofuscinosis (Pardo et al. 1994). The *pmn* mouse, first described in 1991, displays hindlimb weakness as early as 3 weeks postnatal (Schmalbruch et al. 1991). However, unlike MND patients, *pmn* mice have prominent distal motor axonopathy whilst motor neuron somata are relatively preserved (Schmalbruch et al. 1991). Thus, many of these earlier murine models of MND have often failed to reliably encapsulate both the clinical and histopathological features of MND.

2.2 SOD1 Transgenic Model of Amyotrophic Lateral Sclerosis

The recent discovery that a subset of patients with familial ALS express a toxic gain-of-function mutation in the Cu/Zn superoxide dismutase one (SOD1) protein has addressed problems with earlier mouse models and has greatly expanded the range of potential experiments into MND. Under normal physiological conditions, the SOD1 protein encodes a cytosolic isoenzyme which buffers the potentially harmful actions of reactive oxygen species (Valentine et al. 2005). When mutated however, the SOD1 protein confers an autosomal dominant phenotype of ALS which shares identical pathological features to the sporadic form of the disease. In humans, approximately a quarter of all familial ALS cases can be attributed to mutations in SOD1 (Cleveland and Rothstein 2001; Valentine et al. 2005).

In 1994, Gurney and colleagues developed a transgenic mouse which expressed high levels of mutant human SOD1 ($SOD1^{G93A}$), in which the glycine residue at site 93 of the SOD1 protein was replaced with an alanine. The $SOD1^{G93A}$ mouse was found to closely mimic disease progression in humans, with mice exhibiting comparable motor (tremors and eventual paralysis) and histopathological (selective grey matter degeneration) features. The onset and rate of disease progression in SOD1 mice are affected by transgene copy number – mice expressing high levels of mutant protein experience early onset and a more rapid disease course than low transgene expressing mice (Gurney et al. 1994). To date, at least six additional lines of mice expressing variant mutations of SOD1 have been developed, including human $SOD1^{G37R}$ (Wong et al. 1995), $SOD1^{G85R}$ (Ripps et al. 1995), $SOD1^{G127X}$ (Jonsson et al. 2004), $SOD1^{G93R}$ (Friedlander et al. 1997), $SOD1^{D90A}$ (Jonsson et al. 2006) and mouse $SOD1^{G86R}$ (Ripps et al. 1995), all of which display differences in the onset and rate of disease progression.

More recently, the development of human $SOD1^{G93A}$ rats has permitted for a greater range of experiments due to their larger size, for example, studies that require intrathecal administration of drugs or the administration of neural stem cells (Howland et al. 2002). Compared to mouse models, the $SOD1^{G93A}$ rat displays a more rapid disease progression and a transient appearance of neuronal vacuoles (Howland et al. 2002). In addition, $SOD1^{G93A}$ rats on a Sprague-Dawley background display variability in disease severity dependent on the location of initial paralysis. Rats that develop forelimb paralysis prior to hindlimb paralysis exhibit faster disease progression, a phenomenon which reflects the phenotypic variability seen in human ALS patients (Abe et al. 1996). The mechanisms behind this variable phenotype remain unknown, however variable SOD1 expression, modifier genes, and differences in the stability of SOD1 protein aggregates have all be proposed to explain this disparity (Matsumoto et al. 2006).

SOD1 knockout mice and mice overexpressing human wildtype SOD1 ($SOD1^{WT}$) do not develop ALS (Wong et al. 1995; Reaume et al. 1996). Hence the ALS phenotype is caused by gain of a novel, toxic property of the SOD1 mutant enzyme, presumably unrelated to the SOD function. Further, introducing recombinant vectors that restrict SOD1 expression to either neurons or astrocytes does not result in neurodegeneration (Gong et al. 2000; Pramatarova et al. 2001). Rather, expression of the transgene in neurons and astrocytes is required for ALS pathogenesis, highlighting the importance of the immediate environment in neuronal injury (Wang et al. 2005). The precise mechanism by which mutant SOD1 exerts its toxicity remains elusive.

SOD1 transgenic rodent models most accurately recapitulate the major clinical and histopathological hallmarks of ALS and thus have been the standard model in which to evaluate novel drugs (Ripps et al. 1995; Lee et al. 1996). Further, these animals carry a genetic mutation demonstrated in a group of individuals with familial ALS. However, studies which appear to alleviate ALS in rodents do not always translate into successful human therapeutics. For example, creatine, a nitrogenous organic acid shown to improve mitochondrial function, was found to extend survival in SOD1 mice (Klivenyi et al. 1999) but had no beneficial effect in human trials (Groeneveld et al. 2003). Results should hence be interpreted with caution, taking into consideration both the faster rate of degeneration of motor neurons in high expressing SOD1 mice and the relative uncommonness of SOD1 mutations in ALS patients (1–2%).

3 Clinical Evidence for Complement Involvement

A study conducted in 1973 by Whittaker and colleagues, measured immunoglobulins and C3 in the serum of 38 MND patients in an attempt to identify an immune component of the disease. Although this initial study found no change in serum C3 levels, numerous studies have subsequently detected elevated complement factors in patients with MND (Table 1). In particular, components of the classical (or lectin) activation pathway have been frequently identified. The first of these studies examined C3 immunofluorescence in postmortem tissue obtained from 16 ALS patients (Donnenfeld et al. 1984). The study found significant C3 deposition in the spinal cord of six patients and in the motor cortex of five patients, which appeared to be on astrocyte-like cells, with no apparent neuronal staining (Donnenfeld et al. 1984). Subsequently Annunziata and Volpi (1985) measured C3c, C4, C1 inactivator, and C3 activator fractions in the serum and cerebrospinal fluid (CSF) of 13 ALS patients. They detected an increase in C3c in the CSF of these diseased patients compared to normal subjects, and no change in any of the other factors. Apostolski and colleagues (1991) measured serum C4, C3 and Factor B levels in 33 patients with ALS, and found an increase in C4 compared to normal patients. A separate study examined C3d and C4d in the motor cortex and spinal cord of eight ALS and five normal patients (Kawamata et al. 1992). They found clusters of C3d- and C4d-coated fibers particularly on oligodendroglia, but also on degenerating neurites in these regions. Staining for these components was very rare in non-diseased control patients (Kawamata et al. 1992). Tsuboi and Yamada (1994) investigated serum and CSF levels of C1q and C4d in 15 ALS patients and found significantly increased C4d levels in the CSF, which correlated with disease severity. Using a new sensitive method of protein detection (laser nephelometry), Trbojevic-Cepe and co-workers (1998) detected an upregulation of C4 in the CSF of 4 of the 12 MND patients examined. In 1999, Grewel and colleagues examined post-mortem tissue from six ALS and four control patients. They showed an overall 40% increase in C1q (subcomponent B) and clusterin mRNA in the motor cortex of ALS brains, with no change in the temporal cortex compared to non-diseased brains. An increase in mRNA for these complement factors was also seen in the spinal cord, particularly in the medial motor nucleus of the anterior grey horn (Grewal et al. 1999). Elevated mRNA for C2 has also been seen in ALS patients (Jiang et al. 2005). In this study, microarray analysis was performed on ventral horn spinal cord homogenates of five ALS patients and an increase in only 0.7% of genes examined (37/4,845) was found, with C2 mRNA shown to be elevated ~fourfold in MND patients compared with control patients (Jiang et al. 2005). Finally, Goldknopf and co-workers (2006) obtained serum from 422 patients with various neurological pathologies, including MND, and used 2D-gel electrophoresis to isolate proteins significantly increased in these diseased patients. Of the 34 proteins elevated compared to non-diseased control patients, nine were related to the complement system, suggesting a common link in complement activation amongst numerous neurodegenerative disorders. Specifically, in MND patients, C3c, C3dg, and Factor H were elevated compared to control individuals (Goldknopf et al. 2006).

These combined findings of increased complement factors in the serum, CSF and neurological tissue in MND patients strongly suggests an involvement of classical pathway complement activation in the disease process. Where these complement factors originate, and what triggers their activation is not currently known. Complement factors can be produced by all cells of the central nervous system (Gasque et al. 2000) and thus these factors may be produced locally, although there is blood-brain barrier (BBB) breakdown in the end stages of MND (Apostolski et al. 1991; Garbuzova-Davis et al. 2007), so the circulation could also be a source of these factors. Immunoglobulin deposits (Donnenfeld et al. 1984) and auto-antibodies (Niebroj-Dobosz et al. 2006) have been detected in the central nervous system of ALS patients, which could potentially initiate classical pathway activation on neuronal tissue. Whether complement activation is intimately involved in the initiation of MND or simply occurs as a secondary consequence of motor neuron insult is not known. Regardless, the overall evidence from these clinical studies is that complement system activation occurs in patients with MND, and may play a role in disease pathology. Evidence that this is also true in experimental models of MND is reviewed below.

Table 1 Clinical evidence of complement factors in patients with MND

Complement factor(s) elevated	Sample	Methods	Author/Study
C3	Spinal cord, motor cortex	Direct immunofluorescence	Donnenfeld et al. (1984)
C3c	Serum and CSF	Single radial immunodiffusion	Annunziata and Volpi (1985)
C4	Serum	Single radial immunodiffusion	Apostolski, et al. (1991)
C3d, C4d	Spinal cord, motor cortex	Immunohistochemistry	Kawamata et al. (1992)
C4d	CSF	Sandwich ELISA	Tsuboi and Yamada (1994)
C4	CSF	Laser nephelometry	Trbojevic-Cepe et al. (1998)
C1q, clusterin (mRNA)	Spinal cord, motor cortex	Northern blot, in situ hybridization	Grewal et al. (1999)
C2 (mRNA)	Spinal cord	Microarray	Jiang et al. (2005)
C3c, C3dg, Factor H	Serum	2D gel electrophoresis	Goldknopf et al. (2006)

4 Experimental Evidence for Complement Involvement

The advent of high capacity molecular methods such as microarray technologies has allowed investigators to examine widespread changes in gene expression in various disease models. This technique has been used by several investigators to examine changes in the expression of various genes in the transgenic mouse SOD1 model of ALS (Table 2). As described above, the SOD1 transgenic mouse model relies on the over-expression of mutant SOD1 enzymes, which leads to progressive symptoms similar to the human condition. An important consideration when interpreting microarray data in these studies is the use of appropriate controls to differentiate between gene expression changes due to potential disease-inducing effects of the mutant SOD1 enzyme, as compared to changes due purely to the over-expression of a foreign protein. Microarray studies using appropriate controls have found, with universal agreement, an involvement of the classical complement pathway, with elevated levels of C1q in particular (Table 2). The experimental evidence for these findings is described further below.

The first study to demonstrate experimentally an involvement of complement products in a SOD1 transgenic mouse model was performed by Perrin and colleagues in 2005. This group used laser-capture microdissection techniques to specifically isolate ventral motor neurons from the lumbar spinal cord of $SOD1^{G93A}$ mice. Using microarray analysis they detected a significant upregulation of all polypeptide subcomponents of C1q in early symptomatic (P90) and end-stage (P120) diseased mice, compared to motor neurons from non-transgenic, wild-type mice (~fivefold and eightfold respectively) (Perrin et al. 2005).

In 2007, Lobsiger and colleagues published their findings in two separate SOD1 transgenic mouse models, indicating a strong involvement of C1q in these ALS models. In these studies, the authors were careful to select an appropriate control, aware of previous studies' lack of specificity towards gene identification. Transgenic mice overexpressing human $SOD1^{WT}$ were used as a comparison strain. Complement factors consisted of one of the major groups identified using this method (Lobsiger et al. 2007). Laser-capture microdissection was again used to isolate lumbar motor neurons initially from $SOD1^{G37R}$ mice, prior to disease onset. Motor neurons taken from $SOD1^{G37R}$ mice, 2 months prior to clinical onset (P105) had upregulated genes for all three C1q subcomponents, compared to $SOD1^{WT}$ mice. Furthermore the complement regulatory molecule, decay accelerating factor (DAF) was found to be downregulated at this point. A second transgenic strain expressing an inactive mutant SOD1 enzyme ($SOD1^{G85R}$), which develops delayed and prolonged disease, also demonstrated upregulation in C1q subcomponents in presymptomatic mice. This indicates a common mechanism of toxicity from mutant SOD1 enzymes, divergent of the biochemical character. In this study, the authors further went on to demonstrate that protein for C1q was also expressed by motor neurons using immunohistochemical techniques on spinal cord sections of both $SOD1^{G37R}$ and $SOD1^{G85R}$ mice.

Table 2 Experimental evidence of complement factors in transgenic mouse ALS models

Complement factor(s) elevated	Sample	Transgenic model	Author/Study
C1q (mRNA)	Ventral motor neurons	Mouse SOD1 ^{G93A}	Perrin et al. (2005)
C1q, DAF (mRNA, protein)	Ventral motor neurons	Mouse SOD1^{G37R} and SOD1^{G85R}	Lobsiger et al. (2007)
C1q, C4 (mRNA)	Ventral motor neurons	Mouse SOD1^{G93A}	Ferraiuolo et al. (2007)
C1q (mRNA)	Lumbar spinal cord	Mouse SOD1^{L126delTT}	Fukada et al. (2007)

A separate study, also published in 2007, used very similar techniques to those above to further demonstrate complement involvement in SOD1^{G93A} mice (Ferraiuolo et al. 2007). This study again used SOD1^{WT} mice, as well as non-transgenic, wild-type mice as appropriate controls. Using laser-capture microdissection and microarray techniques, lumbar motor neurons from mice in the late, symptomatic stage of disease (P120) had significant upregulation in genes for C1q (subcomponent B) and C4. Microarray analysis revealed that increases in these factors were not seen in presymptomatic (P60) mice or at disease onset (P90). Using semi-quantitative reverse transcription-PCR however, C1q (subcomponent B) mRNA was found at disease onset (P90) as well as late-stage (P120) mice (~sevenfold and eightfold respectively above controls).

The above three studies all demonstrate that motor neurons from different SOD1 transgenic mice express elevated C1q mRNA (and protein) at varying stages in the disease process. A very recent study published in 2007 also used microarray analysis in a separate SOD1 transgenic mouse model, using whole lumbar spinal cord homogenates (Fukada et al. 2007). This study used SOD1^{L126delTT} transgenic mice and found elevated C1q (subcomponent B) mRNA in post-symptomatic (P154) mice compared to non-transgenic, wild-type mice. Semi-quantitative real-time PCR also demonstrated a ~tenfold increase in these C1q mRNA levels compared to pre-symptomatic (P98) levels. Although the precise cells producing this mRNA was not be determined, this study is in harmony with the other studies mentioned above, strongly indicating complement activation in SOD1 rodents.

Our laboratory has performed preliminary experiments in the rat SOD1^{G93A} transgenic model of ALS. Using specific C5a receptor antagonists developed in our laboratories, we dosed animals prior to the onset of disease (P70) and found a delay in the onset of motor symptoms and an extension in survival in drug-treated rats (Woodruff and Taylor 2005). We have also recently shown an up regulation of C5aR in motor neurons and astrocytes during disease progression in these transgenic rats (Denny et al. 2006). Together, these preliminary experiments demonstrate a potential pathogenic role for C5a in the rat $SOD1^{G93A}$ transgenic model of ALS.

5 Therapeutic Possibilities

To date, riluzole (Rilutek; Aventis Pharmaceuticals Inc), is the only approved therapeutic to treat MND. One of the major actions of riluzole is to prevent the pre-synaptic release of glutamate (Miller et al. 2007). In clinical trials, it has been shown to extend survival by around 2–3 months and delay the onset of ventilator-dependence or tracheostomy (Miller et al. 2007). Given this modest extension in survival, there is an urgent need to develop new therapeutics which significantly extend life span and improve morbidity in MND.

There have been numerous studies conducted using different therapeutic agents and strategies, which have had varying success in treating animals with experimental ALS and MND. Some of the more promising candidates include minocylcin, recombinant growth factors, anti-glutamatergic agents and stem cell therapies (Mitchell and Borasio 2007). However, there is a distinct lack of translation of laboratory evidence to successful clinical trials in MND (Benatar 2007). This indicates either an inability for current animal models to properly model human MND, or differential effects of these therapies in rodents vs. humans; no single animal model can ensure clinical success. Despite this lack of translation to the clinic, studies and trials still need to be conducted to identify new therapeutic targets. We have a primitive understanding of MND, and new drug targets are essential to improve therapeutic outcomes. The complement system would appear to be a viable pathway to target, given the clinical evidence of complement involvement in this disease (McGeer and McGeer 2002). This sentiment is supported by our preliminary work which indicates C5a is involved in the development of pathology in transgenic $SOD1^{G93A}$ rats (Woodruff and Taylor 2005).

Our laboratory has developed a series of cyclic peptide C5a receptor antagonists which are potent inhibitors of C5a receptors on human inflammatory cells (Finch et al. 1999; March et al. 2004). These compounds also appear to bind with high affinity to polymorphonuclear cells (PMNs) isolated from rats and dogs, but not guinea-pig, rabbit, pig and mouse cells (Woodruff et al. 2001), however evidence exists of high binding affinity in mice neutrophils using a different assay (Huber-Lang et al. 2002). The lead compound (PMX53) in these series of C5a antagonists of structure Ac-Phe-[Orn-Pro-(D-Cyclohexylalanine)-Trp-Arg], has been demonstrated to display therapeutic efficacy in a range of inflammatory disease models. These include models of: arthritis (Woodruff et al. 2002); ischemia-reperfusion injuries of the gut (Arumugam et al. 2002), kidney

(Arumugam et al. 2003), liver (Arumugam et al. 2004) and limb (Woodruff et al. 2004); inflammatory bowel disease (Woodruff et al. 2003); sepsis (Strachan et al. 2000; Huber-Lang et al. 2002); ruptured abdominal aortic aneurysm (Harkin et al. 2004); immune-complex mediated reactions of the peritoneum (Strachan et al. 2000) and dermis (Strachan et al. 2001); renal (Boor et al. 2007) and liver fibrosis (Hillebrandt et al. 2005); experimental lupus nephritis (Bao et al. 2005); antibody-dependent (Girardi et al. 2003) and independent (Girardi et al. 2006) fetal injury and loss; traumatic brain cryoinjury (Sewell et al. 2004) and incisional (Clark et al. 2006) and neuropathic (Griffin et al. 2007) pain. In regards to central nervous system diseases, our group has also recently shown a therapeutic effect of this drug in a short-term model of neurodegeneration (Woodruff et al. 2006a). In this study, we also tested another analogue of PMX53 (PMX205) structure hydrocinnamate-[Orn-Pro-(D-Cyclohexylalanine)-Trp-Arg]. We found this compound to display similar effects to PMX53, however with a trend towards increased efficacy in this model, associated with an apparent increased BBB penetration of PMX205 over PMX53 (Woodruff et al. 2006a). In previous studies, we also showed an increased potency of PMX205 over PMX53 in a rat model of inflammatory bowel disease (Woodruff et al. 2006). Given the potential increase in potency and efficacy of PMX205 and its apparent increased BBB permeability, it would be this particular compound we would promote for any future clinical trialing in MND (Woodruff and Taylor 2005).

In addition to inhibiting C5a receptors, targeting other factors in the complement cascade may also prove to be a viable therapeutic option to treat MND. Several complement inhibitors have been developed over recent years, and compounds such as sCR1, C5 antibodies, compstatin or others (Mollnes 2004) could be used as potential therapies for MND. The need to chronically administer a drug in MND however promotes the use of the C5a receptor antagonists such as PMX53 and PMX205 given their small size, oral activity, BBB permeability and selective nature. Their selectivity allows for the production of all other complement factors, including the terminal membrane attack complex, thus reducing immune suppression, a likely side effect of other inhibitors of complement which act higher in the cascade. PMX53 also appears to be safe when administered to humans, successfully completing three Phase I/IIa clinical trials (Woodruff et al. 2006b).

Combined therapies will most likely be needed to effectively treat MND – indeed most trials require patients to remain on riluzole on ethical grounds. Using a compound which has different mechanism of action to riluzole (for example the anti-inflammatory activity of C5a receptor antagonists) may be more effective when administered as a combined therapy. Extensive controlled clinical trials will need to be conducted in order to ascertain any potential therapeutic benefit of a complement inhibitor. The devastating and intractable nature of MND allows for urgent consideration of anti-complement agents in these trials. Further experimental work in other models of MND using different doses and time-frames for treatment should be conducted before these human trials are undertaken (Ludolph et al. 2007).

6 Summary

There are several compelling lines of evidence to implicate the involvement of classical complement pathway activation in MND. These include the detection of elevated levels of complement activation fragments in the serum, CSF, spinal cord and motor cortex of patients with this condition. Recently it was discovered that in transgenic SOD1 mice, motor neurons express high levels of C1q mRNA and its protein implicating complement activity in these ALS disease models. Our laboratory has preliminary data indicating that blockade of the C5a receptor using a specific C5a receptor antagonist ameliorates disease symptoms in a rat model of ALS. Collectively, these studies suggest that complement activation and subsequent production of bioactive factors plays a role in the development of pathology of MND. Inhibitors to target these complement factors could be a novel therapeutic option for the treatment of MND.

Acknowledgements

We would like to thank the Motor Neuron Disease Research Institute of Australia and the National Health and Medical Research Council of Australia for grants supporting our research.

References

Abe, K., Aoki, M., Ikeda, M., Watanabe, M., Hirai, S. and Itoyama, Y. (1996) Clinical characteristics of familial amyotrophic lateral sclerosis with Cu/Zn superoxide dismutase gene mutations. *J Neurol Sci*, 136, 108–116

Andrews, J.M., Gardner, M.B., Wolfgram, F.J., Ellison, G.W., Porter, D.D. and Brandkamp, W.W. (1974) Studies on a murine form of spontaneous lower motor neuron degeneration – the wobbler (wa) mouse. *Am J Pathol*, 76, 63–78

Annunziata, P. and Volpi, N. (1985) High-levels of C3c in the cerebrospinal-fluid from amyotrophic lateral sclerosis patients. *Acta Neurol Scand*, 72, 61–64

Apostolski, S., Nikolic, J., Bugarski-Prokopljevic, C., Miletic, V., Pavlovic, S. and Filipovic, S. (1991) Serum and CSF immunological findings in ALS. *Acta Neurol Scand*, 83, 96–98

Arumugam, T.V., Shiels, I.A., Woodruff, T.M., Reid, R.C., Fairlie, D.P. and Taylor, S.M. (2002) Protective effect of a new C5a receptor antagonist against ischemia-reperfusion injury in the rat small intestine. *J Surg Res*, 103, 260–267

Arumugam, T.V., Shiels, I.A., Strachan, A.J., Abbenante, G., Fairlie, D.P. and Taylor, S.M. (2003) A small molecule C5a receptor antagonist protects kidneys from ischemia/reperfusion injury in rats. *Kidney Int*, 63, 134–142

Arumugam, T.V., Woodruff, T.M., Stocks, S.Z., Proctor, L.M., Pollitt, S., Shiels, I.A., Reid, R.C., Fairlie, D.P. and Taylor, S.M. (2004) Protective effect of a human C5a receptor antagonist against hepatic ischaemia-reperfusion injury in rats. *J Hepatol*, 40, 934–941

Bao, L., Osawe, I., Puri, T., Lambris, J.D., Haas, M. and Quigg, R.J. (2005) C5a promotes development of experimental lupus nephritis which can be blocked with a specific receptor antagonist. *Eur J Immunol*, 35, 2496–2506

Benatar, M. (2007) Lost in translation: Treatment trials in the SOD1 mouse and in human ALS. *Neurobiol Dis*, 26, 1–13

Boor, P., Konieczny, A., Villa, L., Schult, A.L., Bucher, E., Rong, S., Kunter, U., van Roeyen, C.R.C., Polakowski, T., Hawlisch, H., Hillebrandt, S., Lammert, F., Eitner, F., Floege, J. and Ostendorf, T. (2007) Complement C5 mediates experimental tubulointerstitial fibrosis. *J Am Soc Nephrol*, 18, 1508–1515

Brown, R.H., Jr. (1995) Amyotrophic lateral sclerosis: recent insights from genetics and transgenic mice. *Cell*, 80, 687–692

Chio, A., Benzi, G., Dossena, M., Mutani, R. and Mora, G. (2005) Severely increased risk of amyotrophic lateral sclerosis among Italian professional football players. *Brain*, 128, 472–476

Clark, J.D., Qiao, Y.L., Li, X.Q., Shi, X.Y., Angst, M.S. and Yeomans, D.C. (2006) Blockade of the complement C5a receptor reduces incisional allodynia, edema, and cytokine expression. *Anesthesiology*, 104, 1274–1282

Cleveland, D.W. and Rothstein, J.D. (2001) From Charcot to Lou Gehrig: deciphering selective motor neuron death in ALS. *Nat Rev Neurosci,* 2, 806–819

Denny, K.J., Crane, J.W., Taylor, S.M. and Noakes, P.G. (2006) Differential localization and expression of complement in a rat model of motor neuron disease. *Proc Aust Soc Biochem Mol Biol*, 38, 58

Donnenfeld, H., Kascsak, R.J. and Bartfeld, H. (1984) Deposits of IgG and C3 in the spinal cord and motor cortex of ALS patients. *J Neuroimmunol*, 6, 51–57

Dupuis, L., di Scala, F., Rene, F., de Tapia, M., Oudart, H., Pradat, P.F., Meininger, V. and Loeffler, J.P. (2003) Up-regulation of mitochondrial uncoupling protein 3 reveals an early muscular metabolic defect in amyotrophic lateral sclerosis. *FASEB J*, 17, 2091–2093

Ferraiuolo, L., Heath, P.R., Holden, H., Kasher, P., Kirby, J. and Shaw, P.J. (2007) Microarray analysis of the cellular pathways involved in the adaptation to and progression of motor neuron injury in the SOD1 G93A mouse model of familial ALS. *J Neurosci*, 27, 9201–9219

Finch, A.M., Wong, A.K., Paczkowski, N.J., Wadi, S.K., Craik, D.J., Fairlie, D.P. and Taylor, S.M. (1999) Low-molecular-weight peptidic and cyclic antagonists of the receptor for the complement factor C5a. *J Med Chem*, 42, 1965–1974

Friedlander, R.M., Brown, R.H., Gagliardini, V., Wang, J. and Yuan, J. (1997) Inhibition of ICE slows ALS in mice. *Nature*, 388, 31

Fukada, Y., Yasui, K., Kitayama, M., Doi, K., Nakano, T., Watanabe, Y. and Nakashima, K. (2007) Gene expression analysis of the murine model of amyotrophic lateral sclerosis: Studies of the Leu126delTT mutation in SOD1. *Brain Res*, 1160, 1–10

Friedlander, R.M., Brown, R.H., Gagliardini, V., Wang, J. and Yuan, J. (1997) Inhibition of ICE slows ALS in mice. *Nature*, 388, 31

Garbuzova-Davis, S., Haller, E., Saporta, S., Kolomey, I., Nicosia, S.V. and Sanberg, P.R. (2007) Ultrastructure of blood-brain barrier and blood-spinal cord barrier in SOD1 mice modeling ALS. *Brain Res*, 1157, 126–137

Gasque, P., Dean, Y.D., McGreal, E.P., VanBeek, J. and Morgan, B.P. (2000) Complement components of the innate immune system in health and disease in the CNS. *Immunopharmacology*, 49, 171–186

Girardi, G., Berman, J., Redecha, P., Spruce, L., Thurman, J.M., Kraus, D., Hollmann, T.J., Casali, P., Caroll, M.C., Wetsel, R.A., Lambris, J.D., Holers, V.M. and Salmon, J.E. (2003) Complement C5a receptors and neutrophils mediate fetal injury in the antiphospholipid syndrome. *J Clin Invest*, 112, 1644–1654

Girardi, G., Yarilin, D., Thurman, J.M., Holers, V.M. and Salmon, J.E. (2006) Complement activation induces dysregulation of angiogenic factors and causes fetal rejection and growth restriction. *J Exp Med*, 203, 2165–2175

Goldknopf, I.L., Sheta, E.A., Bryson, J., Folsom, B., Wilson, C., Duty, J., Yen, A.A. and Appel, S.H. (2006) Complement C3c and related protein biomarkers in amyotrophic lateral sclerosis and Parkinson's disease. *Biochem Biophys Res Commun*, 342, 1034–1039

Gong, Y.H., Parsadanian, A.S., Andreeva, A., Snider, W.D. and Elliott, J.L. (2000) Restricted expression of G86R Cu/Zn superoxide dismutase in astrocytes results in astrocytosis but does not cause motoneuron degeneration. *J Neurosci*, 20, 660–665

Grewal, R.P., Morgan, T.E. and Finch, C.E. (1999) C1qB and clusterin mRNA increase in association with neurodegeneration in sporadic amyotrophic lateral sclerosis. *Neurosci Lett*, 271, 65–67

Griffin, R.S., Costigan, M., Brenner, G.J., Ma, C.H.E., Scholz, J., Moss, A., Allchorne, A.J., Stahl, G.L. and Woolf, C.J. (2007) Complement induction in spinal cord microglia results in anaphylatoxin C5a-mediated pain hypersensitivity. *J Neurosci*, 27, 8699–8708

Groeneveld, G.J., Veldink, J.H., van der Tweel, I., Kalmijn, S., Beijer, C., de Visser, M., Wokke, J.H., Franssen, H. and van den Berg, L.H. (2003) A randomized sequential trial of creatine in amyotrophic lateral sclerosis. *Ann Neurol*, 53, 437–445

Gurney, M.E., Pu, H., Chiu, A.Y., Dal Canto, M.C., Polchow, C.Y., Alexander, D.D., Caliendo, J., Hentati, A., Kwon, Y.W., Deng, H.X. and et al. (1994) Motor neuron degeneration in mice that express a human Cu,Zn superoxide dismutase mutation. *Science*, 264, 1772–1775

Harkin, D.W., Romaschin, A., Taylor, S.M., Rubin, B.B. and Lindsay, T.F. (2004) Complement C5a receptor antagonist attenuates multiple organ injury in a model of ruptured abdominal aortic aneurysm. *J Vasc Surg*, 39, 196–206

Hillebrandt, S., Wasmuth, H.E., Weiskirchen, R., Hellerbrand, C., Keppeler, H., Werth, A., Schirin-Sokhan, R., Wilkens, G., Geier, A., Lorenzen, J., Kohl, J., Gressner, A.M., Matern, S. and Lammert, F. (2005) Complement factor 5 is a quantitative trait gene that modifies liver fibrogenesis in mice and humans. *Nat Genet*, 37, 835–843

Howland, D.S., Liu, J., She, Y.J., Goad, B., Maragakis, N.J., Kim, B., Erickson, J., Kulik, J., DeVito, L., Psaltis, G., DeGennaro, L.J., Cleveland, D.W. and Rothstein, J.D. (2002) Focal loss of the glutamate transporter EAAT2 in a transgenic rat model of SOD1 mutant-mediated amyotrophic lateral sclerosis (ALS). *Proc Natl Acad Sci U S A*, 99, 1604–1609

Huber-Lang, M.S., Riedeman, N.C., Sarma, J.V., Younkin, E.M., McGuire, S.R., Laudes, I.J., Lu, K.T., Guo, R.F., Neff, T.A., Padgaonkar, V.A., Lambris, J.D., Spruce, L., Mastellos, D., Zetoune, F.S. and Ward, P.A. (2002) Protection of innate immunity by C5aR antagonist in septic mice. *FASEB J*, 16, 1567–1574

Jiang, Y.M., Yamamoto, M., Kobayashi, Y., Yoshihara, T., Liang, Y., Terao, S., Takeuchi, H., Ishigaki, S., Katsuno, M., Adachi, H., Niwa, J., Tanaka, F., Doyu, M., Yoshida, M., Hashizume, Y. and Sobue, G. (2005) Gene expression profile of spinal motor neurons in sporadic amyotrophic lateral sclerosis. *Ann Neurol*, 57, 236–251

Jonsson, P.A., Ernhill, K., Andersen, P.M., Bergemalm, D., Brannstrom, T., Gredal, O., Nilsson, P. and Marklund, S.L. (2004) Minute quantities of misfolded mutant superoxide dismutase-1 cause amyotrophic lateral sclerosis. *Brain*, 127, 73–88

Jonsson, P.A., Graffmo, K.S., Brannstrom, T., Nilsson, P., Andersen, P.M. and Marklund, S.L. (2006) Motor neuron disease in mice expressing the wild type-like D90A mutant superoxide dismutase-1. *J Neuropathol Exp Neurol*, 65, 1126–1136

Kawamata, T., Akiyama, H., Yamada, T. and McGeer, P.L. (1992) Immunologic reactions in amyotrophic lateral sclerosis brain and spinal cord tissue. *Am J Pathol*, 140, 691–707

Klivenyi, P., Ferrante, R.J., Matthews, R.T., Bogdanov, M.B., Klein, A.M., Andreassen, O.A., Mueller, G., Wermer, M., Kaddurah-Daouk, R. and Beal, M.F. (1999) Neuroprotective effects of creatine in a transgenic animal model of amyotrophic lateral sclerosis. *Nat Med*, 5, 347–350

Krieger, C., Perry, T.L., Hansen, S. and Mitsumoto, H. (1991) The wobbler mouse: amino acid contents in brain and spinal cord. *Brain Res*, 551, 142–144

Lee, M.K., Borchelt, D.R., Wong, P.C., Sisodia, S.S. and Price, D.L. (1996) Transgenic models of neurodegenerative diseases. *Curr Opin Neurobiol*, 6, 651–660

Lobsiger, C.S., Boillee, S. and Cleveland, D.W. (2007) Toxicity from different SOD1 mutants dysregulates the complement system and the neuronal regenerative response in ALS motor neurons. *Proc Natl Acad Sci U S A*, 104, 7319–7326

Ludolph, A.C., Bendotti, C., Blaugrund, E., Hengerer, B., Loffler, J.P., Martin, J., Meininger, V., Meyer, T., Moussaoui, S., Robberecht, W., Scott, S., Silani, V. and Van Den Berg, L.H. (2007) Guidelines for the preclinical in vivo evaluation of pharmacological active drugs for ALS/MND: report on the 142nd ENMC international workshop. *Amyotroph Lateral Scler*, 8, 217–223

March, D.R., Proctor, L.M., Stoermer, M.J., Sbaglia, R., Abbenante, G., Reid, R.C., Woodruff, T.M., Wadi, K., Paczkowski, N., Tyndall, J.D., Taylor, S.M. and Fairlie, D.P. (2004) Potent cyclic antagonists of the complement C5a receptor on human polymorphonuclear leukocytes. Relationships between structures and activity. *Mol Pharmacol*, 65, 868–879

Matsumoto, A., Okada, Y., Nakamichi, M., Nakamura, M., Toyama, Y., Sobue, G., Nagai, M., Aoki, M., Itoyama, Y. and Okano, H. (2006) Disease progression of human SOD1 (G93A) transgenic ALS model rats. *J Neurosci Res*, 83, 119–133

McGeer, P.L. and McGeer, E.G. (2002) Inflammatory processes in amyotrophic lateral sclerosis. *Muscle Nerve*, 26, 459–470

Messer, A. and Flaherty, L. (1986) Autosomal dominance in a late-onset motor neuron disease in the mouse. *J Neurogenet*, 3, 345–355

Miller, R.G., Mitchell, J.D., Lyon, M. and Moore, D.H. (2007) Riluzole for amyotrophic lateral sclerosis (ALS)/motor neuron disease (MND). *Cochrane Database Syst Rev* Jan 24;(1):CD001447

Mitchell, J.D. and Borasio, G.D. (2007) Amyotrophic lateral sclerosis. *Lancet*, 369, 2031–2041

Mollnes, T.E. (2004) Therapeutic manipulation of the complement system. In: *The Complement System: Novel Roles in Health and Disease*, Szebeni, J. (ed.). Kluwer Academic Publishers, Boston, pp. 483–516

Moreau, C., Devos, D., Brunaud-Danel, V., Defebvre, L., Perez, T., Destee, A., Tonnel, A.B., Lassalle, P. and Just, N. (2005) Elevated IL-6 and TNF-alpha levels in patients with ALS: inflammation or hypoxia? *Neurology*, 65, 1958–1960

Niebroj-Dobosz, I., Dziewulska, D. and Janik, P. (2006) Auto-antibodies against proteins of spinal cord cells in cerebrospinal fluid of patients with amyotrophic lateral sclerosis (ALS). *Folia Neuropathol*, 44, 191–196

Pardo, C.A., Rabin, B.A., Palmer, D.N. and Price, D.L. (1994) Accumulation of the adenosine triphosphate synthase subunit C in the mnd mutant mouse. A model for neuronal ceroid lipofuscinosis. *Am J Pathol*, 144, 829–835

Perrin, F.E., Boisset, G., Docquier, M., Schaad, O., Descombes, P. and Kato, A.C. (2005) No widespread induction of cell death genes occurs in pure motoneurons in an amyotrophic lateral sclerosis mouse model. *Hum Mol Genet*, 14, 3309–3320

Pramatarova, A., Laganiere, J., Roussel, J., Brisebois, K. and Rouleau, G.A. (2001) Neuron-specific expression of mutant superoxide dismutase 1 in transgenic mice does not lead to motor impairment. *J Neurosci*, 21, 3369–3374

Reaume, A.G., Elliott, J.L., Hoffman, E.K., Kowall, N.W., Ferrante, R.J., Siwek, D.F., Wilcox, H.M., Flood, D.G., Beal, M.F., Brown, R.H., Jr., Scott, R.W. and Snider, W.D. (1996) Motor neurons in Cu/Zn superoxide dismutase-deficient mice develop normally but exhibit enhanced cell death after axonal injury. *Nat Genet*, 13, 43–47

Ripps, M.E., Huntley, G.W., Hof, P.R., Morrison, J.H. and Gordon, J.W. (1995) Transgenic mice expressing an altered murine superoxide dismutase gene provide an animal model of amyotrophic lateral sclerosis. *Proc Natl Acad Sci U S A*, 92, 689–693

Schmalbruch, H., Jensen, H.J., Bjaerg, M., Kamieniecka, Z. and Kurland, L. (1991) A new mouse mutant with progressive motor neuronopathy. *J Neuropathol Exp Neurol*, 50, 192–204

Sewell, D.L., Nacewicz, B., Liu, F., Macvilay, S., Erdei, A., Lambris, J.D., Sandor, M. and Fabry, Z. (2004) Complement C3 and C5 play critical roles in traumatic brain cryoinjury: blocking effects on neutrophil extravasation by C5a receptor antagonist. *J Neuroimmunol*, 155, 55–63

Shaw, B.F. and Valentine, J.S. (2007) How do ALS-associated mutations in superoxide dismutase 1 promote aggregation of the protein? *Trends Biochem Sci*, 32, 78–85

Spalloni, A., Albo, F., Ferrari, F., Mercuri, N., Bernardi, G., Zona, C. and Longone, P. (2004) Cu/Zn-superoxide dismutase (GLY93→ALA) mutation alters AMPA receptor subunit expression and function and potentiates kainate-mediated toxicity in motor neurons in culture. *Neurobiol Dis*, 15, 340–350

Strachan, A.J., Woodruff, T.M., Haaima, G., Fairlie, D.P. and Taylor, S.M. (2000) A new small molecule C5a receptor antagonist inhibits the reverse-passive Arthus reaction and endotoxic shock in rats. *J Immunol*, 164, 6560–6565

Strachan, A.J., Shiels, I.A., Reid, R.C., Fairlie, D.P. and Taylor, S.M. (2001) Inhibition of immune-complex mediated dermal inflammation in rats following either oral or topical administration of a small molecule C5a receptor antagonist. *Br J Pharmacol*, 134, 1778–1786

Tomiyama, M., Kannari, K., Nunomura, J., Oyama, Y., Takebe, K. and Matsunaga, M. (1994) Quantitative autoradiographic distribution of glutamate receptors in the cervical segment of the spinal cord of the wobbler mouse. *Brain Res*, 650, 353–357

Trbojevic-Cepe, M., Brinar, V., Pauro, M., Vogrinc, Z. and Stambuk, N. (1998) Cerebrospinal fluid complement activation in neurological diseases. *J Neurol Sci*, 154, 173–181

Tsuboi, Y. and Yamada, T. (1994) Increased concentration of C4d complement protein in CSF in amyotrophic lateral sclerosis. *J Neurol Neurosurg Psychiatry*, 57, 859–861

Valentine, J.S., Doucette, P.A. and Zittin Potter, S. (2005) Copper-zinc superoxide dismutase and amyotrophic lateral sclerosis. *Annu Rev Biochem*, 74, 563–593

Wang, J., Xu, G., Slunt, H.H., Gonzales, V., Coonfield, M., Fromholt, D., Copeland, N.G., Jenkins, N.A., and Borchelt, D.R. (2005) Coincident thresholds of mutant protein for paralytic disease and protein aggregation caused by restrictively expressed superoxide dismutase cDNA. *Neurobiol Dis*, 20, 943–952

Whitaker, J.N., Sciabbarrasi, J., Engel, W.K., Warmolts, J.R. and Strober, W. (1973) Serum immunoglobulin and complement (C3) levels: a study in adults with idiopathic, chronic polyneuropathies and motor neuron diseases. *Neurology*, 23, 1164–1173

Wong, P.C., Pardo, C.A., Borchelt, D.R., Lee, M.K., Copeland, N.G., Jenkins, N.A., Sisodia, S.S., Cleveland, D.W. and Price, D.L. (1995) An adverse property of a familial ALS-Linked SOD1 mutation causes motor-neuron disease characterized by vacuolar degeneration of mitochondria. *Neuron*, 14, 1105–1116

Woodruff, T.M. and Taylor, S.M. (2005) Promics Pty Ltd. Use of a C5a receptor inhibitor to treat neurological or neurodegenerative condition (e.g. Huntington's disease, spinocerebellar ataxia, dentatorubral pallidoluysian atrophy, ischemic damage, motor neuron disease) involving inflammation. *International Patent: WO2005092366-A1*

Woodruff, T.M., Strachan, A.J., Sanderson, S.D., Monk, P.N., Wong, A.K., Fairlie, D.P. and Taylor, S.M. (2001) Species dependence for binding of small molecule agonist and antagonists to the C5a receptor on polymorphonuclear leukocytes. *Inflammation*, 25, 171–177

Woodruff, T.M., Strachan, A.J., Dryburgh, N., Shiels, I.A., Reid, R.C., Fairlie, D.P. and Taylor, S.M. (2002) Antiarthritic activity of an orally active C5a receptor antagonist against antigen-induced monarticular arthritis in the rat. *Arthritis Rheum*, 46, 2476–2485

Woodruff, T.M., Arumugam, T.V., Shiels, I.A., Reid, R.C., Fairlie, D.P. and Taylor, S.M. (2003) A potent human C5a receptor antagonist protects against disease pathology in a rat model of inflammatory bowel disease. *J Immunol*, 171, 5514–5520

Woodruff, T.M., Arumugam, T.V., Shiels, I.A., Reid, R.C., Fairlie, D.P. and Taylor, S.M. (2004) Protective effects of a potent C5a receptor antagonist on experimental acute limb ischemia-reperfusion in rats. *J Surg Res*, 116, 81–90

Woodruff, T.M., Pollitt, S., Proctor, L.M., Stocks, S.Z., Manthey, H.D., Williams, H.M., Mahadevan, I.B., Shiels, I.A. and Taylor, S.M. (2005) Increased potency of a novel complement factor 5a receptor antagonist in a rat model of inflammatory bowel disease. *J Pharmacol Exp Ther*, 314, 811–817

Woodruff, T.M., Crane, J.W., Proctor, L.M., Buller, K.M., Shek, A.B., de Vos, K., Pollitt, S., Williams, H.M., Shiels, I.A., Monk, P.N. and Taylor, S.M. (2006a) Therapeutic activity of C5a receptor antagonists in a rat model of neurodegeneration. *FASEB J*, 20, 1407–1417

Woodruff, T.M., Proctor, L.M., Strachan, A.J. and Taylor, S.M. (2006b) Complement factor 5a as a therapeutic target. *Drugs Future*, 31, 325–334

12. The Role of Membrane Complement Regulatory Proteins in Cancer Immunotherapy

Jun Yan[1], Daniel J. Allendorf[1], Bing Li[1], Ruowan Yan[1], Richard Hansen[1], and Rossen Donev[2]

[1]Tumor Immunobiology Program of the James Graham Brown Cancer Center, Department of Medicine, University of Louisville School of Medicine, Louisville, KY, USA, jun.yan@louisville.edu
[2]Complement Biology Group, Department of Medical Biochemistry and Immunology, School of Medicine, Cardiff University, UK

Abstract. Anti-tumor monoclonal antibody therapy represents one of the earliest targeted therapies in clinical cancer care and has achieved great clinical promise. Complement activation mediated by anti-tumor mAbs can result in direct tumor lysis or enhancement of antibody-dependent cellular cytotoxicy. Chemotaxis of phagocytic cells by complement activation products C5a is also required for certain cancer immunotherapy such as combined β-glucan with anti-tumor mAb therapy. However, high expression levels of membrane-bound complement regulatory proteins (mCRPs) such as CD46, CD55 and CD59 on tumors significantly limit the anti-tumor mAb therapeutic efficacy. In addition, mCRPs have been shown to directly or indirectly down-regulate adaptive T cell responses. Therefore, it is desirable to combine anti-tumor mAb therapy or tumor vaccines with the blockade of mCRPs. Such strategies so far include the utilization of neutralizing mAbs for mCRPs, small interfering RNAs or anti-sense oligos for mCRPs, and chemotherapeutic drugs or cytokines. In vitro studies have demonstrated the feasibility and efficacy of such methods, although concerns have been raised about the utilization of neutralizing mAbs in vivo due to widespread expression of mCRPs on normal cells and tissues. Strategies have been developed to address these issues and more in vivo studies are needed to further validate these combination approaches.

1 Complement System and its Activation

The complement system is a major component of the innate immunity. It is composed of many distinct plasma proteins that react with each other to opsonize pathogens and induce a series of inflammatory responses that help to fight infection. It efficiently protects the host from pathogenic microorganisms, contributes to immune regulation, and links both innate and adaptive immunity (Mastellos and Lambris 2002; Walport 2001a,b). Tumor cells and apoptotic cells may also be potential targets of complement attack, as demonstrated by the findings that activated complement components are deposited on tumor masses such as breast (Niculescu et al. 1992; Vetvicka et al. 1997) and thyroid carcinoma (Lucas et al. 1996) and apoptotic cells (Zwart et al. 2004).

J.D. Lambris (ed.), *Current Topics in Complement II.*
doi: 10:1007/978-0-387-78952-1_12,

There are three activation pathways leading to complement activation: the classical pathway, alternative pathway, and mannose-binding lectin pathway. Although these three pathways have different initiation mechanisms, C3 activation is a central step for all pathways. The final step for complement activation is formation of the terminal C5b-9 membrane-attack complex (MAC), which creates a pore in the cell membranes of certain pathogens or cells that can lead to their death (Muller-Eberhard 1986). A recent study indicated a new pathway of complement activation in which C5 can be directly activated by the coagulation system, therefore bypassing C3 activation (Huber-Lang et al. 2006). Interestingly, the complement system can be also directly activated by tumor cells through the alternative (Budzko et al. 1976; Matsumoto et al. 1997) or the lectin pathway (Ma et al. 1999).

MAC formation leading to complement-dependent cytotoxicity (CDC) is one of weapons of the complement system in the fight against pathogens or cancer cells. In addition, complement activation links to antibody-dependent cellular cytotoxicity (ADCC) through the interaction of complement activation product iC3b with complement receptor 3 (CR3, CD11b/CD18, $\alpha_M\beta_2$-integrin on phagocytic cells and NK cells (Ross 2000). Both in vitro and in vivo data indicate that the adhesion of iC3b-opsonized tumor cells to CR3 on phagocytic cells or NK cells results in enhanced Fcγ receptor (FCγR)-mediated ADCC (Gelderman et al. 2004b). Although CR3 fails to trigger the killing of tumor cells following their interaction with ligand iC3b, our previous studies demonstrate that CR3-dependent cellular cytotoxicity (CR3-DCC) can be elicited for iC3b-opsonized tumor cells if exogenous polysaccharide β-glucan is provided (Hong et al. 2003; Yan et al. 2005). CR3 has two binding domains, the inserted-(I) domain and the lectin-like domain (LLD). Dual ligation of CR3 on neutrophils leads to cytotoxicity and degranulation (Li et al. 2006).

Complement activation also elicits a number of other biological effects, such as recruitment of inflammatory phagocytic cells, smooth muscle contraction, and increase of vascular permeability (Morgan 2000). Such reaction is mediated by the smaller complement activation fragments C3a and C5a. C5a and C3a also act on the endothelial cells lining blood vessels to induce adhesion molecules. Therefore, complement activation is a double-edged sword for the host immune system. The uncontrolled complement activation could lead to host tissue damage, thereby causing pathological consequences such as the development of autoimmune diseases (Linton and Morgan 1999; Manderson et al. 2004).

Given that complement activation could potentially lead to destructive effects and the way in which its activation is rapidly amplified via a triggered-enzyme cascade, it is necessary for the host to establish regulatory mechanisms preventing self-destruction. There are several soluble regulatory proteins such as C1 inhibitor, C4b binding protein (C4bBP), factors H, B, D, and I, as well as membrane-bound regulatory proteins CD35, CD46, CD55, and CD59 to control complement activation at critical stages. Those mechanisms are evolutionarily designed to protect host cells or tissues from undesirable complement activation (Ruiz-Arguelles and Llorente 2007; Sohn et al. 2007; Song 2006). Unexpectedly, overexpression of membrane-bound complement regulatory proteins on tumor cells is also part of the mechanisms that allow tumor cells to escape immune surveillance (Fishelson et al. 2003; Macor and Tedesco 2007). In this review, we will focus on the membrane-bound complement regulatory proteins and their role in the cancer immunotherapy.

2 Membrane-Bound Complement Regulatory Proteins and their Expression on Tumors

Membrane-bound complement regulatory proteins (mCRPs) include CD35 (complement receptor 1, CR1), CD46 (membrane cofactor protein, MCP), CD55 (decay accelerating factor, DAF), and CD59 (Fishelson et al. 2003). mCRPs act on different stages of the complement activation cascade, thereby interrupting the subsequent reaction. CD35 acts as a cofactor for factor I in the cleavage of C3b to iC3b and subsequently into C3c and C3dg (Medof et al. 1982). CD35 also binds to C4b, promoting its degradation into C4c and C4d. In addition, CD35 accelerates the decay of the C3/C5 convertase. CD35 mainly expresses on erythrocytes and most types of leukocytes. Expression of CD35 on tumors is rare and is only observed in follicular dendritic cell tumors, malignant endometrial tissue and leukemic blasts (Seya et al. 1994).

The mCRPs related to tumors are mainly CD46, CD55 and CD59. CD46 is a transmembrane co-factor protein and has cofactor activation for factor I-mediated degradation of C3b and C4b, thereby disrupting the subsequent C3 convertase formation (Seya et al. 1986). It has been reported that CD46 plays a more critical role in the alternative pathway rather than in the classical pathway of complement activation (Barilla-LaBarca et al. 2002). As compare to CD46, decay accelerating factor CD55 is a glycosyl phosphatidylinositol (GPI)-anchored protein and functions by dissociating C3 convertase (C4b2b and C3bBb) and C5 convertase (C4b2b3b and C3bBbC3b) (Spendlove et al. 2006). Unlike CD46, CD55 does not displace C3b or C4b; it accelerates the decay of C3 and C5 convertases. It preferentially acts on the classical pathway. CD59 is also called protectin and binds to C8 and C9 to prevent C9 polymerization. Essentially, CD59 prevents MAC formation and thus protects host cells or tissues from CDC-mediated lysis (Fonsatti et al. 2000).

mCRPs are virtually expressed on all cell types and most of tissues. Strikingly, mCRPs are also overexpressed on many tumor cells, including adenoma (Koretz et al. 1992), breast cancer (Madjd et al. 2004), cervical cancer (Simpson et al. 1997), colorectal cancer (Inoue et al. 1994), gastric cancer (Kiso et al. 2002), glicoblastoma (Junnikkala et al. 2000), hepatoma (Spiller et al. 2000), kidney cancer (Magyarlaki et al. 1996), leukemia (Jurianz et al. 2001), lymphoma (Takei et al. 2006), lung cancer (Sakuma et al. 1993), malignant endometrial tissue (Murray et al. 2000), malignant glioma (Shinoura et al. 1994), melanoma (Weichenthal et al. 1999), neuroblastoma (Chen et al. 2000), ovarian cancer (Bjorge et al. 1997), osteosarcoma (Pritchard-Jones et al. 2005), pancreatic carcinoma (Schmitt et al. 1999), prostate cancer (Babiker et al. 2005), and thyroid carcinoma (Yamakawa et al. 1994). These studies suggest that almost all cancers express at least one of the mCRPs and many express two or three. Upregulation of mCRPs is considered as one of critical mechanisms for tumor cells to escape immunological attack. Moreover, it was demonstrated that overexpression of CD59 in human neuroblastoma enhances tumor growth (Chen et al. 2000). The mechanisms leading to overexpression of mCRPs in tumor cells are not well understood. It has been demonstrated that vascular endothelial growth factor (VEGF) secreted by tumor cells or tumor stromal cells induces upregulation of mCRPs (Mason et al. 2004). In addition, cytokines such as TNF-α, IL-1β, IL-6 can upregulate CD55 and CD59 expression on heptocarcinoma (Spiller et al. 2000). In colon cancers, prostaglandin E2 (PEG2) and epidermal growth factor (EGF) are capable of upregulating CD55 expression (Holla et al. 2005). A recent study has demonstrated that CD46 mRNA expression is induced by IL-6 and by activation of the signal transducers and activators of transcription 3 (STAT-3) (Buettner et al. 2007). Previous studies have indicated that STAT-3 is persistently activated in most of cancer cells and primary tumor tissues as compared to normal tissues or cells (Buettner et al. 2002; Yu and Jove 2004). Therefore, activation of STAT-3 causes CD46 overexpression on tumors and protects human cancer cells from complement-mediated cytotoxicity (Buettner et al. 2007). In addition, p53, a broadly distributed tumor suppressor protein, has been demonstrated a role in regulating CD59 expression on tumor cells (Donev et al. 2006).

3 Antitumor mAb Therapy and mCRPs on Tumors

The application of humanized anti-tumor mAbs as a targeted therapy holds great clinical promise and has become more widely used in clinical practice (Adams and Weiner 2005). For example, anti-CD20 mAb (rituximab) has been used for B-cell non-hodgkin's lymphoma (Weiner and Link 2004). Anti-52 mAb (alemtuzumab) is approved for the treatment of chronic lymphocytic leukemia (Faderl et al. 2005). Anti-her-2/neu mAb (trastuzumab) is a humanized mAb against human epidermal growth factor receptor 2 and has been used in metastatic breast cancer patients (Leyland-Jones 2002). Anti-epidermal growth factor receptor mAb (anti-EGFR mAb, cetuximab) has been approved for colorectal and lung cancer (Ross et al. 2004). In addition, numerous humanized mAbs are in different phases of clinical trials, including anti-carcinoembryonic antigen (CEA) for gastrointenstinal cancer (Stein et al. 2005), anti-epithelial cell adhesion molecules (EpCAM) for colorectal cancer and others (Liljefors et al. 2005; Ruf et al. 2007).

The mechanisms by which anti-tumor mAbs inhibit or kill tumor cells are diverse and include inhibition of growth factor receptors such as trastuzumab and cetuximab, ADCC, and CDC. Anti-tumor mAbs can also effect the delivery of cytotoxic payloads such as radioisotope (Adams and Weiner 2005). Interestingly, most therapeutic chimeric or humanized mAbs are of human immunoglobulin G1 (IgG1) framework and can potently activate the complement, resulting in C3b deposition and subsequently formation of the opsonin iC3b on tumor cells.

However, complement mediated cytotoxicity such as CDC or CR3-ADCC or CR3-DCC has not been thought to play a critical role in the antitumor effect elicited by most anti-tumor mAbs due to overexpression of mCRPs on most tumor cells (Gelderman et al. 2004b; Macor and Tedesco 2007). For example, overexpression of a particular mCRP has been reported for each tumor type to resist therapeutic mAbs (Niehans et al. 1996). The CDC killing activity mediated by anti-Her2/neu-mAb was significantly increased following inhibition of mCRP function on tumor cells (Jurianz et al. 1999). Further studies showed that the upregulation of CD55 and CD59 limited CDC and diminished the anti-tumor mAb therapeutic efficacy in vitro (Gelderman et al. 2004a,b). Overexpression of mCRPs on tumors not only directly impairs complement-mediated killing activity but also abrogates efficient complement activation and dampens the release of potent chemoattractants such as C3a and C5a. In the combined anti-tumor mAb with β-glucan therapy, β-glucan-mediated CR3-DCC requires primed neutrophils trafficking into the tumor microenvironment (Allendorf et al. 2005; Li et al. 2007). Upregulation of mCRPs such as CD55 prevented sufficient complement activation and subsequent C5a release resulting in the paucity of primed neutrophils trafficking into tumors (Li et al. 2007).

Although many in vitro studies have clearly demonstrated that the therapeutic potential of anti-tumor mAbs is significantly impaired in tumor cells due to overexpression of mCRPs, it is intriguing whether the upregulation of mCRPs irreversibly correlates with anti-tumor mAb therapeutic efficacy in vivo, particularly in the clinical patient setting. Studies with anti-CD20 mAb therapy suggest that upregulation of mCRPs, CD59 in particular, on various B-cell tumors is associated with rituximab resistance in patients (Bannerji et al. 2003; Treon et al. 2001). There is an ongoing retrospective clinical study to investigate mCRP expression on colonrectal carcinoma and anti-EGFR mAb therapeutic efficacy (Christoph Strey personal communication).

4 mCRPs and Adaptive T-cell Responses

Several recent studies suggest that mCRPs also regulate adaptive T cell responses (Longhi et al. 2006; Marie et al. 2002). Activation of human CD4 T cells by anti-CD3 and anti-CD46 crosslinking leads to induction of T-regulatory cell 1 (Tr-1) phenotype (Barchet et al. 2006; Kemper et al. 2003). These cells secrete large amounts of IL-10 and inhibit CD4 T cell proliferation. In addition, ligation of CD46 with a physiological relevant ligand such as C3b or pathogen also induces Tr-1 cell generation (Price et al. 2005). These data suggest that CD46 may downregulate effector CD4 T cell activation via Tr-1 regulatory cells.

The role of CD55 on T cell immunity is highlighted by the studies with CD55-deficient mice (Heeger et al. 2005; Liu et al. 2005). Works from Heeger and Medof's group have demonstrated that during primary T cell activation, the absence of CD55 on antigen-presenting cells (APCs) and on T cells enhances T cell proliferation and augments the induced IFN-γ-producing cells (Heeger et al. 2005).

The effect is factor D- and, at least in part, C5-dependent, indicating that local alternative pathway activation is essential. Furthermore, APCs deficient of CD55 produce more IL-12 and C5a and promote more IFN-γ-producing T cells. This process is dependent on C5a receptor expressed on APCs. It appears that C5a release regulated by CD55 interacts with C5aR on APCs and regulate IL-12 production and T cell differentiation (Lalli et al. 2007). Studies from Song's group have shown that T cell recall responses using splenocytes from mice immunized with surrogate antigen ovalbumin (OVA) or an MHC class II restricted myelin oligodendrocyte glycoprotein (MOG)-derived peptide are more profound in CD55-deficient mice than in wildtype mice (Liu et al. 2005). T cells from CD55-deficient mice secrete more IFN-γ and IL-2 upon antigen restimulation. This effect is dependent on a functional complement system since the enhanced cytokine release is abrogated in C3-deficient CD55-double deficient mice. A recent study has also demonstrated that CD55 plays a critical role in virus-specific CD8 T cell proliferation and expansion (Fang et al. 2007). This study is particularly important since most tumor vaccines are designed to elicit potent tumor-antigen specific CD8 T cell responses.

Studies from CD59-deficient mice also suggest that CD59 can down-regulate T cell activity (Longhi et al. 2005). For example, virus-specific CD4 T cells were significantly augmented in CD59-deficient mice with respect to wildtype counterparts. Further study indicates that CD59 on T cells, but not on APCs, is critical in triggering enhanced proliferation in CD59-deficient mice (Longhi et al. 2005).

Given the importance of mCRPs on adaptive T cell immunity, it is therefore vitial to integrate current tumor vaccines with approaches of blocking or suppression of mCRPs, thereby eliciting maximum anti-tumor T cell responses.

5 Modulation of mCRPs for Immunotherapy

Since mCRPs can prevent efficient complement activation, inhibit complement-mediated killing mechanisms such as CDC, ADCC, CR3-DCC, and also down-regulate effector T cell responses, it is therefore hypothesized that blockade or neutralizing mCRPs would significantly improve anti-tumor mAb-based tumor immunotherapy or vaccine-mediated anti-tumor immune responses. These strategies include neutralizing mAbs against mCRPs, small interfering RNAs or anti-sense oligos to knockdown mCRPs, utilization of chemotherapeutic drugs or cytokines to downregulate mCRPs, and a recently proposed new approach for suppression of expression of membrane-bound complement regulator (mCR) genes.

5.1 Neutralizing mAbs

Specific inhibition of mCRP activity has been achieved with blocking mAbs against CD46, CD55, and CD59. In most of in vitro studies, anti-mCRP blocking mAbs have successfully demonstrated the enhancement of susceptibility of tumor cells to complement-mediated killing mechanisms. For example, neutralization of CD55 with blocking mAb in Burkitt lymphoma cells (Kuraya et al. 1992), leukemia cells (Jurianz et al. 2001), melanoma cells (Cheung et al. 1988), and breast cancer cells (Madjd et al. 2004) can significantly increase their sensitivity to complement-mediated killing. Similarly, blockade of CD59 with neutralizing mAb significantly enhances efficiency of complement-mediated lysis to neutroblastoma cells (Chen et al. 2000), leukemia cells (Jurianz et al. 2001), breast (Ellison et al. 2007), ovarian (Donin et al. 2003), renal (Gorter et al. 1996), and prostate carcinoma cells (Jarvis et al. 1997). Blocking mAb for CD46 is controversial in the in vitro studies. In renal carcinoma, blocking CD46 did not significantly affect complement sensitivity (Ajona et al. 2007). This may be related to a particular blocking mAb since inhibition of CD46 mRNA expression significantly increases complement-mediated lysis (Buettner et al. 2007).

Inhibition of mCRPs with neutralizing mAbs may also enhance ADCC effect via iC3b-CR3 interaction. Our recent study indicated that anti-CD55 blocking mAb, but not anti-CD46 blocking mAb, significantly enhanced iC3b deposition on tumors mediated by anti-her-2/neu mAb (Li et al. 2007). Although blocking anti-CD55 itself does not significantly increase the iC3b-CR3-mediated ADCC, enhanced iC3b deposition on tumors synergizes with yeast-derived β-glucan to elicit enhanced CR3-DCC in vitro. More importantly, in vivo administration with anti-CD55 mAb with β-glucan plus anti-her-2/neu mAb elicited tumor regression and long survival in animals bearing the previously resistant SKOV-3 human ovarian carcinoma. In addition, blocking anti-CD55 significantly led to C5a release and massive neutrophil influx within tumors.

However, one concern regarding use of anti-mCRP mAb blockade in vivo is widespread expression of mCRPs on normal tissues or cells such as red blood cells (Lublin and Atkinson 1989). This could potentially lead to hemolytic or vascular disease as a result of increased complement activation on normal cells or targeting by ADCC. This drawback may be overcome by using bi-specific mAb against tumor Ag with higher affinity and CD55 or CD59 with lower affinity (Gelderman et al. 2002a,b; Harris et al. 1997). A previous study has demonstrated that this strategy could specifically target tumor cells with minimally binding to normal cells and increase β-glucan mediated CR3-DCC (Gelderman et al. 2006). Indeed, bi-specific mAb to epithelial cell adhesion molecule (Ep-CAM) and Crry in rat has demonstrated a significant therapeutic efficacy for a rat colorectal cancer lung metastases model in vivo (Gelderman et al. 2004a). Moreover, a recent study showed that CD55 is highly expressed on tumor cells but not on non-neoplastic epithelia, suggesting that it might predominately target the tumor (Ravindranath and Shuler 2006).

5.2 Small Interfering RNAs or Anti-sense Oligos

Since the in vivo utilization of blocking mCRP mAbs could potentially cause undesirable adverse effects, novel strategies to block mCRPs on tumors have been developed. For example, using small interfering RNA (siRNA) technology, CD55 expression levels can be significantly downregulated in prostate cancer cells leading to a profound attenuation of overall tumor burden in vivo (Loberg et al. 2006). Similarly, CD46 siRNA downregulates CD46 expression on prostate cancer resulting in enhanced CDC in vitro (Buettner et al. 2007). Our recent study using CD59 siRNA showed that downregulation of CD59 on human ovarian carcinoma SKOV-3, non-small cell lung carcinoma NCI-H23, and breast carcinoma ZR-75-1 significantly enhanced their susceptibility to anti-tumor mAb and complement-mediated cell lysis (Yan R., et al. unpublished observations).

In addition to siRNAs, anti-sense phosphorothioate oligonucleotides (S-ODNs) are also used to knockdown mCRP expression on tumor cells (Zell et al. 2007). Using S-ODNs for CD46 and CD55, the expression levels of these two molecules were significantly decreased in breast, lung, and prostate carcinoma. The inhibition of mCRPs on tumors led to enhanced CDC both for CD46 and CD55. In addition, C3 opsonization on CD46/CD55-deficient tumor cells was also significantly enhanced. Further in vivo study is needed to test the efficiency and potency of this strategy.

RNA interference (RNAi) can be induced by synthetic siRNA or by vector-driven expression of shRNA. Vectors are usually delivered by viruses resulting in incorporation of the vector into the host genome and a long-term gene silencing. However, this induces unwanted immune response and possible toxic effects. In contrast, siRNA provides a transient gene silencing solving the drawbacks with possible insertional mutagenesis and immune response induction. However, the major challenge is its delivery into cells in vivo and the faded silencing effect due to the high proliferation rate of tumors.

5.3 Chemotherapeutic Drugs

Interestingly, the chemotherapeutic drug fludarabine down-regulates CD55 expression on tumor cells (Di Gaetano et al. 2001). This may well explain the synergistic cytotoxicity of fludarabine and anti-CD20 mAb (rituximab) in a follicular lymphoma cell line (Di Gaetano et al. 2001). We also showed that Paclitaxel could significantly downregulate CD59 on human ovarian carcinoma and synergize with anti-her-2/neu mAb for tumor cytotoxicity (Yan et al. unpublished observation). Study with other chemodrugs is underway. This may be very important since many anti-tumor mAbs are used in combination with chemotherapeutic drugs. The right combination may lead to the maximum therapeutic outcomes.

5.4 Peptide Inhibitors of mCR Gene Expression

Recently we have proposed a new strategy for decreasing expression of mCRPs on tumor surface by downmodulating mCR gene expression (Donev et al. 2006). This can be achieved by targeting transcriptional regulators of the mCR genes. We showed that p53 is a potential target for modulation of expression of CD59 in neuroblastoma (Donev et al. 2006), a tumor type in which mutations in p53 are rare (Valsesia-Wittmann et al. 2004). However, in most other tumors, the DNA-binding domain of p53 is usually mutated (Greenblatt et al. 1994). Hence, p53 is unlikely to be involved in regulation of CD59 expression. Recently we identified another transcription factor involved in overexpression of CD59 in neuroblastoma. This is the neural-restrictive silencer factor (NRSF, REST), which is expressed as a truncated protein not only in neuroblastoma (Palm et al. 1999), but also in small cell lung carcinoma (Coulson et al. 2000) and colorectal cancer (Westbrook et al. 2005). We showed that the expression of this truncated isoform of REST is related to everexpression of CD59 in neuroblastoma and it can be targeted with peptides to sensitize tumor to CDC killing (Donev et al. unpublished data).

We believe that targeting both the mCR gene expression and the stability of synthesized RNA with peptide inhibitors and RNAi, respectively, will significantly decrease the number of mCRPs on tumor surface, resulting in effective CDC killing.

6 Concluding Remarks

It is becoming clear that the evolutionarily ancient complement system can be manipulated to substantially contribute to our current state of the art oncology treatment, particularly to anti-tumor mAb therapy. However, upregulation of mCRPs on tumors imposes an obstacle to maximize the therapeutic efficacy mediated by anti-tumor mAbs or tumor vaccines. Such obstacles may be overcome by the co-administration of neutralizing anti-mCRP mAbs or siRNAs or anti-sense Oliges to achieve this goal. Indeed, many in vitro studies have demonstrated the synergistic effect when anti-tumor mAb is used in combination with blocking mAbs for mCRPs or other approaches. However, their in vivo efficacy needs to be further investigated.

Acknowledgements

This work was supported by NIH/NCI RO1 CA86412, the Kentucky Lung Cancer Research Board, the James Graham Brown Cancer Center Pilot Project Program to J.Y. and by the MRC New Investigator Grant 81345 to R.D.

References

Adams, G. P. and Weiner, L. M. (2005). Monoclonal antibody therapy of cancer. *Nat Biotechnol* 23, 1147–1157

Ajona, D., Hsu, Y. F., Corrales, L., Montuenga, L. M., and Pio, R. (2007). Down-regulation of human complement factor H sensitizes non-small cell lung cancer cells to complement attack and reduces in vivo tumor growth. *J Immunol* 178, 5991–5998

Allendorf, D. J., Yan, J., Ross, G. D., Hansen, R. D., Baran, J. T., Subbarao, K., Wang, L., and Haribabu, B. (2005). C5a-mediated leukotriene B4-amplified neutrophil chemotaxis is essential in tumor immunotherapy facilitated by anti-tumor monoclonal antibody and {beta}-glucan. *J Immunol* 174, 7050–7056

Babiker, A. A., Nilsson, B., Ronquist, G., Carlsson, L., and Ekdahl, K. N. (2005). Transfer of functional prostasomal CD59 of metastatic prostatic cancer cell origin protects cells against complement attack. *Prostate* 62, 105–114

Bannerji, R., Kitada, S., Flinn, I. W., Pearson, M., Young, D., Reed, J. C., and Byrd, J. C. (2003). Apoptotic-regulatory and complement-protecting protein expression in chronic lymphocytic leukemia: relationship to in vivo rituximab resistance. *J Clin Oncol* 21, 1466–1471

Barchet, W., Price, J. D., Cella, M., Colonna, M., MacMillan, S. K., Cobb, J. P., Thompson, P. A., Murphy, K. M., Atkinson, J. P., and Kemper, C. (2006). Complement-induced regulatory T cells suppress T-cell responses but allow for dendritic-cell maturation. *Blood* 107, 1497–1504

Barilla-LaBarca, M. L., Liszewski, M. K., Lambris, J. D., Hourcade, D., and Atkinson, J. P. (2002). Role of membrane cofactor protein (CD46) in regulation of C4b and C3b deposited on cells. *J Immunol* 168, 6298–6304

Bjorge, L., Hakulinen, J., Wahlstrom, T., Matre, R., and Meri, S. (1997). Complement-regulatory proteins in ovarian malignancies. *Int J Cancer* 70, 14–25

Budzko, D. B., Lachmann, P. J., and McConnell, I. (1976). Activation of the alternative complement pathway by lymphoblastoid cell lines derived from patients with Burkitt's lymphoma and infectious mononucleosis. *Cell Immunol* 22, 98–109

Buettner, R., Mora, L. B., and Jove, R. (2002). Activated STAT signaling in human tumors provides novel molecular targets for therapeutic intervention. *Clin Cancer Res* 8, 945–954

Buettner, R., Huang, M., Gritsko, T., Karras, J., Enkemann, S., Mesa, T., Nam, S., Yu, H., and Jove, R. (2007). Activated signal transducers and activators of transcription 3 signaling induces CD46 expression and protects human cancer cells from complement-dependent cytotoxicity. *Mol Cancer Res* 5, 823–832

Chen, S., Caragine, T., Cheung, N. K., and Tomlinson, S. (2000). CD59 expressed on a tumor cell surface modulates decay-accelerating factor expression and enhances tumor growth in a rat model of human neuroblastoma. *Cancer Res* 60, 3013–3018

Cheung, N. K., Walter, E. I., Smith-Mensah, W. H., Ratnoff, W. D., Tykocinski, M. L., and Medof, M. E. (1988). Decay-accelerating factor protects human tumor cells from complement-mediated cytotoxicity in vitro. *J Clin Invest* 81, 1122–1128

Coulson, J. M., Edgson, J. L., Woll, P. J., and Quinn, J. P. (2000). A splice variant of the neuron-restrictive silencer factor repressor is expressed in small cell lung cancer: a potential role in derepression of neuroendocrine genes and a useful clinical marker. *Cancer Res* 60, 1840–1844

Di Gaetano, N., Xiao, Y., Erba, E., Bassan, R., Rambaldi, A., Golay, J., and Introna, M. (2001). Synergism between fludarabine and rituximab revealed in a follicular lymphoma cell line resistant to the cytotoxic activity of either drug alone. *Br J Haematol* 114, 800–809

Donev, R. M., Cole, D. S., Sivasankar, B., Hughes, T. R., and Morgan, B. P. (2006). p53 regulates cellular resistance to complement lysis through enhanced expression of CD59. *Cancer Res* 66, 2451–2458

Donin, N., Jurianz, K., Ziporen, L., Schultz, S., Kirschfink, M., and Fishelson, Z. (2003). Complement resistance of human carcinoma cells depends on membrane regulatory proteins, protein kinases and sialic acid. *Clin Exp Immunol* 131, 254–263

Ellison, B. S., Zanin, M. K., and Boackle, R. J. (2007). Complement susceptibility in glutamine deprived breast cancer cells. *Cell Div* 2, 20

Faderl, S., Coutre, S., Byrd, J. C., Dearden, C., Denes, A., Dyer, M. J., Gregory, S. A., Gribben, J. G., Hillmen, P., Keating, M., Rosen, S., Venugopal, P., and Rai, K. (2005). The evolving role of alemtuzumab in management of patients with CLL. *Leukemia* 19, 2147–2152

Fang, C., Miwa, T., Shen, H., and Song, W. C. (2007). Complement-dependent enhancement of CD8+ T cell immunity to lymphocytic choriomeningitis virus infection in decay-accelerating factor-deficient mice. *J Immunol* 179, 3178–3186

Fishelson, Z., Donin, N., Zell, S., Schultz, S., and Kirschfink, M. (2003). Obstacles to cancer immunotherapy: expression of membrane complement regulatory proteins (mCRPs) in tumors. *Mol Immunol* 40, 109–123

Fonsatti, E., Altomonte, M., Coral, S., De Nardo, C., Lamaj, E., Sigalotti, L., Natali, P. G., and Maio, M. (2000). Emerging role of protectin (CD59) in humoral immunotherapy of solid malignancies. *Clin Ter* 151, 187–193

Gelderman, K. A., Blok, V. T., Fleuren, G. J., and Gorter, A. (2002a). The inhibitory effect of CD46, CD55, and CD59 on complement activation after immunotherapeutic treatment of cervical carcinoma cells with monoclonal antibodies or bispecific monoclonal antibodies. *Lab Invest* 82, 483–493

Gelderman, K. A., Kuppen, P. J., Bruin, W., Fleuren, G. J., and Gorter, A. (2002b). Enhancement of the complement activating capacity of 17-1A mAb to overcome the effect of membrane-bound complement regulatory proteins on colorectal carcinoma. *Eur J Immunol* 32, 128–135

Gelderman, K. A., Kuppen, P. J., Okada, N., Fleuren, G. J., and Gorter, A. (2004a). Tumor-specific inhibition of membrane-bound complement regulatory protein Crry with bispecific monoclonal antibodies prevents tumor outgrowth in a rat colorectal cancer lung metastases model. *Cancer Res* 64, 4366–4372

Gelderman, K. A., Tomlinson, S., Ross, G. D., and Gorter, A. (2004b). Complement function in mAb-mediated cancer immunotherapy. *Trends Immunol* 25, 158–164

Gelderman, K. A., Lam, S., Sier, C. F., and Gorter, A. (2006). Cross-linking tumor cells with effector cells via CD55 with a bispecific mAb induces beta-glucan-dependent CR3-dependent cellular cytotoxicity. *Eur J Immunol* 36, 977–984

Gorter, A., Blok, V. T., Haasnoot, W. H., Ensink, N. G., Daha, M. R., and Fleuren, G. J. (1996). Expression of CD46, CD55, and CD59 on renal tumor cell lines and their role in preventing complement-mediated tumor cell lysis. *Lab Invest* 74, 1039–1049

Greenblatt, M. S., Bennett, W. P., Hollstein, M., and Harris, C. C. (1994). Mutations in the p53 tumor suppressor gene: clues to cancer etiology and molecular pathogenesis. *Cancer Res* 54, 4855–4878

Harris, C. L., Kan, K. S., Stevenson, G. T., and Morgan, B. P. (1997). Tumour cell killing using chemically engineered antibody constructs specific for tumour cells and the complement inhibitor CD59. *Clin Exp Immunol* 107, 364–371

Heeger, P. S., Lalli, P. N., Lin, F., Valujskikh, A., Liu, J., Muqim, N., Xu, Y., and Medof, M. E. (2005). Decay-accelerating factor modulates induction of T cell immunity. *J Exp Med* 201, 1523–1530

Holla, V. R., Wang, D., Brown, J. R., Mann, J. R., Katkuri, S., and DuBois, R. N. (2005). Prostaglandin E2 regulates the complement inhibitor CD55/decay-accelerating factor in colorectal cancer. *J Biol Chem* 280, 476–483

Hong, F., Hansen, R. D., Yan, J., Allendorf, D. J., Baran, J. T., Ostroff, G. R., and Ross, G. D. (2003). Beta-glucan functions as an adjuvant for monoclonal antibody immunotherapy by recruiting tumoricidal granulocytes as killer cells. *Cancer Res* 63, 9023–9031

Huber-Lang, M., Sarma, J. V., Zetoune, F. S., Rittirsch, D., Neff, T. A., McGuire, S. R., Lambris, J. D., Warner, R. L., Flierl, M. A., Hoesel, L. M., Gebhard, F., Younger, J. G., Drouin, S. M., Wetsel, R. A., and Ward, P. A. (2006). Generation of C5a in the absence of C3: a new complement activation pathway. *Nat Med* 12, 682–687

Inoue, H., Mizuno, M., Uesu, T., Ueki, T., and Tsuji, T. (1994). Distribution of complement regulatory proteins, decay-accelerating factor, CD59/homologous restriction factor 20 and membrane cofactor protein in human colorectal adenoma and cancer. *Acta Med Okayama* 48, 271–277

Jarvis, G. A., Li, J., Hakulinen, J., Brady, K. A., Nordling, S., Dahiya, R., and Meri, S. (1997). Expression and function of the complement membrane attack complex inhibitor protectin (CD59) in human prostate cancer. *Int J Cancer* 71, 1049–1055

Junnikkala, S., Jokiranta, T. S., Friese, M. A., Jarva, H., Zipfel, P. F., and Meri, S. (2000). Exceptional resistance of human H2 glioblastoma cells to complement-mediated killing by expression and utilization of factor H and factor H-like protein 1. *J Immunol* 164, 6075–6081

Jurianz, K., Maslak, S., Garcia-Schuler, H., Fishelson, Z., and Kirschfink, M. (1999). Neutralization of complement regulatory proteins augments lysis of breast carcinoma cells targeted with rhumAb anti-HER2. *Immunopharmacology* 42, 209–218

Jurianz, K., Ziegler, S., Donin, N., Reiter, Y., Fishelson, Z., and Kirschfink, M. (2001). K562 erythroleukemic cells are equipped with multiple mechanisms of resistance to lysis by complement. *Int J Cancer* 93, 848–854

Kemper, C., Chan, A. C., Green, J. M., Brett, K. A., Murphy, K. M., and Atkinson, J. P. (2003). Activation of human CD4+ cells with CD3 and CD46 induces a T-regulatory cell 1 phenotype. *Nature* 421, 388–392

Kiso, T., Mizuno, M., Nasu, J., Shimo, K., Uesu, T., Yamamoto, K., Okada, H., Fujita, T., and Tsuji, T. (2002). Enhanced expression of decay-accelerating factor and CD59/homologous restriction factor 20 in intestinal metaplasia, gastric adenomas and intestinal-type gastric carcinomas but not in diffuse-type carcinomas. *Histopathology* 40, 339–347

Koretz, K., Bruderlein, S., Henne, C., and Moller, P. (1992). Decay-accelerating factor (DAF, CD55) in normal colorectal mucosa, adenomas and carcinomas. *Br J Cancer* 66, 810–814

Kuraya, M., Yefenof, E., Klein, G., and Klein, E. (1992). Expression of the complement regulatory proteins CD21, CD55 and CD59 on Burkitt lymphoma lines: their role in sensitivity to human serum-mediated lysis. *Eur J Immunol* 22, 1871–1876

Lalli, P. N., Strainic, M. G., Lin, F., Medof, M. E., and Heeger, P. S. (2007). Decay accelerating factor can control T cell differentiation into IFN-{gamma}-producing effector cells via regulating local C5a-induced IL-12 production. *J Immunol* 179, 5793–5802

Leyland-Jones, B. (2002). Trastuzumab: hopes and realities. *Lancet Oncol* 3, 137–144

Li, B., Allendorf, D. J., Hansen, R., Marroquin, J., Ding, C., Cramer, D. E., and Yan, J. (2006). Yeast beta-glucan amplifies phagocyte killing of iC3b-opsonized tumor cells via complement receptor 3-Syk-phosphatidylinositol 3-kinase pathway. *J Immunol* 177, 1661–1669

Li, B., Allendorf, D. J., Hansen, R., Marroquin, J., Cramer, D. E., Harris, C. L., and Yan, J. (2007). Combined yeast {beta}-glucan and antitumor monoclonal antibody therapy requires C5a-mediated neutrophil chemotaxis via regulation of decay-accelerating factor CD55. *Cancer Res* 67, 7421–7430

Liljefors, M., Nilsson, B., Fagerberg, J., Ragnhammar, P., Mellstedt, H., and Frodin, J. E. (2005). Clinical effects of a chimeric anti-EpCAM monoclonal antibody in combination with granulocyte-macrophage colony-stimulating factor in patients with metastatic colorectal carcinoma. *Int J Oncol* 26, 1581–1589

Linton, S. M. and Morgan, B. P. (1999). Complement activation and inhibition in experimental models of arthritis. *Mol Immunol* 36, 905–914

Liu, J., Miwa, T., Hilliard, B., Chen, Y., Lambris, J. D., Wells, A. D., and Song, W. C. (2005). The complement inhibitory protein DAF (CD55) suppresses T cell immunity in vivo. *J Exp Med* 201, 567–577

Loberg, R. D., Day, L. L., Dunn, R., Kalikin, L. M., and Pienta, K. J. (2006). Inhibition of decay-accelerating factor (CD55) attenuates prostate cancer growth and survival in vivo. *Neoplasia* 8, 69–78

Longhi, M. P., Sivasankar, B., Omidvar, N., Morgan, B. P., and Gallimore, A. (2005). Cutting edge: murine CD59a modulates antiviral CD4+ T cell activity in a complement-independent manner. *J Immunol 175*, 7098–7102

Longhi, M. P., Harris, C. L., Morgan, B. P., and Gallimore, A. (2006). Holding T cells in check – a new role for complement regulators? *Trends Imm*unol 27, 102–108

Lublin, D. M. and Atkinson, J. P. (1989). Decay-accelerating factor: biochemistry, molecular biology, and function. *Annu Rev Immunol* 7, 35–58

Lucas, S. D., Karlsson-Parra, A., Nilsson, B., Grimelius, L., Akerstrom, G., Rastad, J., and Juhlin, C. (1996). Tumor-specific deposition of immunoglobulin G and complement in papillary thyroid carcinoma. *Hum Pathol* 27, 1329–1335

Ma, Y., Uemura, K., Oka, S., Kozutsumi, Y., Kawasaki, N., and Kawasaki, T. (1999). Antitumor activity of mannan-binding protein in vivo as revealed by a virus expression system: mannan-binding proteindependent cell-mediated cytotoxicity. *Proc Natl Acad Sci U S A* 96, 371–375

Macor, P. and Tedesco, F. (2007). Complement as effector system in cancer immunotherapy. *Immunol Lett* 111, 6–13

Madjd, Z., Durrant, L. G., Bradley, R., Spendlove, I., Ellis, I. O., and Pinder, S. E. (2004). Loss of CD55 is associated with aggressive breast tumors. *Clin Cancer Res* 10, 2797–2803

Magyarlaki, T., Mosolits, S., Baranyay, F., and Buzogany, I. (1996). Immunohistochemistry of complement response on human renal cell carcinoma biopsies. *Tumori* 82, 473–479

Manderson, A. P., Botto, M., and Walport, M. J. (2004). The role of complement in the development of systemic lupus erythematosus. *Annu Rev Immunol* 22, 431–456

Marie, J. C., Astier, A. L., Rivailler, P., Rabourdin-Combe, C., Wild, T. F., and Horvat, B. (2002). Linking innate and acquired immunity: divergent role of CD46 cytoplasmic domains in T cell induced inflammation. *Nat Immunol* 3, 659–666

Mason, J. C., Steinberg, R., Lidington, E. A., Kinderlerer, A. R., Ohba, M., and Haskard, D. O. (2004). Decay-accelerating factor induction on vascular endothelium by vascular endothelial growth factor (VEGF) is mediated via a VEGF receptor-2 (VEGF-R2)- and protein kinase C-alpha/epsilon (PKCalpha/epsilon)-dependent cytoprotective signaling pathway and is inhibited by cyclosporin A. *J Biol Chem* 279, 41611–41618

Mastellos, D. and Lambris, J. D. (2002). Complement: more than a 'guard' against invading pathogens? *Trends Immunol* 23, 485–491

Matsumoto, M., Takeda, J., Inoue, N., Hara, T., Hatanaka, M., Takahashi, K., Nagasawa, S., Akedo, H., and Seya, T. (1997). A novel protein that participates in nonself discrimination of malignant cells by homologous complement. *Nat Med* 3, 1266–1270

Medof, M. E., Iida, K., Mold, C., and Nussenzweig, V. (1982). Unique role of the complement receptor CR1 in the degradation of C3b associated with immune complexes. *J Exp Med* 156, 1739–1754

Morgan, B. P. (2000). The complement system: an overview. *Methods Mol Biol* 150, 1–13

Muller-Eberhard, H. J. (1986). The membrane attack complex of complement. *Annu Rev Immunol* 4, 503–528

Murray, K. P., Mathure, S., Kaul, R., Khan, S., Carson, L. F., Twiggs, L. B., Martens, M. G., and Kaul, A. (2000). Expression of complement regulatory proteins-CD 35, CD 46, CD 55, and CD 59-in benign and malignant endometrial tissue. *Gynecol Oncol* 76, 176–182

Niculescu, F., Rus, H. G., Retegan, M., and Vlaicu, R. (1992). Persistent complement activation on tumor cells in breast cancer. *Am J Pathol* 140, 1039–1043

Niehans, G. A., Cherwitz, D. L., Staley, N. A., Knapp, D. J., and Dalmasso, A. P. (1996). Human carcinomas variably express the complement inhibitory proteins CD46 (membrane cofactor protein), CD55 (decay-accelerating factor), and CD59 (protectin). *Am J Pathol* 149, 129–142

Palm, K., Metsis, M., and Timmusk, T. (1999). Neuron-specific splicing of zinc finger transcription factor REST/NRSF/XBR is frequent in neuroblastomas and conserved in human, mouse and rat. *Brain Res Mol Brain* Res 72, 30–39

Price, J. D., Schaumburg, J., Sandin, C., Atkinson, J. P., Lindahl, G., and Kemper, C. (2005). Induction of a regulatory phenotype in human CD4+ T cells by streptococcal M protein. *Journal of Immunology* 175, 677–684

Pritchard-Jones, K., Spendlove, I., Wilton, C., Whelan, J., Weeden, S., Lewis, I., Hale, J., Douglas, C., Pagonis, C., Campbell, B., Alvarez, P., Halbert, G., and Durrant, L. G. (2005). Immune responses to the 105AD7 human anti-idiotypic vaccine after intensive chemotherapy, for osteosarcoma. *Br J Cancer* 92, 1358–1365

Ravindranath, N. M., and Shuler, C. (2006). Expression of complement restriction factors (CD46, CD55 & CD59) in head and neck squamous cell carcinomas. *J Oral Pathol Med* 35, 560–567

Ross, G. D. (2000). Regulation of the adhesion versus cytotoxic functions of the Mac-1/CR3/alphaMbeta2-integrin glycoprotein. *Crit Rev Immunol* 20, 197–222

Ross, J. S., Schenkein, D. P., Pietrusko, R., Rolfe, M., Linette, G. P., Stec, J., Stagliano, N. E., Ginsburg, G. S., Symmans, W. F., Pusztai, L., and Hortobagyi, G. N. (2004). Targeted therapies for cancer 2004. *Am J Clin Pathol* 122, 598–609

Ruf, P., Gires, O., Jager, M., Fellinger, K., Atz, J., and Lindhofer, H. (2007). Characterisation of the new EpCAM-specific antibody HO-3: implications for trifunctional antibody immunotherapy of cancer. *Br J Cancer* 97, 315–321

Ruiz-Arguelles, A. and Llorente, L. (2007). The role of complement regulatory proteins (CD55 and CD59) in the pathogenesis of autoimmune hemocytopenias. *Autoimmun Rev* 6, 155–161

Sakuma, T., Kodama, K., Hara, T., Eshita, Y., Shibata, N., Matsumoto, M., Seya, T., and Mori, Y. (1993). Levels of complement regulatory molecules in lung cancer: disappearance of the D17 epitope of CD55 in small-cell carcinoma. *Jpn J Cancer Res* 84, 753–759

Schmitt, C. A., Schwaeble, W., Wittig, B. M., Meyer zum Buschenfelde, K. H., and Dippold, W. G. (1999). Expression and regulation by interferon-gamma of the membrane-bound complement regulators CD46 (MCP), CD55 (DAF) and CD59 in gastrointestinal tumours. *Eur J Cancer* 35, 117–124

Seya, T., Turner, J. R., and Atkinson, J. P. (1986). Purification and characterization of a membrane protein (gp45-70) that is a cofactor for cleavage of C3b and C4b. *J Exp Med* 163, 837–855

Seya, T., Matsumoto, M., Hara, T., Hatanaka, M., Masaoka, T., and Akedo, H. (1994). Distribution of C3-step regulatory proteins of the complement system, CD35 (CR1), CD46 (MCP), and CD55 (DAF), in hematological malignancies. *Leuk Lymphoma* 12, 395–400

Shinoura, N., Heffelfinger, S. C., Miller, M., Shamraj, O. I., Miura, N. H., Larson, J. J., DeTribolet, N., Warnick, R. E., Tew, J. J., and Menon, A. G. (1994). RNA expression of complement regulatory proteins in human brain tumors. *Cancer Lett* 86, 143–149

Simpson, K. L., Jones, A., Norman, S., and Holmes, C. H. (1997). Expression of the complement regulatory proteins decay accelerating factor (DAF, CD55), membrane cofactor protein (MCP, CD46) and CD59 in the normal human uterine cervix and in premalignant and malignant cervical disease. *Am J Pathol* 151, 1455–1467

Sohn, J. H., Bora, P. S., Jha, P., Tezel, T. H., Kaplan, H. J., and Bora, N. S. (2007). Complement, innate immunity and ocular disease. *Chem Immunol Allergy* 92, 105–114

Song, W. C. (2006). Complement regulatory proteins and autoimmunity. *Autoimmunity* 39, 403–410

Spendlove, I., Ramage, J. M., Bradley, R., Harris, C., and Durrant, L. G. (2006). Complement decay accelerating factor (DAF)/CD55 in cancer. *Cancer Immunol Immunother* 55, 987–995

Spiller, O. B., Criado-Garcia, O., Rodriguez De Cordoba, S., and Morgan, B. P. (2000). Cytokine-mediated up-regulation of CD55 and CD59 protects human hepatoma cells from complement attack. *Clin Exp Immunol* 121, 234–241

Stein, R., Govindan, S. V., Hayes, M., Griffiths, G. L., Hansen, H. J., Horak, I. D., and Goldenberg, D. M. (2005). Advantage of a residualizing iodine radiolabel in the therapy of a colon cancer xenograft targeted with an anticarcinoembryonic antigen monoclonal antibody. *Clin Cancer Res* 11, 2727–2734

Takei, K., Yamazaki, T., Sawada, U., Ishizuka, H., and Aizawa, S. (2006). Analysis of changes in CD20, CD55, and CD59 expression on established rituximab-resistant B-lymphoma cell lines. *Leuk Res* 30, 625–631

Treon, S. P., Mitsiades, C., Mitsiades, N., Young, G., Doss, D., Schlossman, R., and Anderson, K. C. (2001). Tumor cell expression of CD59 is associated with resistance to CD20 serotherapy in patients with B-cell malignancies. *J Immunother* 24, 263–271

Valsesia-Wittmann, S., Magdeleine, M., Dupasquier, S., Garin, E., Jallas, A. C., Combaret, V., Krause, A., Leissner, P., and Puisieux, A. (2004). Oncogenic cooperation between H-Twist and N-Myc overrides failsafe programs in cancer cells. *Cancer Cell* 6, 625–630

Vetvicka, V., Thornton, B. P., Wieman, T. J., and Ross, G. D. (1997). Targeting of natural killer cells to mammary carcinoma via naturally occurring tumor cell-bound iC3b and beta-glucan-primed CR3 (CD11b/CD18). *J Immunol* 159, 599–605

Walport, M. J. (2001a). Complement. First of two parts. *N Engl J Med* 344, 1058–1066

Walport, M. J. (2001b). Complement. Second of two parts. *N Engl J Med* 344, 1140–1144

Weichenthal, M., Siemann, U., Neuber, K., and Breitbart, E. W. (1999). Expression of complement regulator proteins in primary and metastatic malignant melanoma. *J Cutan Pathol* 26, 217–221

Weiner, G. J., and Link, B. K. (2004). Monoclonal antibody therapy of B cell lymphoma. *Expert Opin Biol Ther* 4, 375–385

Westbrook, T. F., Martin, E. S., Schlabach, M. R., Leng, Y., Liang, A. C., Feng, B., Zhao, J. J., Roberts, T. M., Mandel, G., Hannon, G. J., Depinho, R. A., Chin, L., and Elledge, S. J. (2005). A genetic screen for candidate tumor suppressors identifies REST. *Cell* 121, 837–848

Yamakawa, M., Yamada, K., Tsuge, T., Ohrui, H., Ogata, T., Dobashi, M., and Imai, Y. (1994). Protection of thyroid cancer cells by complement-regulatory factors. *Cancer* 73, 2808–2817

Yan, J., Allendorf, D. J., and Brandley, B. (2005). Yeast whole glucan particle (WGP) beta-glucan in conjunction with antitumour monoclonal antibodies to treat cancer. *Expert Opin Biol Ther* 5, 691–702

Yu, H. and Jove, R. (2004). The STATs of cancer – new molecular targets come of age. *Nat Rev Cancer* 4, 97–105

Zell, S., Geis, N., Rutz, R., Schultz, S., Giese, T., and Kirschfink, M. (2007). Down-regulation of CD55 and CD46 expression by anti-sense phosphorothioate oligonucleotides (S-ODNs) sensitizes tumour cells to complement attack. *Clin Exp Immunol* 150, 576–584

Zwart, B., Ciurana, C., Rensink, I., Manoe, R., Hack, C. E., and Aarden, L. A. (2004). Complement activation by apoptotic cells occurs pred et al. ominantly via IgM and is limited to late apoptotic (secondary necrotic) cells. *Autoimmunity* 37, 95–102

13. Role of Complement in Ethanol-Induced Liver Injury

Michele T. Pritchard[1], Megan R. McMullen[1], M. Edward Medof[4], Abram Stavitsky[3], and Laura E. Nagy[1,2]

[1]Department of Pathobiology, Cleveland Clinic, Cleveland, OH, USA, laura.nagy@case.edu
[2]Department of Gastroenterology, Cleveland Clinic, Cleveland, OH, USA, laura.nagy@case.edu
[3]Department of Molecular Biology and Microbiology, Case Western Reserve University, Cleveland, OH, USA
[4]Department of Pathology, Case Western Reserve University, Cleveland, OH, USA

Abstract. The complement cascade is a phylogenetically ancient part of our immune system and is critical to an organism's ability to ward off infection. Interest in a possible role for the complement system in the development of ethanol-induced liver injury was inspired by the large body of data implicating the complement system in the development of acute and chronic inflammatory responses to bacteria/bacterial products, as well as in response to cell injury, both hallmarks of ethanol-induced liver injury. Recent investigations have demonstrated that complement is involved in the pathogenesis of ethanol-induced liver injury. Here we review the available data on the contribution of complement to ethanol-induced liver injury and then discuss the potential mechanisms by which the essential roles of complement in protecting the host from infection and facilitating wound healing may contribute to and/or protect from the pathogenesis of alcohol-induced liver injury.

1. Alcoholic Liver Disease

Alcoholic liver disease (ALD) develops in approximately 20% of all alcoholics with a higher prevalence in females (Lieber 1994). The development of fibrosis and cirrhosis is a complex process involving both parenchymal and non-parenchymal cells resident in the liver, as well as the recruitment of other cell types to the liver in response to damage and inflammation (Gressner and Bachem 1995). The progression of the alcohol-induced liver injury follows a pattern characteristic to all types of liver fibrosis, regardless of the causative agent. This progression is marked by the appearance of fatty liver, hepatocyte necrosis and apoptosis, inflammation, regenerating nodules, fibrosis and cirrhosis (Martinez-Hernandez and Amenta 1993). Fibrosis is thought to be initiated in response to hepatocellular damage, with inflammatory processes contributing to the progression of the disease (Gressner and Bachem 1995). Interestingly, many of the events involved in the development of fibrosis are typical of other tissue responses to injury, such as wound healing in the skin and soft tissues (Raghow 1994). Continued ethanol exposure may disorder the

J.D. Lambris (ed.), *Current Topics in Complement II.*
doi: 10:1007/978-0-387-78952-1_13,

highly regulated "wound healing response", resulting in continued hepatocellular damage, inflammation and fibrosis. Importantly, fibrosis is reversible, while cirrhosis is not, so that therapeutic strategies aimed at decreasing inflammation and fibrosis will likely slow the progression of the disease (Friedman 2000).

1.1 Innate and Adaptive Immunity in Ethanol-Induced Liver Injury

Both innate and adaptive immune systems are thought to be involved in the progression of ALD. While the innate and adaptive immune systems are two distinct branches of the immune response, these two components of immunity are intimately linked at many stages of an organism's response to injury or stress. Components of the innate immune response, including NK and NKT cells (Minagawa et al. 2004), Kupffer cells (resident hepatic macrophages) (Nagy 2003) and the complement system (Markiewski et al. 2004; Strey et al. 2003), as well as T-cells and antibody-dependent adaptive immune responses (Tuma and Casey 2003), are involved in the hepatic response to various types of injury, including bacterial and viral infections, exposure to toxins (including ethanol), partial hepatectomy and ischemia-reperfusion.

Activation of the innate immune response in the liver during chronic ethanol exposure is associated with increased production of pro-inflammatory cytokines and chemokines (Thurman 1998; Tilg and Diehl 2000), as well as reactive oxygen species (ROS) (Hoek and Pastorino 2002; Jaeschke et al. 2002). Kupffer cells, the resident macrophages in the liver, are critical to the onset of ethanol-induced liver injury. Ablation of Kupffer cells prevents the development of fatty liver and inflammation, early stages in the progression of ethanol-induced liver damage, in rats chronically exposed to ethanol (Adachi et al. 1994). Endotoxin/ lipopolysaccharide (LPS), a component of gram-negative bacterial cell walls, is an important activator of Kupffer cells. LPS concentration is increased in the blood of alcoholics (Fukui et al. 1991; Bode et al. 1987) and animals exposed to ethanol (Nanji et al. 1993), probably due to impaired barrier function of the intestinal mucosa (4). Formation of ROS, as well as highly reactive products of ethanol metabolism (acetaldehyde, malondialdehyde, lipid peroxides, etc) can lead to the generation of antibodies to modified proteins during chronic ethanol exposure.

2. Complement and Ethanol-Induced Liver Injury

The complement system is a network of more than 30 proteins, involved in both innate and adaptive immune responses (Gasque 2004). The functions of complement can be generally grouped into three major types of activity: 1) Defense against microbial infection involves the generation of lytic complexes (membrane attack complex, MAC) on the surface of pathogens, formation of opsonins that promote phagocytosis and destruction of bacteria, as well as the regulation of local pro-inflammatory responses; 2) Bridging of the innate and adaptive immune pathways and 3) Removal of immune complexes and cellular debris that can result from inflammatory injury (Gasque 2004). Activation of complement occurs via the classical, lectin or alternative pathways, all three pathways culminate in the activation of C3, which in turn activates the terminal pathway leading to the formation of the C5b-9 membrane attack complex (MAC). While complement is critical to an organism's ability to ward off infection and plays a role in the repair of tissues in response to injury, uncontrolled complement activation may result in direct injury to a variety of target cells.

The liver plays essential functions in the complement system, acting as the primary site of production and secretion of circulating complement proteins. Cells in the liver also express complement receptors and intrinsic regulatory proteins. In healthy liver, Kupffer cells and stellate cells express both C3a and C5a receptors. C5a receptor expression can be induced in hepatocytes in response to inflammatory cytokines (Schieferdecker et al. 2001) or in regenerating hepatocytes (Daveau et al. 2004). While the liver is exposed to complement proteins, as well as activated complexes, it is resistant to complement-induced lysis (Koch et al. 2005). This protection likely involves the activity of intrinsic complement regulatory proteins, as CD55/DAF and CD59 are expressed by hepatocytes, hepatic endothelial cells (Lin et al. 2001) and Kupffer cells (Stavitsky and Nagy unpublished observations).

The complement pathway contributes to inflammatory response in many ways. For example, C3a and C5a stimulate the production of cytokines by various types of cells (Monsinjon et al. 2003; Schieferdecker et al. 2001), either alone or in the presence of other inflammatory mediators, such as LPS (Schieferdecker et al. 2001). C5a is particularly critical in mediating a pro-inflammatory response, as it functions as a chemotactic agent, recruiting neutrophils to the site of infection/injury by regulating the expression of chemokines and adhesion molecules (Jauneau et al. 2003; DiScipio et al. 1999). C3a also has anti-inflammatory activity. C3a suppresses LPS-induced TNFα, IL-1 and IL-6 secretion from isolated peripheral blood mononuclear cells and lymphocytes (Fischer and Hugli 1997; Takabayashi et al. 1996, 1998). Further, C3a receptor knock-out mice show increased LPS-induced pro-inflammatory responses, suggesting important anti-inflammatory function for C3a-mediated activation of the C3a receptor (Kildsgaard et al. 2000).

Activation of the complement pathway can also enhance fibrosis. For example, C5 exacerbates fibrosis in a model of bleomycin-induced lung injury (Addis-Lieser et al. 2005) and C5a receptor antagonists have anti-fibrotic effects in mice (Hillebrandt et al. 2005). Importantly, C5 polymorphisms are associated with advanced hepatic fibrosis in patients with chronic hepatitis C infection (Hillebrandt et al. 2005).

Despite the many parallels between the effectors utilized by the complement pathway and those involved in the pathophysiological progression of liver injury, little is known about the role of complement in mediating chronic ethanol-induced liver injury. Recent investigations, making use of rat and mouse models of acute and chronic ethanol exposure, have addressed two important questions: 1) does ethanol exposure activate the complement pathway and 2) does complement contribute to and/or protect from ethanol-induced liver injury.

2.1 Ethanol and Complement Activation

The three pathways of complement activation culminate in the cleavage of C3 (Walport 2001a). Chronic ethanol feeding to mice increased the cleavage of C3, increasing the concentration of C3a in the plasma (Pritchard et al. 2007), measured using specific monoclonal antibody that recognizes neo-activation epitopes on the C3a peptide (Mastellos et al. 2004; Markiewski et al. 2004). C3 deposition is also increased in the livers of mice after chronic ethanol feeding (Jarvelainen et al. 2002). Jarvelainen et al. (2002) reported that ethanol feeding to rats increased deposition of complement proteins C3 and C8, but not C1, in the liver. Lack of C1 deposition suggests that ethanol feeding activates the complement cascade via the lectin and/or alternative pathways (Jarvelainen et al. 2002). As LPS is a potent activator of the alternative pathway, ethanol-induced activation of the alternative pathway would be consistent with the increased LPS exposure observed in humans after alcohol consumption, as well as in animal models of ethanol exposure (Fukui et al. 1991; Bode et al. 1987; Nanji et al. 1993). However, the sensitivity and/or the timing of these assays for C1 deposition may have precluded the identification of a role for the classical pathway in mediating the effects of ethanol in liver injury. Further studies are required to characterize the specific complement activation pathways stimulated by ethanol, as well as the dynamics of their activation during the pathogenesis of ethanol-induced liver injury.

2.2 Role of Complement in Ethanol-Induced Liver Injury

Activation of complement during ethanol exposure suggests that complement may contribute to the pathogenesis of ethanol-induced liver injury. If the complement system contributes to ethanol-induced liver injury, then mice lacking key proteins in the complement activation cascade should be protected from ethanol-induced liver injury. In contrast, mice lacking decay accelerating factor (CD55/DAF), a complement regulatory protein that inhibits the cleavage-induced activation of C3 and C5, might be more susceptible to ethanol-induced liver injury.

Jarvelainin and colleagues first reported that mice lacking the third component of the complement pathway (*C3–/–*) do not develop hepatic steatosis or increased ALT concentrations in response to acute or chronic ethanol exposure (Bykov et al. 2006). Further studies have revealed that C3 and C5 appear to differentially contribute to the pathogenesis of ethanol-induced liver injury (Pritchard et al. 2007). Mice lacking C3 did not develop steatosis in response to ethanol feeding, but still exhibited a modestly increased ALT in the circulation, a marker of hepatocyte injury, as well as increased expression of inflammatory cytokines in the liver (Pritchard et al. 2007). In contrast, mice lacking C5 had increased hepatic triglycerides, but were completely protected from ethanol-induced increases in ALT and inflammatory cytokines (Pritchard et al. 2007). Mice lacking CD55/DAF exhibited exacerbated liver injury in response to ethanol feeding, suggesting that

CD55/DAF, a complement regulatory protein, serves as a brake on ethanol-induced steatosis and injury to hepatocytes.

3 C3 in the Development of Hepatic Steatosis

Ethanol-induced hepatic steatosis involves the up-regulation of genes regulating fatty acid synthesis and down-regulation of genes involved in fatty acid oxidation, as well as an impairment of VLDL secretion (Nagy 2004). Bykov and colleagues investigated the impact of chronic ethanol on the expression of genes regulating lipid synthesis in wild type and *C3–/–* (Bykov et al. 2007a). Interestingly, ethanol decreased transcripts for lipogenic enzymes in *C3–/–* mice, suggesting that *C3–/–* mice were protected from steatosis by a failure to up-regulate fatty acid synthesis during chronic ethanol feeding (Bykov et al. 2007a).

C3 might also contribute to ethanol-induced steatosis via the regulation of fatty acid oxidation. One of the breakdown products of C3, termed acylation stimulating protein (ASP or C3a*des*Arg) is involved in the regulation of glucose and lipid metabolism (reviewed in (Cianflone et al. 2003). While adipose tissue is the primary target of ASP, in vivo studies suggest that it also regulates fatty acid uptake and oxidation in the liver (Cianflone et al. 2003). While the effects of C3 deficiency on lipid homeostasis are still controversial (Kildsgaard et al. 1999), it is possible that the absence of ASP/C3a*des*Arg, in the *C3–/–* mice contributes to protection from ethanol-induced steatosis via increases in fatty acid oxidation in ethanol-fed mice. Increased fatty acid oxidation would thus decrease the fatty acids available for synthesis of triglyceride and reduce the development of steatosis.

4 Complement and Inflammatory Cytokines in Ethanol-Induced Liver Injury

C3a and C5a, termed the anaphylatoxins, are important regulators of the inflammatory response. Increased expression of inflammatory cytokines in the liver is an important contributor to the pathogenesis of ethanol-induced liver injury (Tilg and Diehl 2000). Expression of inflammatory cytokines in the livers of mice during ethanol exposure was different in *C3–/–* and *C5–/–* mice (Pritchard et al. 2007). *C3–/–* mice, while protected from ethanol-induced steatosis, had increased expression of inflammatory cytokines, including TNF-α, IL-6 and interferon-γ, as well as moderately increased ALT. In contrast, *C5–/–* mice did not increase liver cytokines or ALT, even though they exhibited steatosis (Pritchard et al. 2007). These data suggest that C5, but not C3, contributes to increased production of inflammatory cytokines during ethanol exposure.

While reduced liver cytokines in the *C5–/–* after ethanol exposure clearly point to a pro-inflammatory function of C5a during ethanol exposure, the maintenance of increased inflammatory cytokines in the livers of *C3–/–* mice may involve alternative mechanisms. First, C5 may be activated in a non-complement dependent mechanism, such as a plasmin- (Wetsel and Kolb 1983) and thrombin-mediated cleavage of C5 (Huber-Lang et al. 2006). If this were true, then the presence of C5a, with potent pro-inflammatory functions, in the *C3–/–* mice would be consistent with the increased liver cytokines after ethanol feeding. Secondly, the absence of an important anti-inflammatory function of C3a might also contribute to increased cytokines in *C3–/–* mice during ethanol exposure. Given the key role of LPS-stimulated cytokine production in the pathogenesis of ethanol-induced liver injury (Thurman 1998), the absence of an anti-inflammatory function for C3a in the *C3–/–* mice could allow for the maintenance of LPS-stimulated cytokine production during ethanol exposure. This hypothesis is consistent with the particularly effective ability of C3a to inhibit LPS-stimulated cytokine production (Kildsgaard et al. 2000).

5 Complement Regulatory Proteins in Ethanol-Induced Liver Injury

Given the powerful destructive potential of the complement pathway, several mechanisms have evolved for self-protection of the host organism. A number of soluble complement inhibitors are secreted, including C1 inhibitor, C4b binding protein, factor H and factor I. In addition, cells express membrane-associated inhibitors (e.g. CD46, decay accelerating factor (DAF)/CD55 and CD59). Each of the complement regulatory proteins counters specific steps in the activation pathways or formation of the membrane attack complex. The membrane bound regulatory proteins are effective at protecting self from complement attack by providing localized inhibitory functions. For example, DAF/CD55 inhibits C3 convertase activity, while CD59 prevents the formation of the membrane attack complex (Gasque 2004). Expression of CD55/DAF and CD46 on the surface of all self cells allows for the protection of autologous cells from complement-mediated lysis, particularly relevant because of the spontaneous hydrolysis of the labile thioester bond of C3 (Medzhitov and Janeway 2002). Loss/shedding of membrane associated complement regulatory protein CD46 on apoptotic and necrotic cells has been suggested to enhance the efficiency of removal of these damaged cells (Elward et al. 2005).

Chronic ethanol feeding reduces the expression of Crry (the rat homologue of CD55/DAF) and CD59 in rats after chronic ethanol feeding (Jarvelainen et al. 2002). Further, *CD55/DAF–/–* mice exhibited exacerbated hepatic steatosis and injury in response to chronic ethanol feeding (Pritchard et al. 2007), suggesting that CD55/DAF serves as an important protective element against the hepatotoxic effects of ethanol.

6 Membrane Attack Complex (MAC) and Ethanol-Induced Liver Injury

All three pathways of complement activation converge at the site of C3 cleavage. Subsequent cleavage of C5 is the first step in the further targeted enzymatic cleavage of complement proteins involved in the formation of the membrane attack complex (MAC) which is comprised of C5b, C6, C7, C8 and multiple C9 molecules. These proteins assemble into a pore/hole in the phospholipid bilayer, leading to cell lysis (Gasque 2004). Deficiencies in the ability to form opsonins or generate MAC greatly increase the susceptibility to bacterial infections (Gasque 2004).

While very little data is available on the role of the MAC in ethanol-induced liver injury, there is a suggestion that MAC may play a protective role in this process. For example, chronic ethanol feeding down-regulates the expression of terminal complex components, C6, C8alpha and C9 in liver (Bykov et al. 2007b). Further, rats deficient in C6 (*C6–/–*) exhibited a modest, but significant, worsening of liver steatosis and inflammation induced by 6 weeks of ethanol feeding compared to wild type rats (Bykov et al. 2004). These data suggest that activation of the terminal complement pathway may have a protective function during chronic liver injury.

7 Complement in Hepatocellular Proliferation

The liver has a remarkable capacity to regenerate (Michalopoulos and DeFrances 1997). Regulation of cell proliferation is an additional function of the complement system. While the process is not completely understood, studies have shown that formation of the membrane attack complex stimulates mitogenesis in 3T3-L1 cells, both in the absence and presence of exogenous growth factors (Halperin et al. 1993). This proliferative response may contribute to focal tissue repair in sites of complement-induced injury (Halperin et al. 1993).

Studies from the laboratory of Dr. Lambris demonstrate that both C3 and C5 are required for hepatocyte proliferation in response to toxin-induced injury (CCl_4) and partial hepatectomy (Mastellos et al. 2001; Markiewski et al. 2004; Strey et al. 2003). C3 activation products rapidly increase in the plasma in response to partial hepatectomy, followed by a second phase of activation after 24 h (Markiewski et al. 2004). C3b deposition in liver is observed within 3 h of CCl_4 exposure (Markiewski et al. 2004). The authors suggest that C3 activation at early phases of hepatic regeneration may be involved in priming hepatocytes, while later phases of C3 activation are involved in removal of injured/damaged cells (Markiewski et al. 2004).

While it is well known that chronic ethanol exposure dampens the regenerative response to liver injury (Diehl 2005), there may be stages during the initiation and progression of ethanol-induced liver injury where hepatocyte proliferation is transiently increased. For example, during chronic ethanol feeding to rats, Apte et al. (2004) observed an increase in hepatocyte proliferation during the early stages of steatosis (1–4 weeks of feeding), suggesting that the liver is at least partially capable of repairing ethanol-induced damage at this stage of injury (Apte et al. 2004). Future investigations to determine whether abnormal activation and/or control of complement pathways during chronic ethanol contributes to impaired hepatocellular proliferation may provide important clues into the pathophysiology of ethanol-induced liver injury.

8 Complement and "Waste Disposal": Complement and Eethanol-Induced Apoptosis

The role of complement in the removal of injured and damaged cells after injury may be a host-specific manifestation of the function of complement in opsonization and clearance of bacteria. Indeed, there is a growing appreciation of the role of complement in clearance of apoptotic cells, as well as potentially pathogenic immune complexes (Walport 2002). For example, C1q-deficient mice show impaired clearance of apoptotic cells. In these mice, apoptotic cells accumulate in the kidney and lead to glomerular nephritis with immune deposits (Botto et al. 1998). It has been proposed that C1q can bind directly to surface blebs on apoptotic cells, leading to complement activation (Gasque 2004). Recent studies show that C1q binds to nucleic acids exposed on the surface of apoptotic cells, leading to C3-mediated opsonization and contributing to the efficient removal of apoptotic cells (Elward et al. 2005). Complement can also be activated during ischemia-reperfusion injury (Walport 2001b). C1q is activated during ischemia-reperfusion, likely due to exposure of abnormal phospholipids on the cell surface and/or exposure to mitochondrial proteins (Walport 2001b). While complement is not required for clearance of apoptosis in all tissues (clearance of apoptotic cells is mediated by a number of receptor/ligand mediated systems), evidence indicates that it can play an essential role in "waste-disposal" under certain conditions (Botto and Walport 2002).

Chronic ethanol exposure increases the susceptibility of hepatocytes to apoptosis. While there are not yet data available, we can speculate that specific components of the complement pathway, if activated during chronic ethanol exposure, may have deleterious and/or protective effects in stimulating apoptosis in hepatocyte, as well as other cell types within the liver, and/or the clearance of apoptotic cells from the liver.

9 Conclusions

There is a growing body of evidence to indicate that complement plays a complex role in the initation and progression of ethanol-induced liver injury. Ethanol exposure to mice results in the activation of the complement pathway (Pritchard et al. 2007; Jarvelainen et al. 2002). Importantly, C3 and C5 differentially contribute to the development of steatosis, increased production of pro-inflammatory cytokines and injury to hepatocytes and CD55/DAF, a complement regulatory protein, serves as a brake on ethanol-induced steatosis and injury to hepatocytes (Pritchard et al. 2007). Taken together, these data demonstrate an important role for complement activation in ethanol-induced liver injury and suggest that therapeutic interventions targeted at specific components in the complement pathway might be useful in treatment and/or prevention of alcoholic liver disease.

Acknowledgements

This work was supported by NIH grants AA013868 and AA11975 to LEN and AI23598 MEM.

References

Adachi, Y, Bradford, BU, Gao, W, Bojes, HK, and Thurman, RG, (1994) Inactivation of Kupffer cells prevents early alcohol-induced liver injury, *Hepatology* 20, 453–460

Addis-Lieser, E, Kohl, J, and Chiaramonte, MG, (2005) Opposing regulatory roles of complement factor 5 in the development of bleomycin-induced pulmonary fibrosis, *Journal of Immunology* 175, 1894–1902

Apte, UM, McRee, R, and Ramaiah, SK, (2004) Hepatocyte proliferation is the possible mechanism for the transient decrease in liver injury during steatosis stage of alcoholic liver disease, *Toxicologic Pathology* 32, 567–576

Bode, C, Kugler, V, and Bode, JC, (1987) Endotoxemia in patients with alcoholic and non-alcoholic cirrhosis and in subjects with no evidence of chronic liver disease following acute alcohol excess, *Journal of Hepatology* 4, 8–14

Botto, M and Walport, MJ, (2002) C1q, autoimmunity and apoptosis, *Immunobiology* 205, 395–406

Botto, M, Dell'Agnola, C, Bygrave, AE, Thompson, EM, Cook, HT, Petry, F, Loos, M, Pandolfi, PP, and Walport, MJ, (1998) Homozygous C1q deficiency causes glomerulonephritis associated with multiple apoptotic bodies, *Nature Genetics* 19, 56–59

Bykov, IL, Vakeva, A, Jarvelainen, HA, Meri, S, and Lindros, KO, (2004) Protective function of complement against alcohol-induced rat liver damage, *International Immunopharmacology* 4, 1445–1454

Bykov, I, Junnikkala, S, Pekna, M, Lindros, KO, and Meri, S, (2006) Complement C3 contributes to ethanol-induced liver steatosis in mice, *Annals of Medicine* 38, 280–286

Bykov, I, Jauhiainen, M, Olkkonen, VM, Saarikoski, ST, Ehnholm, C, Junnikkala, S, Vakeva, A, Lindros, KO, and Meri, S, (2007a) Hepatic gene expression and lipid parameters in complement C3(–/–) mice that do not develop ethanol-induced steatosis, *Journal of Hepatology* 46, 907–914

Bykov, I, Junnikkala, S, Pekna, M, Lindros, KO, and Meri, S, (2007b) Effect of chronic ethanol consumption on the expression of complement components and acute-phase proteins in liver, *Clinical Immunology* 124, 213–220

Cianflone, K, Xia, Z, and Chen, LY, (2003) Critical review of acylation-stimulating protein physiology in humans and rodents, *Biochimica et Biophysica Acta* 1609, 127–143

Daveau, M, Benard, M, Scotte, M, Schouft, MT, Hiron, M, Francois, A, Salier, JP, and Fontaine, M, (2004) Expression of a functional C5a receptor in regenerating hepatocytes and its involvement in a proliferative signaling pathway in rat, *Journal of Immunology* 173, 3418–3424

Diehl, AM, (2005) Recent events in alcoholic liver disease V. effects of ethanol on liver regeneration, *American Journal of Physiology Gastrointestinal and Liver Physiology* 288, G1–G6

DiScipio, RG, Daffern, PJ, Jagels, MA, Broide, DH, and Sriramarao, P, (1999) A comparison of C3a and C5a-mediated stable adhesion of rolling eosinophils in postcapillary venules and transendothelial migration in vitro and in vivo, *Journal of Immunology* 162, 1127–1136

Elward, K, Griffiths, M, Mizuno, M, Harris, CL, Neal, JW, Morgan, BP, and Gasque, P, (2005) CD46 plays a key role in tailoring innate immune recognition of apoptotic and necrotic cells, *Journal of Biological Chemistry* 280, 36342–36354

Fischer, WH and Hugli, TE, (1997) Regulation of B cell functions by C3a and C3a(desArg): suppression of TNF-alpha, IL-6, and the polyclonal immune response, *Journal of Immunology* 159, 4279–4286

Friedman, SL, (2000) Molecular regulation of hepatic fibrosis, an integrated cellular response to tissue injury, *Journal of Biological Chemistry* 275, 2247–2250

Fukui, H, Brauner, B, Bode, J, and Bode, C, (1991) Plasma endotoxin concentrations in patients with alcoholic and nonalcoholic liver disease: reevaluation with an improved chromogenic assay, *Journal of Hepatology* 12, 162–169

Gasque, P, (2004) Complement: a unique innate immune sensor for danger signals, *Molecular Immunology* 41, 1089–1098

Gressner, AM and Bachem, MG, (1995) Molecular mechanisms of liver fibrogenesis-a homage to the role of activated fat-storing cells, *Digestion* 56, 335–346

Halperin, JA, Taratuska, A, and Nicholson-Weller, A, (1993) Terminal complement complex C5b-9 stimulates mitogenesis in 3T3 cells, *Journal of Clinical Investigation* 91, 1974–1978

Hillebrandt, S, Wasmuth, HE, Weiskirchen, R, Hellerbrand, C, Keppeler, H, Werth, A, Schirin-Sokhan, R, Wilkens, G, Geier, A, Lorenzen, J, Kohl, J, Gressner, AM, Matern, S, and Lammert, F, (2005) Complement factor 5 is a quantitative trait gene that modifies liver fibrogenesis in mice and humans, *Nature Genetics* 37, 835–843

Hoek, JB and Pastorino, JG, (2002) Ethanol, oxidative stress, and cytokine-induced liver cell injury, *Alcohol* 27, 63–68

Huber-Lang, M, Sarma, JV, Zetoune, FS, Rittirsch, D, Neff, TA, McGuire, SR, Lambris, JD, Warner, RL, Flierl, MA, Hoesel, LM, Gebhard, F, Younger, JG, Drouin, SM, Wetsel, RA, and Ward, PA, (2006) Generation of C5a in the absence of C3: a new complement activation pathway, *Nature Medicine* 12, 682–687

Jaeschke, H, Gores, GJ, Cederbaum, AI, Hinson, JA, Pessayre, D, and Lemasters, JJ, (2002) Mechanisms of hepatotoxicity, *Toxicological Sciences* 65, 166–176

Jarvelainen, HA, Vakeva, A, Lindros, KO, and Meri, S, (2002) Activation of complement components and reduced regulator expression in alcohol-induced liver injury in the rat, *Clinical Immunology* 105, 57–63

Jauneau, AC, Ischenko, A, Chan, P, and Fontaine, M, (2003) Complement component anaphylatoxins upregulate chemokine expression by human astrocytes, *FEBS Letters* 537, 17–22

Kildsgaard, J, Zsigmond, E, Chan, L, and Wetsel, RA, (1999) A critical evaluation of the putative role of C3adesArg (ASP) in lipid metabolism and hyperapobetalipoproteinemia, *Molecular Immunology* 36, 869–876

Kildsgaard, J, Hollmann, TJ, Matthews, KW, Bian, K, Murad, F, and Wetsel, RA, (2000) Cutting edge: targeted disruption of the C3a receptor gene demonstrates a novel protective anti-inflammatory role for C3a in endotoxin-shock, *Journal of Immunology* 165, 5406–5409

Koch, CA, Kanazawa, A, Nishitai, R, Knudsen, BE, Ogata, K, Plummer, TB, Butters, K, and Platt, JL, (2005) Intrinsic resistance of hepatocytes to complement-mediated injury, *Journal of Immunology* 174, 7302–7309

Lieber, CS, (1994) Alcohol and the liver: 1994 update, *Gastroenterology* 106, 1085–1105

Lin, F, Fukuoka, Y, Spicer, A, Ohta, R, Okada, N, Harris, CL, Emancipator, SN, and Medof, ME, (2001) Tissue distribution of products of the mouse decay-accelerating factor (DAF) genes. Exploitation of a Dafl knock-out mouse and site-specific monoclonal antibodies, *Immunology* 104, 215–225

Markiewski, MM, Mastellos, D, Tudoran, R, DeAngelis, RA, Strey, CW, Franchini, S, Wetsel, RA, Erdei, A, and Lambris, JD, (2004) C3a and C3b activation products of the third component of complement (C3) are critical for normal liver recovery after toxic injury, *Journal of Immunology* 173, 747–754

Martinez-Hernandez, A and Amenta, PS, (1993) The hepatic extracellular matrix II. Ontogenesis, regeneration and cirrhosis, *Virchows Archiv A Pathological Anatomy* 423, 77–84

Mastellos, D, Papadimitriou, JC, Franchini, S, Tsonis, PA, and Lambris, JD, (2001) A novel role of complement: mice deficient in the fifth component of complement (C5) exhibit impaired liver regeneration, *Journal of Immunology* 166, 2479–2486

Mastellos, D, Prechl, J, Laszlo, G, Papp, K, Olah, E, Argyropoulos, E, Franchini, S, Tudoran, R, Markiewski, M, Lambris, JD, and Erdei, A, (2004) Novel monoclonal antibodies against mouse C3 interfering with complement activation: description of fine specificity and applications to various immunoassays, *Molecular Immunology* 40, 1213–1221

Medzhitov, R and Janeway, CA Jr, (2002) Decoding the patterns of self and nonself by the innate immune system, *Science* 296, 298–300

Michalopoulos, GK and DeFrances, MC, (1997) Liver regeneration, *Science* 276, 60–66

Minagawa, M, Deng, Q, Liu, ZX, Tsukamoto, H, and Dennert, G, (2004) Activated natural killer T cells induce liver injury by Fas and tumor necrosis factor-alpha during alcohol consumption, *Gastroenterology* 126, 1387–1399

Monsinjon, T, Gasque, P, Chan, P, Ischenko, A, Brady, JJ, and Fontaine, MC, (2003) Regulation by complement C3a and C5a anaphylatoxins of cytokine production in human umbilical vein endothelial cells, *FASEB Journal* 17, 1003–1014

Nagy, LE, (2003) New insights into the role of the innate immune response in the development of alcoholic liver disease. *Experimantal Biology and Medicine* 228, 882–890

Nagy, LE, (2004) Molecular aspects of alcohol metabolism. *Annual Review Nutrition* 24, 55–78

Nanji, AA, Khettry, U, Sadrzadeh, SMH, and Yamanaka, T, (1993) Severity of liver injury in experimental alcoholic liver disease: Correlation with plasma endotoxin, prostaglandin E2, leukotriene B4 and thromboxane B2, *American Journal of Pathology* 142, 367–373

Pritchard, MT, McMullen, MR, Stavitsky, AB, Cohen, JI, Lin, F, Medof, ME, and Nagy, LE, (2007) Differential contributions of C3, C5, and decay-accelerating factor to ethanol-induced fatty liver in mice, *Gastroenterology* 132, 1117–1126

Raghow, R, (1994) The role of extracellular matrix in postinflammatory wound healing and fibrosis, *FASEB Journal* 8, 823–831

Schieferdecker, HL, Schlaf, G, Jungermann, K, and Gotze, O, (2001) Functions of anaphylatoxin C5a in rat liver: direct and indirect actions on nonparenchymal and parenchymal cells, *International Immunopharmacology* 1, 469–481

Strey, CW, Markiewski, M, Mastellos, D, Tudoran, R, Spruce, LA, Greenbaum, LE, and Lambris, JD, (2003) The proinflammatory mediators C3a and C5a are essential for liver regeneration, *Journal of Experimental Medicine* 198, 913–923

Takabayashi, T, Vannier, E, Clark, BD, Margolis, NH, Dinarello, CA, Burke, JF, and Gelfand, JA, (1996) A new biologic role for C3a and C3a desArg: regulation of TNF-alpha and IL-1 beta synthesis, *Journal of Immunology* 156, 3455–3460

Takabayashi, T, Vannier, E, Burke, JF, Tompkins, RG, Gelfand, JA, and Clark, BD, (1998) Both C3a and C3a(desArg) regulate interleukin-6 synthesis in human peripheral blood mononuclear cells, *Journal of Infectious Diseases* 177, 1622–1628

Thurman, RG, (1998) Mechanisms of Hepatic Toxicity II. Alcoholic liver injury involves activation of Kupffer cells by endotoxin, *American Journal of Physiology* 275, G605–G611

Tilg, H and Diehl, AM, (2000) Cytokines in alcoholic and nonalcoholic steatohepatitis, *New England Journal of Medicine* 343, 1467–1476

Tuma, DJ and Casey, CA, (2003) Dangerous byproducts of alcohol breakdown – focus on adducts, *Alcohol Research and Health* 27, 285–290

Walport, MJ, (2001a) Complement. First of two parts, *New England Journal of Medicine* 344, 1058–1066

Walport, MJ, (2001b) Complement. Second of two parts, *New England Journal of Medicine* 344, 1140–1144

Walport, MJ, (2002) Complement and systemic lupus erythematosus, *Arthritis Research* 4, S279–S293

Wetsel, RA and Kolb, WP, (1983) Expression of C5a-like biological activities by the fifth component of human complement (C5) upon limited digestion with noncomplement enzymes without release of polypeptide fragments, *Journal of Experimental Medicine* 157, 2029–2048

14. Immune Complex-Mediated Cytokine Production is Regulated by Classical Complement Activation both In Vivo and In Vitro

Johan Rönnelid*, Erik Åhlin, Bo Nilsson, Kristina Nilsson-Ekdahl, and Linda Mathsson

*Unit of Clinical Immunology, Uppsala University, Uppsala, Sweden
johan.ronnelid@klinimm.uu.se

Abstract. Immune complexes (IC) induce a number of cellular functions, including the enhancement of cytokine production from monocytes, macrophages and plasmacytoid dendritic cells. The range and the composition of cytokines induced by IC in vitro is influenced by the availability of an intact classical complement cascade during cell culture, as we have showed in our studies on artificial IC and on cryoglobulins purified from patients with lymphoproliferative diseases. When IC purified from systemic lupus erythematosus sera were used to stimulate in vitro cytokine production, the amount of circulating IC and IC-induced cytokine levels depended both on in vivo classical complement function as well as on the occurrence of anti-SSA, but not on anti-dsDNA or any other autoantibodies. Collectively these findings illustrate that studies on IC-induced cytokine production in vitro requires stringent cell culture conditions with complete control and definition of access to an intact classical complement pathway in the cell cultures. If IC are formed in vivo, the results have to be interpreted in the context of classical complement activation in vivo as well as the occurrence of IC-associated autoantibodies at the time of serum sampling.

1 Introduction

Immune complexes (IC) consist of antibodies associated with their corresponding antigens. Circulating IC are demonstrated in rheumatic diseases as well as malignant and infectious diseases. The formation of IC is also part of the normal immune response in healthy individuals, e.g. in conjunction to vaccination (Tarkowski et al. 1985). Whereas autoantigen- and autoantibody-containing IC can be demonstrated in autoimmune diseases, IC often include antigens from pathogens and the corresponding antibodies in infectious disease states. IC might also consist of self-aggregating IgG as is the case in many cryoglobulins associated with malignancies.

IC induce a number of cellular functions, including phagocytosis, opsonisation and the induction of anaphylatoxin-mediated chemotaxis. One IC-mediated function that recently has acquired increasing interest concerns their effects on cytokine production. A number of cell types respond to IC by cytokine production,

J.D. Lambris (ed.), *Current Topics in Complement II.*
doi: 10:1007/978-0-387-78952-1_14, © Springer Science+Business Media, LLC 2008

including monocytes (Mathsson et al. 2005, 2006; Mullazehi et al. 2006), macrophages (Blom et al. 2003), monocytoid dendritic cells (Radstake et al. 2004) and plasmacytoid dendritic cells (Lövgren et al. 2004). Our research interest concerns IC-induced production of cytokines from peripheral blood mononuclear cells (PBMC) where the IC are obtained from patients with rheumatoid arthritis (RA) (Mathsson et al. 2006), systemic lupus erythematosus (SLE) (Rönnelid et al. 2003; Mathsson et al. 2007), patients with cryoglobulinaemia on the basis of lymphproliferative diseases (Mathsson et al. 2005) and patients infected with the tropical parasite *Leishmania donovani* (Elshafie et al. 2006, 2007). To define the fundamental requirements for IC-induced cytokine production we also utilize totally artificial immune aggregates like heat-aggregated gamma globulin (HAGG) (Tejde et al. 2004). In situations where the IC are supposed to be tissue-bound in vivo, we instead create a parallel in vitro situation utilizing autoantigen and patient-derived autoantibodies as we have done with anti-type II collagen autoantibodies in RA patients (Mullazehi et al. 2006).

In our studies we have found that the IC-induced cytokine responses in vitro are affected by the classical complement pathway. This influence is manifested at two levels. Classical complement activation together with the occurrence of specific autoantibodies in vivo is associated with levels of circulating IC as well as with the levels of IC-induced cytokines (Mathsson et al. 2007). The access of an intact classical complement pathway in vitro during cell stimulation with IC instead influence the composition of cytokines induced, as well as cytokine levels (Tejde et al. 2004; Mathsson et al. 2005).

2 Down-Regulation of IC-Induced IL-12 Production by the Classical Complement Pathway

In a first study we investigated the induction of the T helper type 2 (Th2) cytokines interleukin (IL)–6 and IL-10 by serum IC obtained from SLE patients (Rönnelid et al. 2003). In that paper we had used fetal calf serum (FCS) as growth factor source in our cell cultures. Besides our and other (Park et al. 1998) studies showing the importance of elevated IL-10 levels in SLE, a series of papers from another group had shown that levels of the Th1 cytokine IL-12 were decreased in SLE, and that IL-10 and IL-12 were inversely regulated in SLE (Liu and Jones 1998a,b; Liu et al. 1999). We therefore wanted to investigate whether this was due to reciprocal IC-mediated regulation of IL-10 and IL-12 in vitro, at the same time as we humanized the cell culture system to utilize normal human serum (NHS) instead of FCS.

In this cell culture system we found that addition of NHS to the cell culture system enhanced the IC-induced production of IL-10 whereas at the same time IL-12p40 production was suppressed in a dose-dependent manner (Tejde et al. 2004) (Fig. 1). The suppressive effect on IL-12p40 production by addition of serum was much smaller when the NHS had been heat-inactivated (Fig. 1b). As IC we used

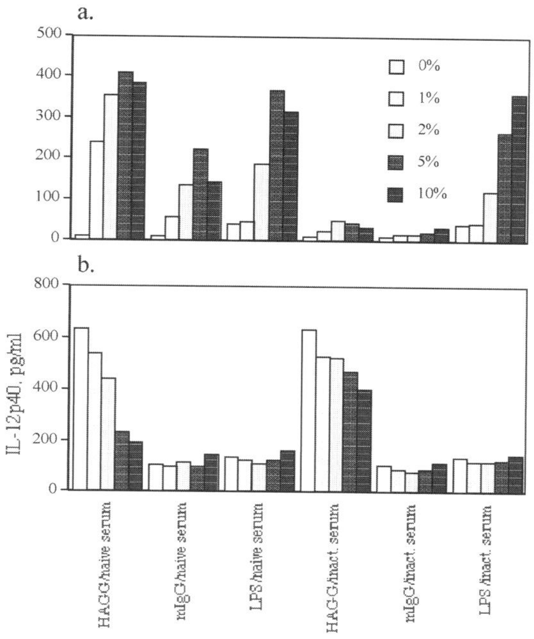

Fig. 1 The effect of gradual addition of human serum to PBMC cultures stimulated with heat-aggregated gamma globulin (HAGG) or the same monomeric non aggregated IgG (mIgG) as negative controls on the production of (**a**) IL-10 and (**b**) IL-12p40. As positive controls, cells were stimulated with 100 ng/mL of lipopolysaccharide (LPS) The human serum was either native (three left sets of bars) or decomplemented by heat inactivation (56°C, 30 min; three right sets of bars)

HAGG with the corresponding non-aggregated monomeric IgG (mIgG) preparation as control. This control preparation showed only minimally different cytokine regulatory effects as compared to cell cultures without any mIgG added (data not shown). Addition of native NHS to control cultures increased IL-10 production somewhat, whereas there was no effect on IL-10 production of addition of heat-inactivated NHS. Neither addition of native or heat-inactivated NHS had any effect on the production of IL-12p40 in control cultures. Addition of either native or heat-inactivated NHS to lipopolysaccharide (LPS) -stimulated cell cultures also increased IL-10 production in a dose dependent manner, but had no effect on the production of IL-12p40 (Fig. 1).

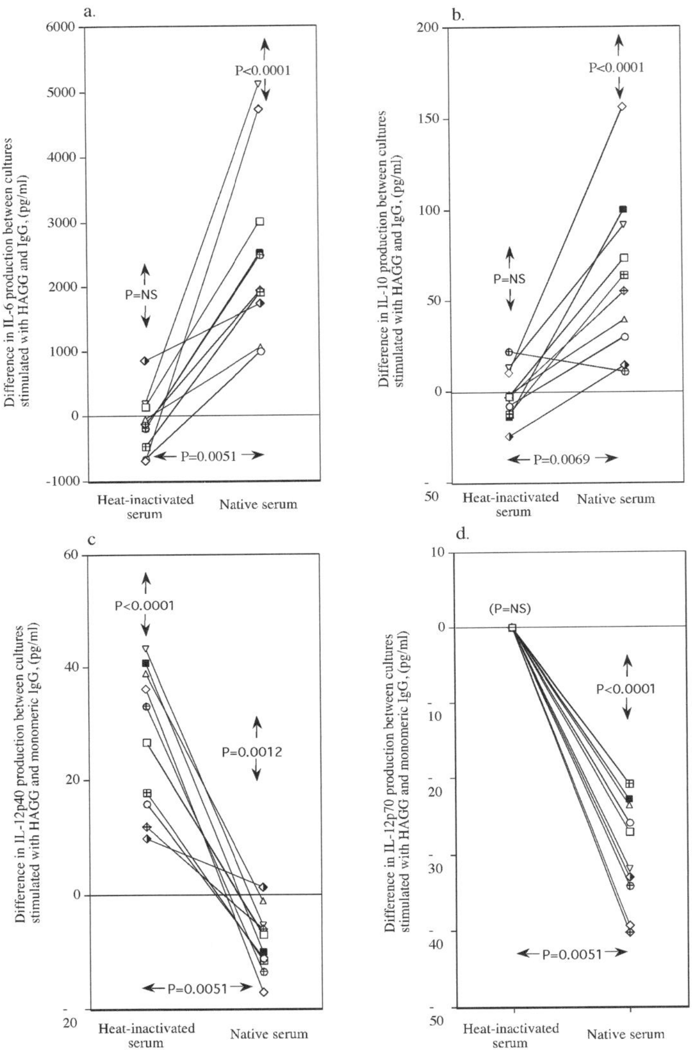

Fig. 2 Effects of native and heat-inactivated NHS on IA-induced cytokine production in ten PBMC donors. PBMC (10^6/ml) were stimulated with HAGG or mIgG (100 μg/ml) in media containing 10% native or heat-inactivated NHS. After 20 h supernatants were collected and levels of (**a**) IL-6, (**b**) IL-10, (**c**) IL-12p40 and (**d**) IL-12p70 analyzed by ELISA. Results are shown as differences in cytokine production between cultures stimulated with HAGG and mIgG, with a positive value signifying a net stimulatory effect of HAGG. P-values within vertical arrows apply to the net differences between HAGG and mIgG-stimulated cultures. P-values within horizontal arrows apply to differences in net HAGG-induced effects between cell cultures with heat-inactivated and native serum. Reprinted from (Tejde et al. 2004) with permission from Blackwell Publishing

Using 10% serum in cell cultures, the net effect of IC on cytokine production (the difference in cytokine production between HAGG- and mIgG-stimulated cultures) was therefore decreased after heat-inactivation for the production of IL-10, but increased for the production of IL-12p40. To investigate whether this effect was responder cell dependent we investigated ten different responder cell populations in parallel. As can be seen in Fig. 2, the effect was similar in each experiment. Whereas there was no net effect on IL-10 production of IC-stimulation in heat-inactivated serum, there was a highly significant stimulatory effect when native serum was used (Fig. 2b). IL-12p40 production on the other hand was significantly enhanced by IC utilizing heat-inactivated serum, but instead suppressed when native NHS was used (Fig 2c). As measurement of IL-12p40 might reflect levels of the p40 homodimer with other biological activity, we repeated the measurement for the IL-12p70 heterodimer. Even if levels were much lower, the general findings for IL-12p40 were corroborated (Fig 2d). IC-induced levels of IL-6 paralleled those for IL-10 (Fig 2a).

We interpreted the results to imply that IC-induced production of IL-12 is regulated by the complement system. In two separate sets of experiments this hypothesis was corroborated. When a C4-deficient serum without any measurable classical complement function was gradually reconstituted with C4, the net IC-induced production of IL-12p40 was suppressed in parallel to regained classical complement function (data not shown). Addition of C4 to a C4-sufficient NHS had on the other hand no effect on net IC-induced production of IL-12p40. Addition of C4 did not convey any consistent effects on net IL-10 production in either type of serum.

In a third set of experiments the complement system was blocked in NHS. Blockade of classical complement function by addition of the complement inhibitor Compstatin (Sahu et al. 2000) to NHS increased net IL-12p40 production in five parallel PBMC cultures (Fig. 3).

The effect on IL-10 production differed from above, as in these cell cultures also net IC-induced production of IL-10 was increased in 4/5 experiments after the addition of Compstatin (Fig. 3).

Collectively these three sets of experiments show that after IC-stimulation the production of IL-12 by PBMC is suppressed by an intact classical complement pathway. IL-10 is instead positively regulated in some but not all of the investigated experimental systems. As IL-10 might down-regulate the production of IL-12 (Liu et al. 1998a) we neutralized in vitro-produced IL-10 in an attempt to overcome the down-regulatory effects on IL-12p40 production in native serum, but without effect. Another indication that the IC-induced effects of the production of IL-10 and IL-12p40 are regulated separately is the fact that IC-induced production of IL-10 but not of IL-12p40 can be blocked by an antibody against FcγRIIa (Rönnelid et al. 2003; Tejde et al. 2004). A number of studies have shown that C5a can suppress the production of IL-12 (Wittmann et al. 1999; Braun et al. 2000) and related cytokines (Hawlisch et al. 2005). Blockade with a C5aR antibody did however not succeed to overcome the suppressive effects of IC in native serum.

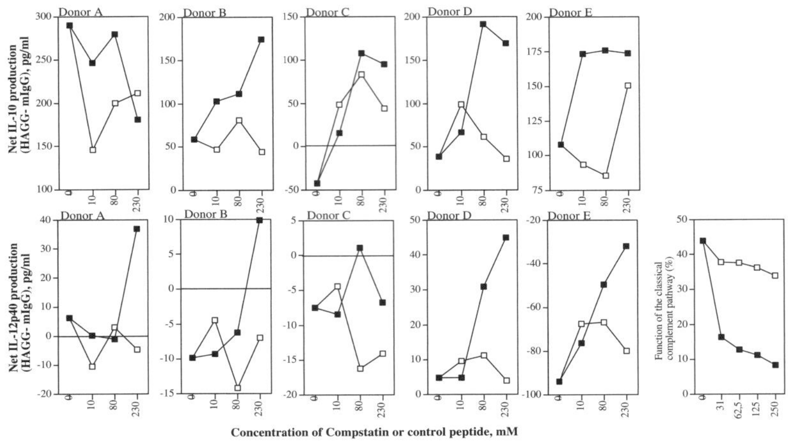

Fig. 3 Effects of gradual suppression of the classical complement pathway by Compstatin. Compstatin (filled symbols) or a linear control peptide (open symbols) were added to 100% normal human serum for 15 min at 37°C, whereupon HAGG or mIgG (100 µg/ml final concentration in the PBMC cultures) were added for additional 20 min at 37°C. Sera were thereafter added to cultures (10% v/v) of PBMC from five healthy donors and incubated for 20 h. Supernatants were then collected and analyzed for IL-10 and IL-12p40 by ELISA. Results are shown as differences in cytokine production between cultures stimulated with HAGG and mIgG. Compstatin concentrations in the graphs indicate final cell culture concentrations. The rightmost graph show the effect on the classical complement function after addition of increasing amounts of Compstatin or control peptide. Reprinted from (Tejde et al. 2004) with permission from Blackwell Publishing

3 Complement Blockade Induce Inverse Regulation of Cryoglobulin-Stimulated TNF-α and IL-10 Production

Cryoglobulins (CG) are IC that form reversibly when serum is cooled below body temperature, but also as a response to changes in pH and ionic strength. CG can be found e.g. in the blood of patients with rheumatic diseases, viral hepatitides and lymphoproliferative diseases. The pathology is dependent on CG deposition in vessel walls with ensuing complement activation (Kallemuchikkal and Gorevic 1999). CG are known as strong complement activators, and complement proteins can be detected in the precipitated CG (Weiner et al. 2001).

We investigated CG-induced production of IL-6, IL-10 and TNF-α in two patients with lymphoproliferative diseases, one patient with a cryoprecipitable IgG3

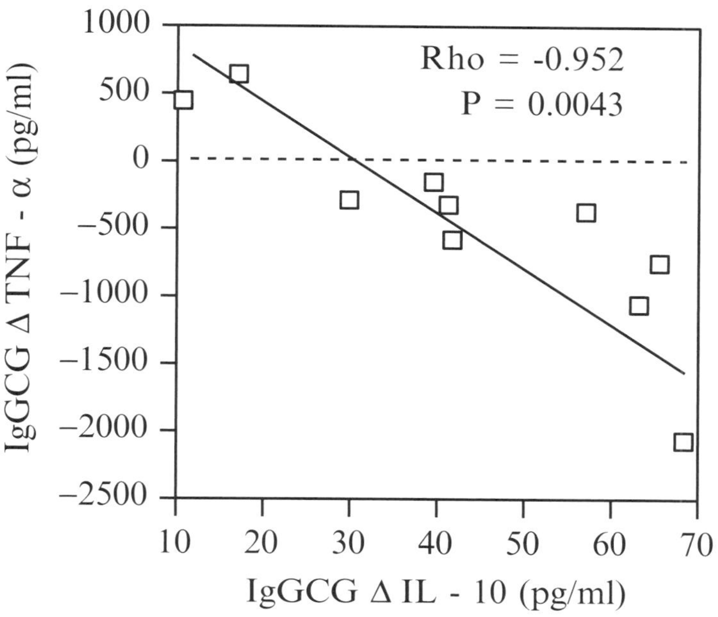

Fig. 4 Correlation between the net effect of complement blockade on the production of IL-10 and TNF-α after stimulation of PBMC cultures from ten healthy donors with IgG3 CG. Δ cytokine production denotes differences between cytokine production in cultures where complement activation had been blocked with optimal doses of Compstatin and cultures treated with a linear control peptide. Statistics were performed with the Spearman Rank Correlation Test. Reprinted from (Mathsson et al. 2005) with permission from Blackwell Publishing

myeloma and one patient with Waldenström's macroglobulinemia and an IgM CG (Mathsson et al. 2005). Both IL-6 (Urbanska-Rys et al. 2000; Lauta 2003) and IL-10 (Lu et al. 1995; Gu et al. 1996) are known to promote the growth of myeloma cells, and TNF-α stimulate myeloma growth by increasing cell adhesion to bone marrow stroma (Hideshima et al. 2001). We could show that changes in temperature and ionic strength mediating CG precipitation also involved corresponding changes in the production of both IL-10 and TNF-α. In those optimal concentrations of Compstatin the effects were opposite on IgG3 CG-induced production of IL-10 and TNF-α. The net effect of addition of Compstatin

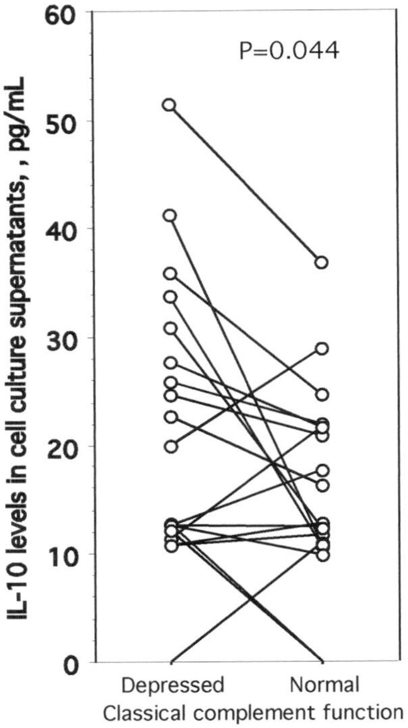

Fig. 5 Activation of the classical complement pathway is associated with increased IC-induced IL-10 production. Paired sera from 19 SLE patients taken on two separate occasions with either depressed (*left*) or normal (*right*) function of the classical complement pathway were PEG-precipitated. PEG-precipitates were added to healthy PBMC cultures and levels of IL-10 were measured in supernatants after 20 h of culture. Rewritten from (Mathsson et al. 2007) with permission from Blackwell Publishing

(as compared to addition of the linear control peptide) to the serum-containing medium before cell culture increased IgG3 CG-induced IL-10 production whereas the production of TNF-α was suppressed. We found a highly correlated inverse relationship between the effects of Compstatin on the IgG3 CG-induced production of IL-10 and TNF-α (Fig. 4). This effect was specific for the IgG3 CG, as modest Compstatin-induced effects of the IgM CG from the Waldenström's macroglobulinemia patient changed the production of IL-10 and TNF-α in the same direction (Mathsson et al. 2005). Even if earlier studies have shown IC-induced TNF-α production to be suppressed by concomitantly produced IL-10 (Berger et al. 1996), our own attempts to neutralize IL-10 were inconclusive.

4 In Vivo Complement Activation and Anti-SSA in SLE

In the two investigations referred to above, in vitro access to an intact complement system was shown to regulate cytokine production. But also in vivo complement activation is associated with IC-induced cytokine responses. To follow up our early findings of IC-induced cytokine production in SLE (Rönnelid et al. 2003), we designed a study to investigate the impact of a) SLE disease activity (expressed as decrease in classical complement activation) and b) occurrence of specific SLE-associated autoantibodies on the levels of IC and on IC-induced cytokines (Mathsson et al. 2007). SLE is associated with a number of specific autoantibodies, where antibodies against double stranded DNA (anti-dsDNA) levels mirror disease activity whereas levels of other antibodies e.g. anti-SSA, anti-SSB, anti-U1snRNP and anti-Sm do not. In a first experiment we compared the cytokine-inducing effect of polyteylene glycol (PEG) precipitated IC from paired SLE sera obtained during periods of normal and depressed classical complement function, respectively. Initially we showed that SLE IC-induced production of IL-10 indeed was significantly increased during periods of high disease activity/depressed classical complement function (Fig. 5). Levels of IL-12p40 were also increased in cell cultures with signs of complement activation. Next we investigated the impact of PEG-precipitated IC from a group of SLE sera with known autoantibody profile obtained during periods of normal complement function. In this investigation, samples containing anti-SSA antibodies and especially samples containing both anti-SSA and anti-SSB antibodies induced higher levels of IL-12p40 than samples without these antibodies ($p = 0.0105$ and $p < 0.0001$, respectively (Mathsson et al. 2007). Also IL-10 production showed a trend towards increase in anti-SSA positive cell cultures. Interestingly only anti-SSA showed this association with cytokine induction, whereas no effect was demonstrated for anti-dsDNA or anti-U1 snRNP antibodies.

In a third experiment we combined a larger set of serum samples. Circulating C1q-binding antibody levels were used as dependent variable and classical complement activation and the occurrence of different specific autoantibodies as independent variables in analysis of variance (ANOVA). The findings were in accordance with the two earlier experiments. Both complement activation and the occurrence of anti-SSA, but not of any other autoantibody type were significantly associated with circulating IC levels (Fig. 6). The ANOVA results also showed a significant degree of interaction, in that the effect of anti-SSA antibodies was visible only in serum samples obtained during periods of classical complement activation (Fig. 6).

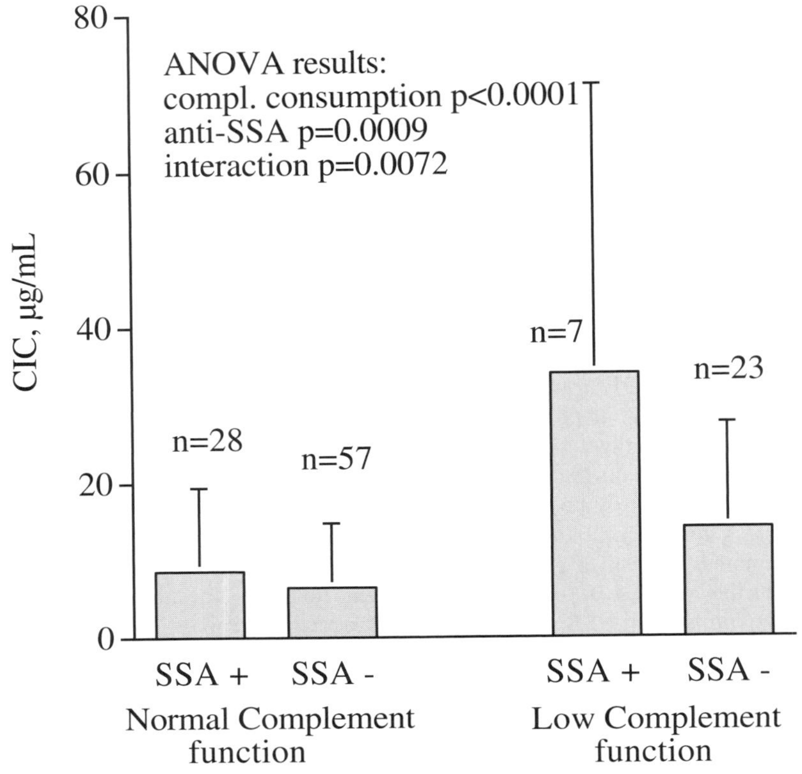

Fig. 6 Levels of cir culating IC (CIC) are associated with both classical complement function and the occurrence of anti-SSA autoantibodies in a synergistic fashion. The graph depicts mean levels + SD of circulating IC in sera with normal and low classical complement function distributed between anti-SSA positive (anti-SSA+) and anti-SSA negative (anti-SSA-) sera. Rewritten from (Mathsson et al. 2007) with permission from Blackwell Publishing

We interpret these findings to imply that during active SLE anti-SSA and possibly anti-SSB form IC together with the corresponding autoantigens released from apoptotic cells. These IC then induce classical complement activation and FcγR-mediated cytokine release by separate pathways. During quiescent periods without SLE flares there is no autoantigen release and the corresponding anti-SSA autoantibodies instead circulate in uncomplexed form.

In this study on PEG-precipitated IC from SLE sera, the induction pattern for IL-12p40 generally followed that for IL-10 and IL-6 in all experiments irrespective of the access to an intact complement cascade in vivo. This imply that whereas the in vivo situation from which the IC are purified determines the amount of IC obtained (Mathsson et al. 2007), it is the access to complement during the subsequent cell culture in vitro which determines the relative amounts of pro- and anti-inflammatory cytokines produced in response to these IC (Tejde et al. 2004; Mathsson et al. 2005).

5 Discussion

The finding that changes in the availability of an intact complement system might direct IC-regulated production of pro- (IL-12, TNF-α) and anti-inflammatory (IL-10, IL-6) cytokines differently can have implications both in inflammatory and oncological diseases: In SLE levels of IL-6 and IL-10 are increased, both cytokines produced by monocytes (Llorente et al. 1993; Lacki et al. 1997). IL-12 production is on the other hand decreased and inversely correlated to disease activity and serum levels of IL-10 (Liu et al. 1998b).

The pathogenesis of joint destruction in RA is intimately coupled to the balance between pro- and anti-inflammatory cytokines within the joint (McInnes and Schett 2007), and whereas exogenously delivered IL-12 may induce disease exacerbation (Peeva et al. 2000), IL-10 suppresses joint inflammation in experimental arthritis models (Kim et al. 2000).

A number of solid tumors and leukemias are associated with circulating IC (Segal-Eiras and Croce 1984), and these IC can be related to immune suppression both in mice (Gorczynski et al. 1975) and humans (Jerry et al. 1976). Furthermore, this immune suppression depends on Fc parts of immunoglobulins, as IC containing F(ab′)2 fragments are without effect (Kilburn et al. 1976) and is also dependent on serum factor(s) undefined at the time of publication (Kilburn et al. 1976). Hypothetically these findings can be interpreted as a tumor-associated immune suppression driven by IC-induced and FcγR-mediated IL-10 production with a concomitant down-regulation of IL-12 in an complement-sufficient environment.

Our findings of increased IgG CG-induced production of IL-10 during complement blockade might also have pathological implications. As the CG activate and deplete the complement system (Weiner et al. 2001), the CG themselves might act to create a complement deficient situation supporting IgG CG-induced production of the cytokine IL-10 that in turn can act as a growth factor for malignant myeloma cells (Lu et al. 1995; Gu et al. 1996).

Our findings also imply a distinct role for anti-SSA in IC formation. Anti-SSA and anti-SSB are often regarded as disease markers with quite stable levels over time (Hassan et al. 2002), and have not been implicated in the SLE disease process. Our recent findings show how these autoantibodies may participate directly in the inflammatory process by enhancing IC formation and IC-induced cytokine production during active SLE. Thereby these autoantibodies are inserted into a new pathophysiological context.

The question of IC-induced cytokine production is complex and include both factors in the IC per se (e.g. immunoglobulin isotypes, antigen-antibody ratio (Berger et al. 1996), IC size, complement activation that might conceal Fc receptors

(Nilsson 2001)), qualities of the responder cells (receptor expression and polymorphisms, maturational level), and factors in serum in vivo or in cell cultures in vitro (pre-existing cytokines, complement components, complement function, immunoglobulins). Ideally each factor should be investigated separately before establishment of standardized cell culture conditions.

6 Concluding Remarks

Taken together our investigations have shown that in vitro studies of IC-induced cytokine production must take into account the availability of an intact classical complement cascade during cell culture in vitro as well as the occurrence of specific autoantibodies in the context of complement activation at the time of sampling in vivo. Once these factors are taken into account, studies of how the complement system might regulate the IC-induced production of cytokines might have impact both in inflammatory and in oncological disease states.

Acknowledgments

We are indebted to professor John D. Lambris for use of Compstatin and the control peptide.

References

Berger, S., Balló, H. and Stutte, H.J. (1996) Immune complex-induced interleukin-6, interleukin-10 and prostaglandin secretion by human monocytes: a network of pro- and antiinflammatory cytokines dependent on the antigen-antibody ratio. *Eur. J. Immunol.* 26, 1297–1301

Blom, A.B., Radstake, T.R., Holthuysen, A.E., Sloetjes, A.W., Pesman, G.J., Sweep, F.G., van de Loo, F.A., Joosten, L.A., Barrera, P., van Lent, P.L. and van den Berg, W.B. (2003) Increased expression of Fcgamma receptors II and III on macrophages of rheumatoid arthritis patients results in higher production of tumor necrosis factor alpha and matrix metalloproteinase. *Arthritis Rheum.* 48, 1002–1014

Braun, M.C., Lahey, E. and Kelsall, B.L. (2000) Selective suppression of IL-12 production by chemoattractants. *J. Immunol.* 164, 3009–3017

Elshafie, A., Åhlin, E., Mathsson, L., Elghazali, G. and Rönnelid, J. (2007) Circulating immune complexes (IC) and IC-induced levels of GM-CSF are increased in Sudanese patients with acute visceral Leishmania donovani infection undergoing sodium stibogluconate treatment: implications for disease pathogenesis. *J. Immunol.* 178, 5383–5389

Elshafie, A., Elghazali, G., Rönnelid, J. and Venge, P. (2006) Cystatin C as a marker of immune complex-associated renal impairment in a Sudanese population with visceral leishmaniasis. *Am. J. Trop. Med. Hyg.* 75, 864–868

Gorczynski, R.M., Kilburn, D.G., Knight, R.A., Norbury, C., Parker, D.C. and Smith, J.B. (1975) Nonspecific and specific immunosuppression in tumour-bearing mice by soluble immune complexes. *Nature* 254, 141–143

Gu, Z.J., Costes, V., Lu, Z.Y., Zhang, X.G., Pitard, V., Moreau, J.F., Bataille, R., Wijdenes, J., Rossi, J.F. and Klein, B. (1996) Interleukin-10 is a growth factor for human myeloma cells by induction of an oncostatin M autocrine loop. *Blood* 88, 3972–3986

Hassan, A.B., Lundberg, I.E., Isenberg, D. and Wahren-Herlenius, M. (2002) Serial analysis of Ro/SSA and La/SSB antibody levels and correlation with clinical disease activity in patients with systemic lupus erythematosus. *Scand. J. Rheumatol.* 31, 133–139

Hawlisch, H., Belkaid, Y., Baelder, R., Hildeman, D., Gerard, C. and Kohl, J. (2005) C5a negatively regulates toll-like receptor 4-induced immune responses. *Immunity* 22, 415–426

Hideshima, T., Chauhan, D., Schlossman, R., Richardson, P. and Anderson, K.C. (2001) The role of tumor necrosis factor alpha in the pathophysiology of human multiple myeloma: therapeutic applications. *Oncogene* 20, 4519–4527

Jerry, L.M., Lewis, M.G. and Cano, P. (1976) Anergy, anti-antibodies and immune complex disease: a syndrome of disordered immune regulation in human cancer. Immuno-cancerology in solid tumors. *M. Martin and L. Dionne. New York, Stratton*, 63–79

Kallemuchikkal, U. and Gorevic, P.D. (1999) Evaluation of cryoglobulins. *Arch. Pathol. Lab. Med.* 123, 119–125

Kilburn, D.G., Fairhurst, M., Levy, J.G. and Whitney, R.B. (1976) Synergism between immune complexes and serum from tumor-bearing mice in the suppression of mitogen responses. *J. Immunol.* 117, 1612–1617

Kim, K.N., Watanabe, S., Ma, Y., Thornton, S., Giannini, E.H. and Hirsch, R. (2000) Viral IL-10 and soluble TNF receptor act synergistically to inhibit collagen-induced arthritis following adenovirus-mediated gene transfer. *J. Immunol.* 164, 1576–1581

Lacki, J.K., Samborski, W. and Mackiewicz, S.H. (1997) Interleukin-10 and interleukin-6 in lupus erythematosus and rheumatoid arthritis, correlations with acute phase proteins. *Clin. Rheumatol.* 16, 275–278

Lauta, V.M. (2003) A review of the cytokine network in multiple myeloma: diagnostic, prognostic, and therapeutic implications. *Cancer* 97, 2440–2452

Liu, T.F. and Jones, B.M. (1998a) Impaired production of IL-12 in system lupus erythematosus. II: IL-12 production in vitro is correlated negatively with serum IL-10, positively with serum IFN-gamma and negatively with disease activity in SLE. *Cytokine* 10, 148–153

Liu, T.F. and Jones, B.M. (1998b) Impaired production of IL-12 in systemic lupus erythematosus. I. Excessive production of IL-10 suppresses production of IL-12 by monocytes. *Cytokine* 10, 140–147

Liu, T.F., Jones, B.M., Wong, R.W. and Srivastava, G. (1999) Impaired production of IL-12 in systemic lupus erythematosus. III: deficient IL-12 p40 gene expression and cross-regulation of IL-12, IL-10 and IFN-gamma gene expression. *Cytokine* 11, 805–811

Llorente, L., Richaud-Patin, Y., Wijdenes, J., Alcocer-Varela, J., Maillot, M.C., Durand-Gasselin, I., Fourrier, B.M., Galanaud, P. and Emilie, D. (1993) Spontaneous production of interleukin-10 by B lymphocytes and monocytes in systemic lupus erythematosus. *Eur. Cytokine Netw.* 4, 421–427

Lövgren, T., Eloranta, M.L., Båve, U., Alm, G.V. and Rönnblom, L. (2004) Induction of interferon-alpha production in plasmacytoid dendritic cells by immune complexes containing nucleic acid released by necrotic or late apoptotic cells and lupus IgG. *Arthritis Rheum.* 50, 1861–1872

Lu, Z.Y., Zhang, X.G., Rodriguez, C., Wijdenes, J., Gu, Z.J., Morel-Fournier, B., Harousseau, J.L., Bataille, R., Rossi, J.F. and Klein, B. (1995) Interleukin-10 is a proliferation factor but not a differentiation factor for human myeloma cells. *Blood* 85, 2521–2527

Mathsson, L., Tejde, A., Carlson, K., Höglund, M., Nilsson, B., Nilsson-Ekdahl, K. and Rönnelid, J. (2005) Cryoglobulin-induced cytokine production via FcgRIIa: inverse effects of complement blockade on the production of TNF-a and IL-10. Implications for the growth of malignant B-cell clones. *Br. J. Haematol.* 129, 830–838

Mathsson, L., Lampa, J., Mullazehi, M. and Rönnelid, J. (2006) Immune complexes from rheumatoid arthritis synovial fluid induce FcgRIIa dependent and rheumatoid factor correlated production of tumour necrosis factor-alpha by peripheral blood mononuclear cells. *Arthritis Res. Ther.* Mar 28;8(3):R64 [Epub ahead of print] doi:10.1186/ar1926

Mathsson, L., Åhlin, E., Sjövall, C., Skogh, T. and Rönnelid, J. (2007) Cytokine induction by circulating immune complexes and signs of in-vivo complement activation in systemic lupus erythematosus are associated with the occurrence of anti-SSA antibodies. *Clin. Exp. Immunol.* 147, 513–520

McInnes, I.B. and Schett, G. (2007) Cytokines in the pathogenesis of rheumatoid arthritis. *Nat. Rev. Immunol.* 7, 429–442

Mullazehi, M., Mathsson, L., Lampa, J. and Rönnelid, J. (2006) Surface-bound anti-type II collagen containing immune complexes induce production of TNF-a, IL-1ß and IL-8 from peripheral blood monocytes via FcgRIIa. A potential patho-physiological mechanism for humoral anti-collagen type II immunity in RA. *Arthritis Rheum.* 54, 1759–1771

Nilsson, U.R. (2001) Deposition of C3b/iC3b leads to the concealment of antigens, immunoglobulins and bound C1q in complement-activating immune complexes. *Mol. Immunol.* 38, 151–160

Park, Y.B., Lee, S.K., Kim, D.S., Lee, J., Lee, C.H. and Song, C.H. (1998) Elevated interleukin-10 levels correlated with disease activity in systemic lupus erythematosus. *Clin. Exp. Rheumatol.* 16, 283–288

Peeva, E., Fishman, A.D., Goddard, G., Wadler, S. and Barland, P. (2000) Rheumatoid arthritis exacerbation caused by exogenous interleukin-12. *Arthritis Rheum.* 43, 461–463

Radstake, T., van Lent, P.L.E.M., Pesman, G.J., Blom, A.B., Sweep, F.G.J., Rönnelid, J., Figdor, C.G., Adema, G.J., Barrera, P. and van den Berg, W.B. (2004) High production of proinflammatory and Th1 cytokines by dendritic cells from patients with rheumatoid arthritis, and down regulation upon Fc{gamma}R triggering. *Ann. Rheum. Dis.* 63, 696–702

Rönnelid, J., Tejde, A., Mathsson, L., Nilsson-Ekdahl, K. and Nilsson, B. (2003) Immune complexes from SLE sera induce IL10 production from normal peripheral blood mononuclear cells by an FcgammaRII dependent mechanism: implications for a possible vicious cycle maintaining B cell hyperactivity in SLE. *Ann. Rheum. Dis*. 62, 37–42

Sahu, A., Soulika, A.M., Morikis, D., Spruce, L., Moore, W.T. and Lambris, J.D. (2000) Binding kinetics, structure-activity relationship, and biotransformation of the complement inhibitor compstatin. *J. Immunol.* 165, 2491–2499

Segal-Eiras, A. and Croce, M.V. (1984) Immune complexes in human malignant tumours. A review. *Allergol. Immunopathol.* (Madr) 12, 225–232

Tarkowski, A., Czerkinsky, C. and Nilsson, L.A. (1985) Simultaneous induction of rheumatoid factor- and antigen-specific antibody-secreting cells during the secondary immune response in man. *Clin. Exp. Immunol.* 61, 379–387

Tejde, A., Mathsson, L., Nilsson Ekdahl, K., Nilsson, B. and Rönnelid, J. (2004) Immune complex stimulation of peripheral blood mononuclear cells result in enhancement or suppression of IL-12 production dependent on serum factors. Role of complement activation. *Clin. Exp. Immunol.* 137, 521–528

Urbanska-Rys, H., Wiersbowska, A., Stepien, H. and Robak, T. (2000) Relationship between circulating interleukin-10 (IL-10) with interleukin-6 (IL-6) type cytokines (IL-6, interleukin-11 (IL-11), oncostatin M (OSM)) and soluble interleukin-6 (IL-6) receptor (sIL-6R) in patients with multiple myeloma. *Eur. Cytokine Netw.* 11, 443–451

Weiner, S.M., Prasauskas, V., Lebrecht, D., Weber, S., Peter, H.H. and Vaith, P. (2001) Occurrence of C-reactive protein in cryoglobulins. *Clin. Exp. Immunol.* 125, 316–322

Wittmann, M., Zwirner, J., Larsson, V.A., Kirchhoff, K., Begemann, G., Kapp, A., Gotze, O. and Werfel, T. (1999) C5a suppresses the production of IL-12 by IFN-gamma-primed and lipopolysaccharide-challenged human monocytes. *J. Immunol.* 162, 6763–6769

15. Subversion of Innate Immunity by Periodontopathic Bacteria via Exploitation of Complement Receptor-3

George Hajishengallis[1], Min Wang[1], Shuang Liang[1], Muhamad-Ali K. Shakhatreh[1], Deanna James[1], So-ichiro Nishiyama[2], Fuminobu Yoshimura[2], and Donald R. Demuth[1]

[1]Department of Periodontics/Oral Health and Systemic Disease, University of Louisville School of Dentistry, Louisville, KY 40292, USA
[2]Department of Microbiology, Aichi-Gakuin University School of Dentistry, Nagoya464-8650, Japan

Abstract. The capacity of certain pathogens to exploit innate immune receptors enables them to undermine immune clearance and persist in their host, often causing disease. Here we review subversive interactions of *Porphyromonas gingivalis*, a major periodontal pathogen, with the complement receptor-3 (CR3; CD11b/CD18) in monocytes/macrophages. Through its cell surface fimbriae, *P. gingivalis* stimulates Toll-like receptor-2 (TLR2) inside-out signaling which induces the high-affinity conformation of CR3. Although this activates CR3-dependent monocyte adhesion and transendothelial migration, *P. gingivalis* has co-opted this TLR2 proadhesive pathway for CR3 binding and intracellular entry. In CR3-deficient macrophages, the internalization of *P. gingivalis* is reduced twofold but its ability to survive intracellularly is reduced 1,000-fold, indicating that CR3 is exploited by the pathogen as a relatively safe portal of entry. The interaction of *P. gingivalis* fimbriae with CR3 additionally inhibits production of bioactive (p70) interleukin-12, which mediates immune clearance. In vivo blockade of CR3 leads to reduced persistence of *P. gingivalis* in the mouse host and diminished ability to cause periodontal bone loss, the hallmark of periodontal disease. Strikingly, the ability of *P. gingivalis* to interact with and exploit CR3 depends upon quantitatively minor components (FimCDE) of its fimbrial structure, which predominantly consists of polymerized fimbrillin (FimA). Indeed, isogenic mutants lacking FimCDE but expressing FimA are dramatically less persistent and virulent than the wild-type organism both in vitro and in vivo. This model of immune evasion through CR3 exploitation by *P. gingivalis* supports the concept that pathogens evolved to manipulate innate immune function for promoting their adaptive fitness.

1 Introduction

Porphyromonas gingivalis is a predominant pathogen associated with periodontitis, an infection-driven chronic inflammatory disease that leads to destruction of the tooth-supporting tissues (Pihlstrom et al. 2005). This gram-negative anaerobic oral organism is moreover implicated as a contributory factor in several systemic conditions, including atherosclerosis (Gibson et al. 2006). The capacity of *P. gingivalis* to cause disease has been attributed to several virulence factors, such

J.D. Lambris (ed.), *Current Topics in Complement II.*
doi: 10:1007/978-0-387-78952-1_15, © Springer Science+Business Media, LLC 2008

as fimbriae, LPS, hemagglutinins, and cysteine proteinases (gingipains), which in concert enable the pathogen to colonize and secure critical nutrients (Lamont and Jenkinson 1998). However, a pathogen's ability to find a niche and establish chronic infection requires more than simple expression of appropriate adhesins or other factors for nutrient procurement. To persist in a hostile host environment, pathogens should be able to evade or subvert the host immune defense system aiming at controlling or eliminating them. Most of those successful pathogens which disable host defenses target preferentially innate immunity (Rosenberger and Finlay 2003). This is not surprising given that innate defenses are the ones first encountered by the pathogens and, furthermore, subversion of innate immunity may additionally undermine the overall host defense. The latter is due to the significant instructive role of the innate response in the development of adaptive immunity (Pasare and Medzhitov 2005). Although *P. gingivalis* is a successful pathogen that can also be found in systemic tissues (Kozarov et al. 2005), the mechanism(s) whereby *P. gingivalis* resists immune elimination are poorly understood. In this regard, the pathogen's in vitro ability to inhibit production of the IL-8 chemokine by gingival epithelial cells (Darveau et al. 1998) or to degrade secreted cytokines (Calkins et al. 1998) is thought to suppress host defenses. Here we summarize recent in vitro and in vivo work from our group indicating that at least one of the mechanisms whereby *P. gingivalis* escapes immunosurveillance involves exploitation of complement receptor 3 (CR3). The utilization of CR3 by *P. gingivalis* depends greatly upon expression of surface fimbriae, comprising polymerized fimbrillin (FimA) associated with quantitatively minor proteins (FimCDE) (Nishiyama et al. 2007; Wang et al. 2007).

2 CR3 in Innate Immunity

CR3 is a β_2 integrin (CD11b/CD18) that is abundantly expressed by phagocytes, such as monocytes and neutrophils (Bhat et al. 1999; Yakubenko et al. 2002). This receptor recognizes a wide variety of structurally unrelated molecules from either the host (e.g., complement C3 fragment [iC3b], intercellular adhesion molecule-1 [ICAM-1], or fibrinogen) or pathogens (e.g., *Bordetella pertussis* filamentous hemagglutinin and *Leishmania* gp63) (Diamond et al. 1993; McGuirk and Mills 2000; Russell and Wright 1988). Similarly to other integrins, CR3 integrates the intracellular and extracellular environments through inside-out and outside-in bidirectional signaling. Inside-out signaling refers to the activation of the CR3 adhesive capacity from within the cell by means of signals generated by other receptors. Upon activation, CR3 binds ligands (or counter-receptors) resulting in stimulation of downstream signaling pathways, referred to as outside-in signaling (Shimaoka et al. 2002; Ginsberg et al. 2005). The ligand binding promiscuity of CR3 suggests that it may possess pattern recognition capabilities. In this context, CR3 has been shown to cluster with other pattern-recognition receptors (PRRs), such as CD14 and Toll-like receptors (TLRs), in membrane lipid rafts of activated cells (Triantafilou et al. 2004; Hajishengallis et al. 2006a).

CR3 plays diverse roles in immunity and inflammation, including iC3b-mediated phagocytosis, promotion of leukocyte transmigration to sites of extravascular inflammation, and induction of cytokine responses (Ehlers 2000). Given these potentially protective roles in innate immune defense, it is puzzling, if not paradoxical, that CR3 appears to be a "preferred" receptor used by certain phylogenetically unrelated pathogens for promoting their survival in the mammalian host (Romani et al. 2002; Mosser and Edelson 1987; Payne and Horwitz 1987; Hellwig et al. 2001). In this chapter, we will present and discuss evidence that the interaction of CR3 with *P. gingivalis* leads to suppression of interleukin (IL)-12 in vitro and in vivo, increased intracellular survival of the pathogen in macrophages, as well as enhanced in vivo persistence and induction of periodontal bone loss.

3 *P. gingivalis* Interacts with CR3 Through its Cell Surface Fimbriae

The Hanazawa group was the first to show that *P. gingivalis* interacts with CR3 though its cell surface fimbriae. Specifically, these investigators showed that the fimbriae of *P. gingivalis* bind mouse macrophage CR3 resulting in induction of tumor necrosis factor-α (TNF-α) and IL-1β production (Takeshita et al. 1998). Subsequently, our group established that mouse or human CR3 does not work in isolation for innate recognition of *P. gingivalis*. Rather, CR3 engages in cooperative interactions with the CD14/TLR2 signaling complex in response to *P. gingivalis* fimbriae. (Hajishengallis et al. 2005, 2006a,b; Harokopakis and Hajishengallis 2005). According to this model, *P. gingivalis* fimbriae bind CD14 and activate TLR2- and phosphatidylinositol 3-kinase (PI3K)-mediated inside-out signaling leading to activation of the ligand-binding capacity of CR3 (Harokopakis and Hajishengallis 2005) (Fig. 1). These interactions take place in membrane lipid rafts where CD14 is constitutively found, whereas TLR2 and CR3 are recruited there following cell stimulation with *P. gingivalis* fimbriae (Hajishengallis et al. 2006a,b). Unlike wild-type *P. gingivalis*, the ability of nonfimbriated mutants to activate and interact with CR3 is relatively poor (Hajishengallis et al. 2006b; Harokopakis et al. 2006).

4 Biological Significance of CR3-*P. gingivalis* Interactions: In Vitro Mechanistic Studies

We investigated the biological significance of *P. gingivalis* interactions with CR3 as they relate to three distinct functions of this integrin, i.e., as an adhesion molecule involved in transmigration, as a phagocytic receptor, and as an outside-in signaling receptor for modulation of cytokine responses. These findings are summarized and discussed below.

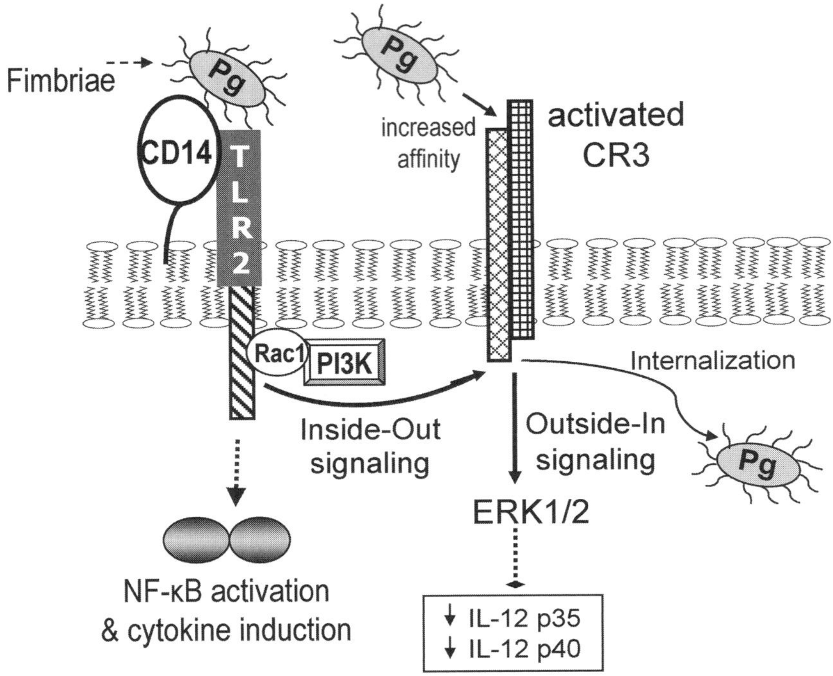

Fig. 1 *P. gingivalis* exploits TLR2 and CR3 and undermines innate immunity. *P. gingivalis* exploits its fimbrial-mediated interaction with macrophage TLR2 to activate the high-affinity conformation of CR3. This pathway proceeds through Rac1/PI3K-mediated inside-out signaling and requires CD14 for facilitating the fimbria-TLR2 interaction (Harokopakis and Hajishengallis 2005; Harokopakis et al. 2006). TLR2 stimulation by *P. gingivalis* also induces NF-κB activation and cytokine production (Hajishengallis et al. 2005, 2006a). Upon CR3 activation, fimbriated *P. gingivalis* can readily interact with CR3 leading to outside-in signaling, which via ERK1/2 downregulates IL-12 p35 and p40 and consequently inhibits production of bioactive (p70) IL-12 (Hajishengallis et al. 2007). This in turn undermines IL-12-mediated immune clearance, leading to persistence of the pathogen and enhanced virulence in the mouse periodontitis model (Hajishengallis et al. 2007). Moreover, CR3 uptake of *P. gingivalis* promotes its intracellular persistence in macrophages (Wang et al. 2007)

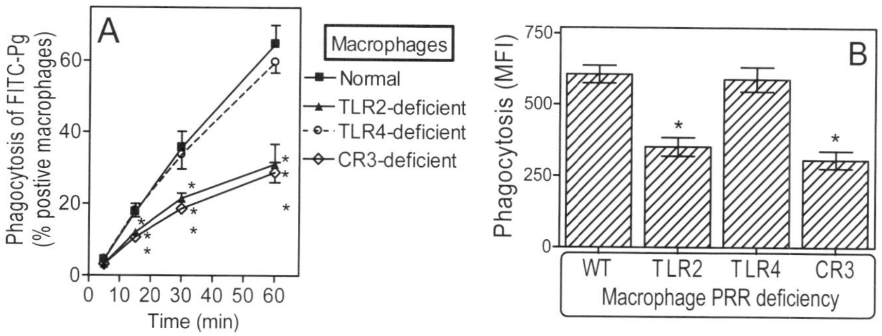

Fig. 2 *P. gingivalis* uptake by macrophages depends on TLR2 and CR3. Peritoneal macrophages from wild-type mice or mice deficient in TLR2, TLR4 (control), or CR3 (CD11b), were incubated with FITC-labeled *P. gingivalis* at a MOI of 10:1 for the indicated times at 37°C. Internalization was assessed by flow cytometry after washing the macrophages and quenching extracellular fluorescence, and was expressed as % FITC-positive macrophages (**a**). The mean fluorescence intensity (MFI) at the 60-min time-point is also shown (**b**) as a relative measure of the number of internalized bacteria. Results are means ± SD ($n = 3$; for clarity only the upper or lower SD is shown in (**a**). Asterisks indicate statistically significant ($p < 0.05$) differences between PRR deficiencies and wild-type controls. Reproduced from Hajishengallis et al. (2006b). Copyright 2006 American Society for Microbiology

4.1 *P. gingivalis* Stimulates CR3-Dependent Transendothelial Migration of Monocytes

Monocyte transmigration is mediated by interacting sets of cell adhesion molecules, including the CR3 – ICAM-1 pair (Libby 2002). The ability of *P. gingivalis* fimbriae to activate TLR2 inside-out signaling for CR3 activation suggested that this pathogen may stimulate monocyte transmigration. Indeed, we found that fimbriated *P. gingivalis* (or purified fimbriae) stimulate CR3-dependent monocyte adhesion to endothelial ICAM-1 and transmigration across endothelial cell monolayers (Harokopakis et al. 2006). This may represent a potentially protective mechanism which can contribute to monocyte recruitment to sites of *P. gingivalis* infection. However, since the adhesion of monocytes to the arterial endothelium and their subsequent migration into the subendothelial area is a hallmark of early atherogenesis (Libby 2002), *P. gingivalis*-induced CR3 activation may constitute a mechanistic basis linking this pathogen to inflammatory atherosclerotic processes. In this regard, viable *P. gingivalis* has been found in atherosclerotic plaques (Kozarov et al. 2005), although it is uncertain how the pathogen resists immune elimination and relocates there from the oral environment. An interesting hypothesis is that *P. gingivalis* not only stimulates the

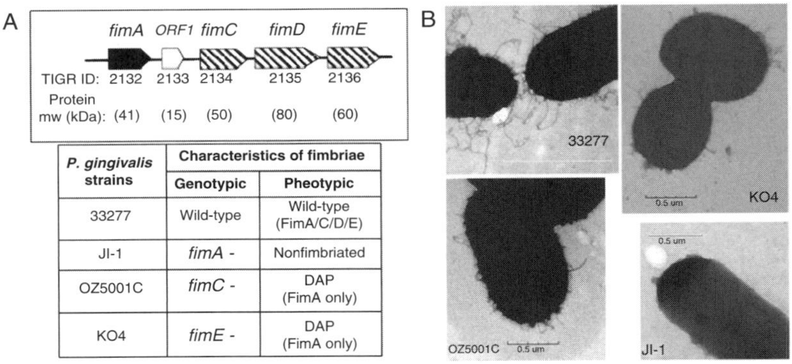

P. gingivalis strains	Characteristics of fimbriae	
	Genotypic	Pheotypic
33277	Wild-type	Wild-type (FimA/C/D/E)
JI-1	*fimA* -	Nonfimbriated
OZ5001C	*fimC* -	DAP (FimA only)
KO4	*fimE* -	DAP (FimA only)

Fig. 3 The *P. gingivalis* fimbrial gene cluster and related mutants. (**a**) The *P. gingivalis* 33,277 fimbrial gene cluster contains *fimA*, encoding the main fimbrillin subunit, and *fimCDE* encoding accessory proteins associated with fimbriae (the role of *ORF1* is unknown, although its product is not associated with fimbriae) (Watanabe et al. 1996; Nishiyama et al. 2007). The direction of transcription for each ORF is shown with the TIGR designation below the ORF. Strains OZ5001C and KO4 lacking *fimC* and *fimE*, respectively, express fimbriae devoid of all accessory proteins (DAP), whereas lack of *fimA* in strain JI-1 abrogates expression of both FimA and FimCDE resulting in a non-fimbriated state (Nishiyama et al. 2007; Wang et al. 2007). (**b**) Wild-type and mutant *P. gingivalis* surface structures visualized by transmission electron microscopy (Wang et al. 2007). Reproduced from Wang et al. (2007). Copyright 2007. The American Association of Immunologists, Inc

transmigratory activity of monocytes/macrophages but also exploits them as "Trojan horses" for disseminating to systemic tissues. Despite the lack of supporting evidence for this intriguing mechanism, additional work by our group has shown that *P. gingivalis* can indeed persist within macrophages if it is taken up through CR3 (below).

4.2 *P. gingivalis* Enters Macrophages via CR3 and Resists Intracellular Killing

Our initial report that *P. gingivalis* fimbriae stimulate TLR2 inside-out signaling for CR3 activation (Harokopakis and Hajishengallis 2005) was published concomitantly with a study by an independent group which showed that mycobacterial lipoarabinomannan also activates this proadhesive pathway (Sendide et al. 2005). Strikingly, mycobacteria exploit the TLR2/CR3 pathway for promoting their entry into monocytes/macrophages (Sendide et al. 2005) where they can parasitize (Ernst 1998). The potential for CR3 exploitation by certain pathogens may, at least partly, be related to the notion that CR3 is not linked to vigorous microbicidal mechanisms, in contrast to most phagocytic receptors (Lowell 2006; Wright and Silverstein 1983; Yamamoto and Johnston 1984; Caron

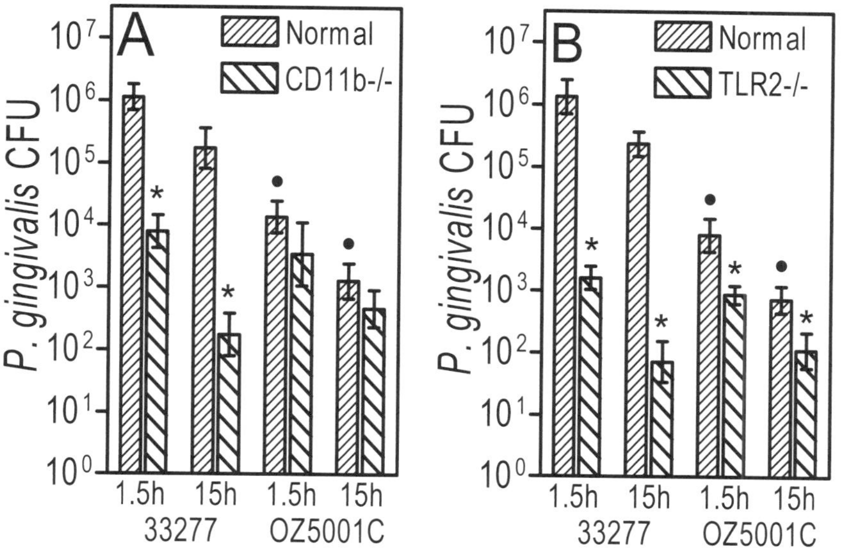

Fig. 4 *P. gingivalis* exploits CR3-mediated internalization to persist in macrophages. The persistence of viable internalized *P. gingivalis* 33,277 or OZ5001C (FimCDE mutant) in normal, CR3-deficient (CD11b–/–) (**a**), or TLR2-deficient (**b**) macrophages was determined by an antibiotic protection-based intracellular survival assay. Data are shown as means ± SD ($n = 3$). Asterisks indicate significant ($p < 0.05$) differences between receptor-deficient and wild-type macrophages, whereas black circles denote significant ($p < 0.05$) differences between 33,277 and OZ5001C. Reproduced from Wang et al. (2007). Copyright 2007 The American Association of Immunologists, Inc

and Hall 1998; Rosenberger and Finlay 2003). Consistent with this concept, the in vivo phagocytic uptake of *Bordetella pertussis* through the Fcγ receptor III (CD16) facilitates its clearance in contrast to CR3-mediated uptake (Hellwig et al. 2001). It is thus intriguing to speculate that pathogen-induced TLR2 inside-out signaling for CR3 activation may be a general pathway exploited by certain pathogens. We therefore investigated whether *P. gingivalis* can similarly induce its uptake through CR3 resulting in intracellular persistence rather than post-phagocytosis killing.

We first determined whether CR3 mediates *P. gingivalis* internalization by macrophages and found that CR3-deficient (CD11b–/–) mouse macrophages display significantly reduced capacity in the uptake of fimbriated *P. gingivalis*, compared to normal macrophages (Hajishengallis et al. 2006b) (Fig. 2). In contrast, no significant differences were observed regarding the uptake of nonfimbriated (FimA-deficient) mutants, suggesting that CR3 preferentially takes up fimbriated *P. gingivalis* (Hajishengallis et al. 2006b). Although TLR2 is not a phagocytic receptor, TLR2 deficiency similarly inhibits the uptake of *P. gingivalis*, consistent with its role in inside-out signaling for CR3 activation (Hajishengallis et al. 2006b) (Fig. 2).

We next followed the fate of internalized *P. gingivalis* in mouse macrophages or human monocytes by monitoring the recovery of viable internalized cells over time. *P. gingivalis* persisted intracellularly in a viable state for at least 72 h, in contrast to *Aggregatibacter* (*Actinobacillus*) *actinomycetemcomitans*, another periodontal pathogen (Socransky et al. 1998), which was readily killed (Wang et al. 2007). Unlike wild-type *P. gingivalis*, a nonfimbriated mutant was not recovered at 72 h but viable counts were obtained after 24 and 48 h, albeit at significantly lower levels. Strikingly, two other isogenic mutants which express a defective form of fimbriae, comprising FimA but lacking the FimCDE components (Fig. 3), were cleared even more rapidly than the nonfimbriated mutant (Wang et al. 2007). Because none of the mutants interact efficiently with CR3, these data imply that CR3 may, at least partly, be responsible for the enhanced persistence of wild-type *P. gingivalis*. Indeed, CR3 deficiency results in dramatic reduction of the intracellular survival of wild-type *P. gingivalis* by a factor of 10^3. In contrast, a FimCDE mutant displays limited intracellular persistence which is not affected CR3 deficiency (Wang et al. 2007) (Fig. 4a).

The implications of the ability of *P. gingivalis* to resist intracellular killing are currently uncertain. However, the differential susceptibility of wild-type *P. gingivalis* and the FimA or FimCDE mutants in intracellular killing correlates with their in vivo virulence in a mouse periodontitis model (Wang et al. 2007) (Fig. 5a). In addition, it is conceivable that the persistence of *P. gingivalis* in macrophages may be sufficient for co-opting the migration potential of these cells, facilitating relocation to systemic tissues, as alluded to above.

4.3 *P. gingivalis* Interaction with CR3 Downregulates IL-12 Induction

Consistent with earlier results (Takeshita et al. 1998), we found that mouse or human CR3 contributes to induction of several proinflammatory cytokines by *P. gingivalis* fimbriae, including TNF-α, IL-1β, and IL-6 (Hajishengallis et al. 2005, 2006a, 2007).

Strikingly, however, the binding of *P. gingivalis* fimbriae to activated CR3 results in reduced production of bioactive (p70) IL-12 (Hajishengallis et al. 2005, 2007), a key cytokine involved in intracellular bacterial clearance (Trinchieri 2003). At the mechanistic level, suppression of IL-12p70 production is mediated by CR3-dependent phosphorylation of ERK1/2 which leads to downregulation of IL-12 p35 and p40 subunits (Hajishengallis et al. 2007) (Fig. 1). Because the ability of mouse macrophages to elicit IL-12p70 in response to *P. gingivalis* fimbriae is upregulated by CR3 deficiency but is abrogated by TLR2 deficiency (Hajishengallis et al. 2005), it can be concluded that CR3 binding of fimbriae inhibits TLR2-dependent induction of IL-12p70. Moreover, *P. gingivalis* fimbriae can block IL-12p70 induction by other bacterial stimuli which activate TLR4. Indeed, the capacity of LPS from *E. coli* or *A. actinomycetemcomitans* to induce

IL-12p70 in IFN-γ-primed monocytes is suppressed by *P. gingivalis* fimbriae, although other proinflammatory cytokines (TNF-α, IL-1β, IL-6, and IL-8) are upregulated (Hajishengallis et al. 2007). Similar downregulation of LPS-induced IL-12p70 is observed when whole cells of *P. gingivalis* are used, provided that the bacteria express fully mature fimbriae containing the FimCDE accessory proteins (Wang et al. 2007). This inhibitory activity is CR3-dependent but is irrelevant to *P. gingivalis* internalization since pretreating cells with cytochalasin D does not reverse the effect (Wang et al. 2007).

IL-12p70 production by macrophages is significant for host defense in that it activates cytotoxic T lymphocytes cells and natural killer cells to produce IFN-γ, which in turn activates the bactericidal function of macrophages (Trinchieri 2003). The ability of *P. gingivalis* to inhibit IL-12 induction may be particularly relevant to oral disease. In this context, *P. gingivalis* readily takes intracellular refuge in permissive cells, such as epithelial cells (Lamont et al. 1995) and endothelial cells (Progulske-Fox et al. 1999), and a reduction in IL-12-dependent stimulation of cell-mediated immunity may compromise the killing of these *P. gingivalis*-infected cells. This may consequently allow the pathogen a window of opportunity to establish infection and create a niche that is appropriate for its survival and growth. Moreover, since *P. gingivalis* inhibits IL-12 induction by other organisms, this mechanism may promote the survival of both *P. gingivalis* and co-habiting organisms in the subgingival pocket.

5 In Vivo Evidence for CR3 Exploitation by *P. gingivalis* and Implications in Periodontitis

Based on the concept that inhibition of IL-12p70 may constitute a microbial tactic to evade immunity, we reasoned that CR3 blockade with a small-molecule antagonist (XVA143; m.w. 585) would upregulate induction of IL-12p70 and IFN-γ in response to *P. gingivalis* and facilitate its clearance by the host. This notion was experimentally confirmed in a peritonitis model of *P. gingivalis* infection (Hajishengallis et al. 2007), suggesting that CR3 antagonists may be used therapeutically for controlling *P. gingivalis* infection. CR3 was conclusively implicated as an exploited receptor in additional experiments demonstrating that CR3-deficient mice elicit higher IL-12p70 and IFN-γ levels and display enhanced clearance of *P. gingivalis* compared to wild-type mice (Hajishengallis et al. 2007).

It seems curious why the host would "allow" a key receptor, such as CR3, become an Achilles' heel to infection by at least some pathogens. From the host point of view, however, CR3-dependent inhibition of IL-12 appears to serve a physiological role. In this regard, the phagocytosis of apoptotic cells by macrophages is heavily dependent upon CR3 and is associated with inhibition of IL-12p70, since apoptotic cells are not normally recognized as danger that would justify induction of cell-mediated immunity (Kim et al. 2004; Mevorach et al. 1998). Moreover, in the case of extracellular pathogens that can readily be controlled with complement activation and humoral immunity, the phagocytosis of iC3b-coated bacteria by CR3 would help control potentially destructive inflammation through IL-12 downregulation. In fact, inhibition of IL-12 would not only suppress T helper type 1 cell-mediated immunity but would also upregulate T helper type 2 responses required for effective humoral (antibody) responses (Trinchieri 1998). It is thus possible that *P. gingivalis* has co-opted a physiological anti-inflammatory CR3-dependent mechanism to evade innate immune clearance. This mechanism may be exploited also by other pathogens. For instance, the interaction of *Bordetella pertussis* filamentous hemagglutinin with CR3 similarly leads to inhibition of IL-12p70 (McGuirk and Mills 2000) and the in vivo phagocytic uptake of *B. pertussis* via CR3 fails to promote its clearance (Hellwig et al. 2001).

In the context of periodontal disease, the virulence of *P. gingivalis* can be measured by its capacity to induce periodontal bone resorption in animal models. Using a validated model of mouse periodontitis (Baker et al. 2000), we demonstrated that CR3 blockade inhibits the ability of *P. gingivalis* to persist in the mouse host and to induce periodontal bone loss (Hajishengallis et al. 2007) (Fig. 5b). Since FimA- or FimCDE-deficient mutants of *P. gingivalis* cannot effectively interact with CR3, they would be expected to display relatively reduced persistence and virulence in the bone loss model. Although our findings confirmed this hypothesis as mentioned above (Fig. 5a) (Wang et al. 2007), additional defects could have contributed to the results. This is based on the notion that the FimCDE accessory proteins mediate binding to certain extracellular matrix proteins which may facilitate optimal *P. gingivalis* colonization in the oral cavity (Nishiyama et al. 2007). Interestingly, however, although the FimCDE-deficient mutants display enhanced adhesive properties compared to the FimA-deficient mutant, the former are less virulent in inducing periodontal bone loss (Fig. 5a) or in resisting intracellular killing by mouse macrophages or human monocytes (Wang et al. 2007). It is possible that expression of FimA devoid of the accessory proteins may elicit robust host responses that could eliminate *P. gingivalis*. In this regard, the FimCDE mutants are stronger inducers of NF-κB activation in vitro than both wild-type and FimA-deficient *P. gingivalis* (Wang et al. 2007). In general, increased microbial immunostimulatory potential correlates with reduced microbial survival in the host, as exemplified by genetically modified *Yersenia pestis* expressing an immunostimulatory version of LPS (Montminy et al. 2006). It is uncertain at the moment why FimCDE-deficient mutants are more proinflammatory than wild-type *P. gingivalis*. However, at least in part, this could be explained by the lack of efficient interactions with CR3. Thus, diminished CR3 outside-in signaling by the FimCDE mutants would not only result in higher IL-12 induction, but also reduced ERK1/2 activation downstream of CR3. Reduced ERK1/2 activation may in turn result in decreased production of the anti-inflammatory IL-10 (Martin et al. 2003).

Although the concept that CR3 plays a role in periodontitis is recent, it is intriguing to speculate that CR3 may, at least partly, be related to the age-related alterations associated with this disease. Periodontitis and other infection-driven chronic inflammatory diseases generally appear rather late in life, but it is not clear whether, or what kind of, age-related alterations in innate immune function are responsible. Interestingly, although phagocytosis generally declines

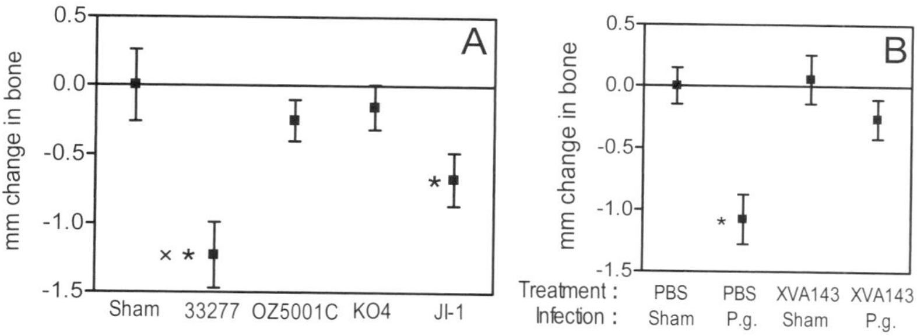

Fig. 5 Involvement of CR3 in induction of periodontal bone loss by *P. gingivalis*. . . Wild-type *P. gingivalis* (strain 33,277) induces significantly higher levels of periodontal bone resorption compared to mutants that do not efficiently interact with CR3 (**a**), while a CR3 antagonist inhibits the ability of *P. gingivalis* 33,277 to cause periodontal bone loss (**b**). In (**a**), BALB/c mice were orally infected or not with *P. gingivalis* 33,277 (wild-type fimbriae), OZ5001C (fimbriae lacking FimCDE), KO4 (fimbriae lacking FimCDE) or JI-1 (nonfimbriated). In (**b**), the mice were pretreated with a CR3 antagonist (XVA143) or PBS control prior to infection with *P. gingivalis* 33,277 (P.g.) or vehicle only (Sham). The mm distance from the cementoenamel junction to the alveolar bone crest was measured at 14 predetermined sites in defleshed maxillae and the data were transformed to indicate bone loss (Hajishengallis et al. 2007; Wang et al. 2007). Results are shown as means ± SD ($n = 5$) and negative values indicate bone loss. In (**a**), asterisks show significant ($p < 0.05$) differences between infected and sham-infected mice. The sign "x" indicates significant difference between 33,277 and JI-1. In (**b**), asterisks denote significant ($p < 0.05$) differences between PBS-treated/*P. gingivalis*-infected mice and the rest of the groups, among which no significant differences were found. Reproduced from Wang et al. (2007) (**a**) and Hajishengallis et al. (2007) (**b**). Copyright 2007 The American Association of Immunologists, Inc

with aging (Sebastian et al. 2005; Butcher et al. 2000), CR3 (CD11b/CD18)-dependent phagocytosis is intact (Butcher et al. 2001). Specifically, unlike FcγRIIIa (CD16)-mediated phagocytosis which declines because of age-related downregulation of CD16 expression, CD11b expression is preserved at old age (Butcher et al. 2001). Interestingly, CD16-mediated phagocytosis readily induces the oxidative burst response, in contrast to phagocytosis through CR3 (Payne and Horwitz 1987; Wright and Silverstein 1983; Lowell 2006). In this respect, studies in macrophages have shown that CD16-derived phagosomes fuse more readily with lysosomes than CR3-derived phagosomes, suggesting association of CD16 with enhanced microbial killing (Vieira et al. 2002). It seems possible, therefore, that CR3-mediated internalization of *P. gingivalis* may stay intact with aging, whereas alternative uptake of the pathogen by strongly microbicidal pathways may decline. In relative terms, this means that CR3-mediated internalization of *P. gingivalis* may increase with aging. It is not known at the moment whether CR3 in advanced age is more readily exploitable by *P. gingivalis* leading to increased disease activity, but it is certainly a testable hypothesis.

6 CR3 Exploitation by *P. gingivalis* Depends on TLR2

CR3 exploitation by *P. gingivalis* is initiated at the level of TLR2, since TLR2 inside-out signaling is required for effective interaction of *P. gingivalis* fimbriae with CR3 (Harokopakis et al. 2006; Harokopakis and Hajishengallis 2005). Consistent with this, we have now found that the intracellular survival of fimbriated *P. gingivalis* is dramatically reduced in TLR2-deficient macrophages relative to normal controls (Wang et al. 2007) (Fig. 4b). These results are in line with a study by an independent group that TLR2-deficient mice are more resistant to *P. gingivalis*-induced periodontal bone loss than wild-type controls (Burns et al. 2006). Although TLR2 deficiency limits efficient activation of CR3, which in turn cannot be readily exploited by *P. gingivalis*, this does not necessarlily rule out the possibility that TLR2 may be exploited by *P. gingivalis* in CR3-independent ways. In this regard, TLR2 signaling has been implicated in immune evasion by *Yersinia enterocolitica* through induction of IL-10-mediated immunosuppression (Sing et al. 2005).

A recent report has presented evidence suggesting that CR3 and TLR4 cooperate for the uptake and intracellular killing of *Salmonella enterica* serovar Typhimurium (van Bruggen et al. 2007). Although this was shown in neutrophils, rather than in macrophages, it could be speculated that CR3 may be exploited in a contextual way, i.e., dependent upon which TLR is predominantly activated by the pathogen. Interestingly, *P. gingivalis* appears to be biased toward preferentially activating TLR2 both in vitro and in vivo (Burns et al. 2006; Hajishengallis et al. 2006a). Although bacterial LPS in general is a strong TLR4 agonist, *P. gingivalis* seems to deviate from the norm in that it expresses a heterogeneous mixture of lipid A species, which can induce cell activation through TLR2 or TLR4 (weakly) or even antagonize TLR4-induced cell activation (Darveau et al. 2004; Dixon and Darveau 2005). Therefore, by altering the proportions of its different lipid A moieties, *P. gingivalis* may increase its virulence through manipulation of the innate response in ways that predominant activation of TLR2 over TLR4, may allow effective exploitation of CR3.

7 Conclusion

The interactions of *P. gingivalis* with CR3 is but one of the ways this pathogen interacts with the complement system in general. Interestingly, *P. gingivalis* is very resistant to killing by complement; this is attributable to the ability of its gingipain proteases to degrade C3 and C5 and thereby prevent deposition of C3b on the bacterial cell surface, which moreover contains a complement-resistant anionic polysaccharide (Popadiak et al. 2007; Slaney et al. 2006). Intriguingly, degradation of C5 by *P. gingivalis* leads to generation of a biologically active C5a-like fragment (Wingrove et al. 1992). This bioactive fragment activates a chemotactic response in neutrophils, presumably through the C5a receptor (C5aR), but the implications for periodontal disease are uncertain. In the light of recent developments that C5aR cross-talks with TLR signaling pathways and inhibits IL-12 (Hawlisch et al. 2005; Zhang et al. 2007), it could be speculated that *P. gingivalis* may use more than one complement-related mechanisms for escaping IL-12-mediated clearance. These considerations along with our findings regarding CR3 exploitation by *P. gingivalis* suggest that this pathogen "prefers" to manipulate the complement response rather than to merely inactivate it. The elucidation of complement-dependent immune evasion strategies of *P. gingivalis* may help control periodontitis, or other systemic conditions associated with *P. gingivalis* infections, through the application of appropriate complement inhibitors.

Acknowledgments

The authors acknowledge support by U.S. Public Health Service Grants DE015254 and DE018292 (to G.H.), and DE14605 (to D.R.D.) from the National Institutes of Health; Grants-in-Aid for Scientific Research (15591957 to F.Y. and 17791318 to S.N.) from the Japan Society for the Promotion of Science; and the AGU High-Tech Research Center Project from the Ministry of Education, Culture, Sports, Science, and Technology, Japan (to F.Y.).

References

Baker, P. J., Dixon, M. and Roopenian, D. C. (2000). Genetic control of susceptibility to *Porphyromonas gingivalis*-induced alveolar bone loss in mice. *Infect Immun* 68, 5864–5868

Bhat, N., Perera, P.-Y., Carboni, J. M., Blanco, J., Golenbock, D. T., Mayadas, T. N. and Vogel, S. N. (1999). Use of a photoactivatable taxol analogue to identify unique cellular targets in murine macrophages: identification of murine CD18 as a major taxol-binding protein and a role for Mac-1 in taxol-induced gene expression. *J Immunol* 162, 7335–7342

Burns, E., Bachrach, G., Shapira, L. and Nussbaum, G. (2006). Cutting edge: TLR2 is required for the innate response to *Porphyromonas gingivalis*: activation leads to bacterial persistence and TLR2 deficiency attenuates induced alveolar bone resorption. *J Immunol* 177, 8296–8300

Butcher, S., Chahel, H. and Lord, J. M. (2000). Ageing and the neutrophil: no appetite for killing? *Immunology* 100, 411–416

Butcher, S. K., Chahal, H., Nayak, L., Sinclair, A., Henriquez, N. V., Sapey, E., O'Mahony, D. and Lord, J. M. (2001). Senescence in innate immune responses: reduced neutrophil phagocytic capacity and CD16 expression in elderly humans. *J Leukoc Biol* 70, 881–886

Calkins, C. C., Platt, K., Potempa, J. and Travis, J. (1998). Inactivation of tumor necrosis factor-a by proteinases (gingipains) from the periodontal pathogen, *Porphyromonas gingivalis*. Implications of immune evasion. *J Biol Chem* 273, 6611–6614

Caron, E. and Hall, A. (1998). Identification of two distinct mechanisms of phagocytosis controlled by different Rho GTPases. *Science* 282, 1717–1721

Darveau, R. P., Belton, C. M., Reife, R. A. and Lamont, R. J. (1998). Local chemokine paralysis, a novel pathogenic mechanism for *Porphyromonas gingivalis*. *Infect Immun* 66, 1660–1665

Darveau, R. P., Pham, T. T., Lemley, K., Reife, R. A., Bainbridge, B. W., Coats, S. R., Howald, W. N., Way, S. S. and Hajjar, A. M. (2004). *Porphyromonas gingivalis* lipopolysaccharide contains multiple lipid A species that functionally interact with both toll-like receptors 2 and 4. *Infect Immun* 72, 5041–5051

Diamond, M. S., Garcia-Aguilar, J., Bickford, J. K., Corbi, A. L. and Springer, T. A. (1993). The I domain is a major recognition site on the leukocyte integrin Mac-1 (CD11b/CD18) for four distinct adhesion ligands. *J Cell Biol* 120, 1031–1043

Dixon, D. R. and Darveau, R. P. (2005). Lipopolysaccharide heterogeneity: innate host responses to bacterial modification of lipid a structure. *J Dent Res* 84, 584–595

Ehlers, M. R. W. (2000). CR3: a general purpose adhesion-recognition receptor essential for innate immunity. *Microbes Infect* 2, 289–294

Ernst, J. D. (1998). Macrophage receptors for *Mycobacterium tuberculosis*. *Infect Immun* 66, 1277–1281

Gibson, F. C., III, Yumoto, H., Takahashi, Y., Chou, H. H. and Genco, C. A. (2006). Innate immune signaling and *Porphyromonas gingivalis*-accelerated atherosclerosis. *J Dent Res* 85, 106–121

Ginsberg, M. H., Partridge, A. and Shattil, S. J. (2005). Integrin regulation. *Curr Opin Cell Biol* 17, 509–516

Hajishengallis, G., Ratti, P. and Harokopakis, E. (2005). Peptide mapping of bacterial fimbrial epitopes interacting with pattern recognition receptors. *J Biol Chem* 280, 38902–38913

Hajishengallis, G., Tapping, R. I., Harokopakis, E., Nishiyama, S.-I., Ratti, P., Schifferle, R. E., Lyle, E. A., Triantafilou, M., Triantafilou, K. and Yoshimura, F. (2006a). Differential interactions of fimbriae and lipopolysaccharide from *Porphyromonas gingivalis* with the toll-like receptor 2-centred pattern recognition apparatus. *Cell Microbiol* 8, 1557–1570

Hajishengallis, G., Wang, M., Harokopakis, E., Triantafilou, M. and Triantafilou, K. (2006b). *Porphyromonas gingivalis* fimbriae proactively modulate b2 integrin adhesive activity and promote binding to and internalization by macrophages. *Infect Immun* 74, 5658–5666

Hajishengallis, G., Shakhatreh, M.-A. K., Wang, M. and Liang, S. (2007). Complement receptor 3 blockade Promotes IL-12-mediated clearance of *Porphyromonas gingivalis* and negates its virulence in vivo. *J Immunol* 179, 2359–2367

Harokopakis, E. and Hajishengallis, G. (2005). Integrin activation by bacterial fimbriae through a pathway involving CD14, toll-like receptor 2, and phosphatidylinositol-3-kinase. *Eur J Immunol* 35, 1201–1210

Harokopakis, E., Albzreh, M. H., Martin, M. H. and Hajishengallis, G. (2006). TLR2 transmodulates monocyte adhesion and transmigration via Rac1- and PI3K-mediated inside-out signaling in response to *Porphyromonas gingivalis* fimbriae. *J Immunol* 176, 7645–7656

Hawlisch, H., Belkaid, Y., Baelder, R., Hildeman, D., Gerard, C. and Kohl, J. (2005). C5a negatively regulates toll-like receptor 4-induced immune responses. *Immunity* 22, 415–426

Hellwig, S. M., van Oirschot, H. F., Hazenbos, W. L., van Spriel, A. B., Mooi, F. R. and van De Winkel, J. G. (2001). Targeting to Fcg receptors, but not CR3 (CD11b/CD18), increases clearance of *Bordetella pertussis*. *J Infect Dis* 183, 871–879

Kim, S., Elkon, K. B. and Ma, X. (2004). Transcriptional suppression of interleukin-12 gene expression following phagocytosis of apoptotic cells. *Immunity* 21, 643–653

Kozarov, E. V., Dorn, B. R., Shelburne, C. E., Dunn, W. A., Jr. and Progulske-Fox, A. (2005). Human atherosclerotic plaque contains viable invasive *Actinobacillus actinomycetemcomitans* and *Porphyromonas gingivalis*. *Arterioscler Thromb Vasc Biol* 25, e17–e18

Lamont, R. J. and Jenkinson, H. F. (1998). Life below the gum line: pathogenic mechanisms of *Porphyromonas gingivalis*. *Microbiol Mol Biol Rev* 62, 1244–1263

Lamont, R. J., Chan, A., Belton, C. M., Izutsu, K. T., Vasel, D. and Weinberg, A. (1995). *Porphyromonas gingivalis* invasion of gingival epithelial cells. *Infect Immun* 63, 3878–3885

Libby, P. (2002). Inflammation in atherosclerosis. *Nature* 420, 868–874

Lowell, C. A. (2006). Rewiring phagocytic signal transduction. *Immunity* 24, 243–245

Martin, M., Schifferle, R. E., Cuesta, N., Vogel, S. N., Katz, J. and Michalek, S. M. (2003). Role of the phosphatidylinositol 3 kinase-Akt pathway in the regulation of IL-10 and IL-12 by *Porphyromonas gingivalis* lipopolysaccharide. *J Immunol* 171, 717–725

McGuirk, P. and Mills, K. H. (2000). Direct anti-inflammatory effect of a bacterial virulence factor: IL-10-dependent suppression of IL-12 production by filamentous hemagglutinin from *Bordetella pertussis*. *Eur J Immunol* 30, 415–422

Mevorach, D., Mascarenhas, J. O., Gershov, D. and Elkon, K. B. (1998). Complement-dependent clearance of apoptotic cells by human macrophages. *J Exp Med* 188, 2313–2320.

Montminy, S. W., Khan, N., McGrath, S., Walkowicz, M. J., Sharp, F., Conlon, J. E., Fukase, K., Kusumoto, S., Sweet, C., Miyake, K., Akira, S., Cotter, R. J., Goguen, J. D. and Lien, E. (2006). Virulence factors of *Yersinia pestis* are overcome by a strong lipopolysaccharide response. *Nat Immunol* 7, 1066–1073

Mosser, D. M. and Edelson, P. J. (1987). The third component of complement (C3) is responsible for the intracellular survival of Leishmania major. *Nature* 327, 329–331

Nishiyama, S.-I., Murakami, Y., Nagata, H., Shizukuishi, S., Kawagishi, I. and Yoshimura, F. (2007). Involvement of minor components associated with the FimA fimbriae of *Porphyromonas gingivalis* in adhesive functions. *Microbiology* 153, 1916–1925

Pasare, C. and Medzhitov, R. (2005). Toll-like receptors: linking innate and adaptive immunity. *Adv Exp Med Biol* 560, 11–18

Payne, N. R. and Horwitz, M. A. (1987). Phagocytosis of *Legionella pneumophila* is mediated by human monocyte complement receptors. *J Exp Med* 166, 1377–1389

Pihlstrom, B. L., Michalowicz, B. S. and Johnson, N. W. (2005). Periodontal diseases. *Lancet* 366, 1809–1820

Popadiak, K., Potempa, J., Riesbeck, K. and Blom, A. M. (2007). Biphasic effect of gingipains from *Porphyromonas gingivalis* on the human complement system. *J Immunol* 178, 7242–7250

Progulske-Fox, A., Kozarov, E., Dorn, B., Dunn, W., Jr., Burks, J. and Wu, Y. (1999). *Porphyromonas gingivalis* virulence factors and invasion of cells of the cardiovascular system. *J Periodontal Res* 34, 393–399

Romani, L., Bistoni, F. and Puccetti, P. (2002). Fungi, dendritic cells and receptors: a host perspective of fungal virulence. *Trends Microbiol* 10, 508–514

Rosenberger, C. M. and Finlay, B. B. (2003). Phagocyte sabotage: disruption of macrophage signalling by bacterial pathogens. *Nat Rev Mol Cell Biol* 4, 385–396

Russell, D. G. and Wright, S. D. (1988). Complement receptor type 3 (CR3) binds to an Arg-Gly-Asp-containing region of the major surface glycoprotein, gp63, of *Leishmania* promastigotes. *J Exp Med* 168, 279–292

Sebastian, C., Espia, M., Serra, M., Celada, A. and Lloberas, J. (2005). MacrophAging: a cellular and molecular review. *Immunobiology* 210, 121–126

Sendide, K., Reiner, N. E., Lee, J. S., Bourgoin, S., Talal, A. and Hmama, Z. (2005). Cross-talk between CD14 and complement receptor 3 promotes phagocytosis of mycobacteria: regulation by phosphatidylinositol 3-kinase and cytohesin-1. *J Immunol* 174, 4210–4219

Shimaoka, M., Takagi, J. and Springer, T. A. (2002). Conformational regulation of integrin structure and function. *Annu Rev Biophys Biomol Struct* 31, 485–516

Sing, A., Reithmeier-Rost, D., Granfors, K., Hill, J., Roggenkamp, A. and Heesemann, J. (2005). A hypervariable N-terminal region of *Yersinia* LcrV determines toll-like receptor 2-mediated IL-10 induction and mouse virulence. *Proc Natl Acad Sci U S A* 102, 16049–16054

Slaney, J. M., Gallagher, A., Aduse-Opoku, J., Pell, K. and Curtis, M. A. (2006). Mechanisms of resistance of *Porphyromonas gingivalis* to killing by serum complement. *Infect Immun* 74, 5352–5361

Socransky, S. S., Haffajee, A. D., Cugini, M. A., Smith, C. and Kent, R. L., Jr. (1998). Microbial complexes in subgingival plaque. *J Clin Periodontol* 25, 134–144

Takeshita, A., Murakami, Y., Yamashita, Y., Ishida, M., Fujisawa, S., Kitano, S. and Hanazawa, S. (1998). *Porphyromonas gingivalis* fimbriae use b2 integrin (CD11/CD18) on mouse peritoneal macrophages as a cellular receptor, and the CD18 b chain plays a functional role in fimbrial signaling. *Infect Immun* 66, 4056–4060

Triantafilou, M., Brandenburg, K., Kusumoto, S., Fukase, K., Mackie, A., Seydel, U. and Triantafilou, K. (2004). Combinational clustering of receptors following stimulation by bacterial products determines lipopolysaccharide responses. *Biochem J* 381, 527–536

Trinchieri, G. (1998). Immunobiology of interleukin-12. Immunol Res 17, 269–278

Trinchieri, G. (2003). Interleukin-12 and the regulation of innate resistance and adaptive immunity. *Nat Rev Immunol* 3, 133–146

van Bruggen, R., Zweers, D., van Diepen, A., van Dissel, J. T., Roos, D., Verhoeven, A. J. and Kuijpers, T. W. (2007). Complement receptor 3 and toll-like receptor 4 act sequentially in uptake and intracellular killing of unopsonized *Salmonella enterica serovar typhimurium* by human neutrophils. *Infect Immun* 75, 2655–2660

Vieira, O. V., Botelho, R. J. and Grinstein, S. (2002). Phagosome maturation: aging gracefully. *Biochem J* 366, 689–704

Wang, M., Shakhatreh, M.-A. K., James, D., Liang, S., Nishiyama, S.-i., Yoshimura, F., Demuth, D. R. and Hajishengallis, G. (2007). Fimbrial proteins of *Porphyromonas gingivalis* mediate in vivo virulence and exploit TLR2 and complement receptor 3 to persist in macrophages. *J Immunol* 179, 2349–2358

Watanabe, K., Onoe, T., Ozeki, M., Shimizu, Y., Sakayori, T., Nakamura, H. and Yoshimura, F. (1996). Sequence and product analyses of the four genes downstream from the fimbrilin gene (fimA) of the oral anaerobe *Porphyromonas gingivalis*. *Microbiol Immunol* 40, 725–734

Wingrove, J. A., DiScipio, R. G., Chen, Z., Potempa, J., Travis, J. and Hugli, T. E. (1992). Activation of complement components C3 and C5 by a cysteine proteinase (gingipain-1) from *Porphyromonas (Bacteroides) gingivalis*. *J Biol Chem* 267, 18902–18907

Wright, S. D. and Silverstein, S. C. (1983). Receptors for C3b and C3bi promote phagocytosis but not the release of toxic oxygen from human phagocytes. *J Exp Med* 158, 2016–2023

Yakubenko, V. P., Lishko, V. K., Lam, S. C. and Ugarova, T. P. (2002). A molecular basis for integrin aMb2 ligand binding promiscuity. *J Biol Chem* 277, 48635–48642

Yamamoto, K. and Johnston, R. B., Jr. (1984). Dissociation of phagocytosis from stimulation of the oxidative metabolic burst in macrophages. *J Exp Med* 159, 405–416

Zhang, X., Kimura, Y., Fang, C., Zhou, L., Sfyroera, G., Lambris, J. D., Wetsel, R. A., Miwa, T. and Song, W. C. (2007). Regulation of toll-like receptor-mediated inflammatory response by complement in vivo. *Blood* 110, 228–2(**a**)36

16. Staphylococcal Complement Inhibitors: Biological Functions, Recognition of Complement Components, and Potential Therapeutic Implications

Brian V. Geisbrecht

School of Biological Sciences, University of Missouri at Kansas City, Kansas City, MO 64110, USA

Abstract. It has been known for quite some time that many pathogenic microorganisms are capable of specifically attenuating or bypassing complement-mediated immune responses. Over the last several years, our understanding of the complement evasion mechanisms utilized by pathogens has increased precipitously through the study of the virulent bacterium *Staphylococcus aureus*. The combination of structural and functional characterization of *S. aureus*-derived complement inhibitors has revealed new mechanisms of complement regulation. Study of these proteins may also hold important clues into the design and optimization of long-awaited therapeutics that specifically and effectively block the complement activation and amplification cascades.

1 *S. aureus* as a Model System for Immune Evasion

Staphylococcus aureus is a prototypic opportunistic pathogen in humans. The organism is a leading cause of nosocomial and community-acquired infections, and is responsible for a remarkably broad range of diseases that span the entire range of severity in clinical presentation (Lowy 1998). Perhaps more so than any other bacterial pathogen, *S. aureus* has evolved the ability to adapt to and persist within a diverse array of physiological microenvironment within its hosts, including skin, bone, and various tissues and structures within the circulatory system. One hallmark of *S. aureus* is its ability to infect and thrive within immune-competent host organisms. As a result, an enormous amount of research has been conducted over the last two decades to arrive at a more complete understanding of (i) how these bacteria survive within the host and (ii) the nature of their effects on host homeostatic, defense, and repair mechanisms. Collectively, this effort has established that *S. aureus* expresses a formidable arsenal of virulence-facilitating proteins and structures that contribute its infectivity.

While the role of various cell surface-retained MSCRAMM adhesins has been firmly established for some time, the identities and functions of a number of anti-inflammatory *S. aureus* molecules have emerged more recently (Reviewed in

J.D. Lambris (ed.), *Current Topics in Complement II.*
doi: 10:1007/978-0-387-78952-1_16,

(Chavakis et al. 2005, 2007; Foster 2005; Rooijakkers et al. 2005b)). Many of these molecules have been characterized structurally as well as functionally, which has revealed critical relationships between these two intimately related aspects of biology. Furthermore, several of these studies have revealed entirely new mechanisms of immune evasion or modulation. As a result, *S. aureus* has become much more than just a different organism to look for new examples of well-documented evasion principles. Rather, it has developed into a model system to study this fascinating interplay of host-pathogen interactions. Indeed, the list of immunosuppressive proteins from *S.aureus* has continued to grow, and so has our insight into their function and overall role in the immune evasive strategy of this highly proficient pathogen.

2 The Anti-Complement Activities of *S. aureus*

As a bacterial cell, *S. aureus* is capable of activating all three complement pathways (Bredius et al. 1992; Kawasaki et al. 1987; Neth et al. 2002; Verbrugh et al. 1979; Wilkinson et al. 1978). Not surprisingly, then, the components of the complement system represent a central target for the numerous immune evasion strategies of *S. aureus*. Depending on the strain examined, however, the bacterium may also synthesize capsular polysaccharides (O'Riordan and Lee 2004) that have been shown to interfere with both opsonization by C3b and antibodies (Cunnion et al. 2001; Peterson et al. 1978) as well as recognition of deposed C3b (Verbrugh et al. 1982), presumably by blocking access to the complement receptors. Additionally, the bacterium is resistant to MAC function owing to the peptidoglycan-rich structure of its gram positive cell wall (Frank 2001). Given this implicit layer of structural protection from complement, there are no reported examples of *S. aureus* proteins that target the function of terminal complement components (C6-C9) or assembly of the MAC. Instead, this bacterium has extensively targeted many of the initial complement components with the apparent goals of (i) specifically preventing opsonization and (ii) suppressing complement-associated inflammatory responses.

2.1 Inhibitors of Complement Activation and Amplification

There are several reported mechanisms through which *S. aureus* may impair the initiation or activation of the classical pathway. Among these, the structure and function of Protein A (SpA) has been characterized most extensively (Fig. 1). This protein recognizes the Fc domain of Ig molecules with extraordinary affinity (Cedergren et al. 1993; Gouda et al. 1992); however, since Protein A is retained on the bacterial cell surface, its activity results in "inverted" opsonization of the bacterial surface, where the Fab regions of the Ig molecules are misoriented. A second *S. aureus* protein with similar Ig binding properties to Protein A has also been described (Zhang et al. 1998). Together, these proteins impede initiation of the classical pathway by preventing recognition of surface-bound Ig via C1q.

Interestingly, it has also been shown that Protein A can bind to gC1qR/p33 that is expressed on the surface of activated platelets (Nguyen et al. 2000). The utility for blocking function of this particular complement receptor on non-phagocytic cells remains to be fully appreciated. Furthermore, it not known whether this property is a conserved feature of Ig-binding proteins, since no such activity has been described for Sbi.

The alternative pathway plays a vital role in the initiation of the complement response against pathogens, as well as in the amplification of both the classical and lectin pathways. The physiological function of C3 lies at the heart of each of these processes, and so it has long been hypothesized as a likely target for immune evasive strategies by pathogens. That *S. aureus* has evolved efficient means of targeting C3 is not surprising, however, the extent to which the bacterium has gone to disturb C3 function is amazing. In fact, there are currently at least five unique *S. aureus* proteins that inhibit various steps of the conversion of native C3 to its activated derivative, C3b.

The Extracellular Fibrinogen-binding Protein (Efb) was the first C3b binding protein produced by *S. aureus* to be identified. In the original study, Efb was shown to inhibit opsonophagocytosis by granulocytes (Lee et al. 2004a). Efb binds to all forms of C3 containing the active-site thioester-containing domain (C3d) via its three-helix bundle C-terminal domain (Hammel et al. 2007b; Lee et al. 2004b) (Fig. 1). Interestingly, Efb exhibits its potent effects on the alternative pathway by inducing an active-like conformation in C3 that cannot be proteolyzed physiologically into C3a and C3b (Hammel et al. 2007b). Recently, an Efb Homologous Protein (Ehp) from *S. aureus* has also been reported (Hammel et al. 2007a). Ehp exhibits a high level of structural similarity to the C3-binding domain of Efb and also appears to induce a similar conformational change in C3 (Fig. 1).

In addition to Efb and Ehp, *S. aureus* also expresses another family of small, helical proteins that block activation of complement activation. These three related proteins, denoted SCIN for Secreted Complement INhibitors, are powerful inhibitors of all three complement pathaways and function by binding and stabilizing C3 convertases in a non-functional state (Rooijakkers et al. 2005a). In contrast to the Ehp family, SCINs do not bind directly to any form of C3. This suggests that the SCIN family members must be specific to substrate conformations or protein interfaces that are found exclusively in the C3 convertase assemblies. That SCINs function through a distinct mechanism from the Efb/Ehp family can be inferred, in part, form the recent crystal structure of SCIN (Rooijakkers et al. 2007). This work revealed that SCIN bares more resemblance to *S. aureus* Protein A modules than it does to either Efb-C or Ehp (Fig. 1). Thus, although additional study is needed to determine the relative contributions of each of these proteins to *S. aureus* complement evasion, it is clear that *S. aureus* can efficiently block the essential functions of C3 through a suite of protein inhibitors that can act concertedly by targeting both the native C3 substrate and its associated convertases.

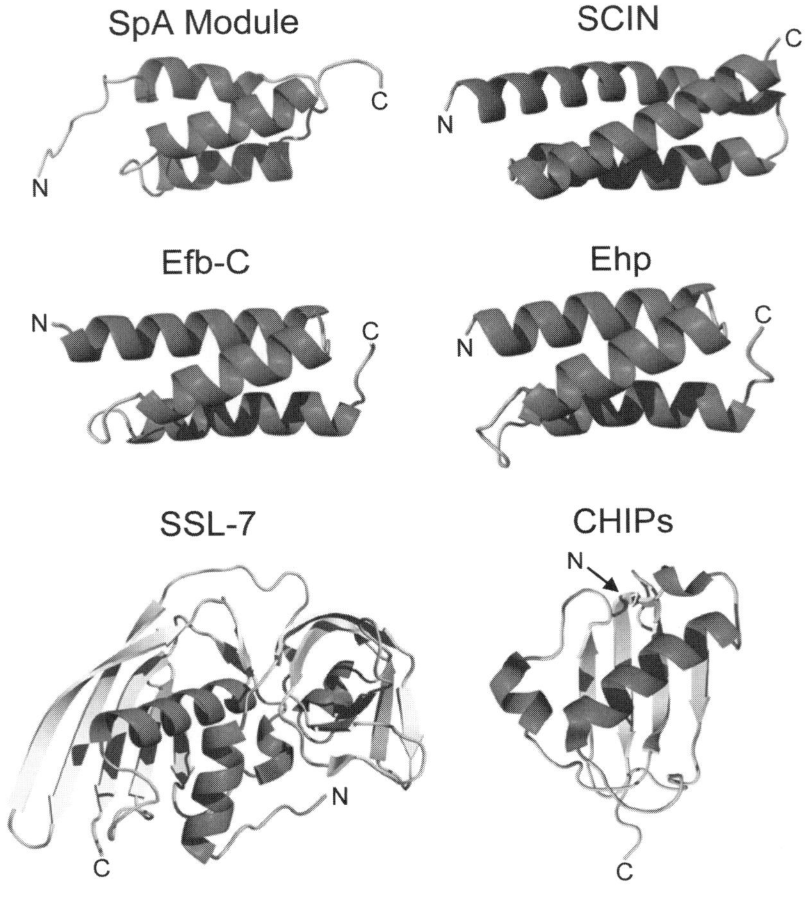

Fig. 1 Three-dimensional structures of complement inhibitory proteins expressed by *Staphylococcus aureus*. Structural representations were reproduced from the following PDB entries: 1BDD (SpA, (Gouda et al. 1992)); 2QFF (SCIN, (Rooijakkers et al. 2007)); 2GOM (Efb-C, (Hammel et al. 2007a)); 2NOJ (Ehp, (Hammel et al. 2007b)); 1V1O (SSL-7, (Al-Shangiti et al. 2004)); and 1XEE (CHIPs, (Haas et al. 2005)). Protein amino (N) and carboxyl (C) termini are indicated

2.2 Inhibitors of Complement-induced Inflammatory Responses

At a functional level, the combined activities of the Efb/Ehp and SCIN families described above serve to limit opsonization of the *S. aureus* cell by C3b. However, *S. aureus* also produces at least two proteins that appear to distrupt downstream inflammatory responses initiated by activation or processing of complement C5. The Superantigen-like Protein-7 (SSL-7) has recently been shown to bind both C5 and human IgA (Al-Shangiti et al. 2004; Langley et al. 2005) (Fig. 1). This interaction blocks binding of IgA to the FcαRI (CD89) both on granulocytes and in vitro, and can occur independently of SSL-7 binding to C5. Presumably, binding to C5 prevents its activation to C5a and C5b, since the presence of SSL-7 markedly reduces the efficiency of human serum in killing of MAC-sensitive *E. coli* cells. Separately, the *S. aureus* CHemotaxis INhibitory Protein (CHIPS) has also been identified as a powerful anti-inflammatory molecule (de Haas et al. 2004). The structure of CHIPs (Fig. 1) is somewhat distantly related to the C-terminal domains found in a large family of bacterial immunomodulatory proteins (Haas et al. 2005), such as the superantigens (including SSL-7) and the extracellular adherence protein (Eap) (Geisbrecht et al. 2005), also from *S. aureus*. However, CHIPs functions by blocking binding of chemotactic ligands to the C5a receptor (C5aR) and Formylated-peptide Receptor (FPR) expressed on neutrophils (Postma et al. 2004, 2005).

2.3 Remaining Questions

When considered as a whole, *S. aureus* has evolved one of the most elaborate, and arguably the most expansive anti-complement arsenal of any pathogen studied thus far. In this respect, it is quite perplexing that this organism appears to have completely ignored the most common complement evasion strategy in existence, namely the recruitment of soluble, host-derived Regulator of Complement Activation (RCA) proteins to the bacterial cell surface (Lambris et al. 2008). As of this writing, no C4BP, fH, or fHL-1-binding proteins from *S. aureus* have been identified. One potential explanation is that the combined effects of the *S. aureus* cellular architecture, Efb and SCIN protein families are sufficiently protective as to make additional RCA-recruiting proteins that affect the conversion of C3 or the stability of C3b irrelevant. On the other hand, it may simply be the case that the RCA recruiting activities of *S. aureus* have yet to be identified. Along these lines, it has recently been reported that certain strains of *S. aureus* can be opsonized by C3b, but that conversion of C3b to iC3b on the bacterial surface can occur by factor I in an fH-independent manner (Cunnion et al. 2004). Moreover, the presence of fI decreased the efficiency of neutrophil phagocytosis of C3b-opsonized *S. aureus* (Cunnion et al. 2005). These results raise the exciting possibility that *S. aureus* may possess one or more molecules that can circumvent the requirement of fH or fHL-1 as a cofactor for fI-mediated degradation of C3b. Such a discovery would clearly add a new layer of complexity to the already substantial anti-complement strategies of this versatile bacterium.

3 Toward a Molecular Understanding of Complement Evasion

In the past several years, there have been tremendous advances in our understanding of the structure of the central complement component C3 (Janssen et al. 2005), as well as the conformational changes that accompany its activation and deactivation cycle (Gros et al. 2008; Janssen et al. 2006). Separately, structure/function studies on the individual complement inhibitory complexes formed between the *S. aureus* proteins Efb (Hammel et al. 2007b) and Ehp (Hammel et al. 2007a) and their cognate C3d domain from human C3 have provided important new insights into the mechanisms through which pathogenic organisms may compromise the efficacy of the complement system (Lambris et al. 2008). When taken together, these independent studies constitute a unique framework for evaluating the vulnerabilities of the complement system, and for the potential utilization of these *S. aureus* proteins as templates for the design of new anti-complement therapeutics. Consequently, this section will summarize our knowledge of C3 recognition and inhibition by *S. aureus* Efb and Ehp.

3.1 Recognition of C3 by *S. aureus* Efb

The complement-inhibitory properties of Efb reside solely within the ordered, carboxy-terminal region of the protein (Hammel et al. 2007b; Lee et al. 2004a,b). Consistent with this observation, the entirety of the C3-binding ability of Efb lies within the same fragment. To better define the contributions of Efb protein to *Staphylococcal* complement evasion, a protease-stable, carboxyl-terminal fragment of *S. aureus* Efb (denoted Efb-C) was crystallized both free and bound to a recombinant form of human C3d (Hammel et al. 2007b). These structures were refined to 1.25 Å (Fig.1) and 2.2 Å (Fig. 2) limiting resolution, respectively. Comparison of the Efb-C protein in both structures revealed no substantive changes in the bacterial component upon binding. Specifically, 61 of the 65 residue Cα positions that these structures share in common superimpose within 2.5 Å and an r.m.s. deviation of 0.47 Å. This observation suggests that the C3d binding site on the Efb surface is preformed and does not require much structural rearrangement to interact with the complement component. At the time, the Efb-C structure was the first example of a soluble, all α-helical regulatory protein that interacts directly with a component of the complement cascade; however, the recent crystal structures of other proteins shown in Fig. 1. have expanded upon this observation.

The bimolecular complex formed between Efb-C and C3(d) is characterized by a low-nanomolar dissociation constant ($K_d \approx 2$ nM) and a notable kinetic stability ($t_{1/2} \approx 1$ h) (Hammel et al. 2007b). In the absence of structural information, these parameters in and of themselves are consistent with a protein-protein interaction formed by extensive charge-charge interactions. Indeed, as is shown in Fig. 2, the

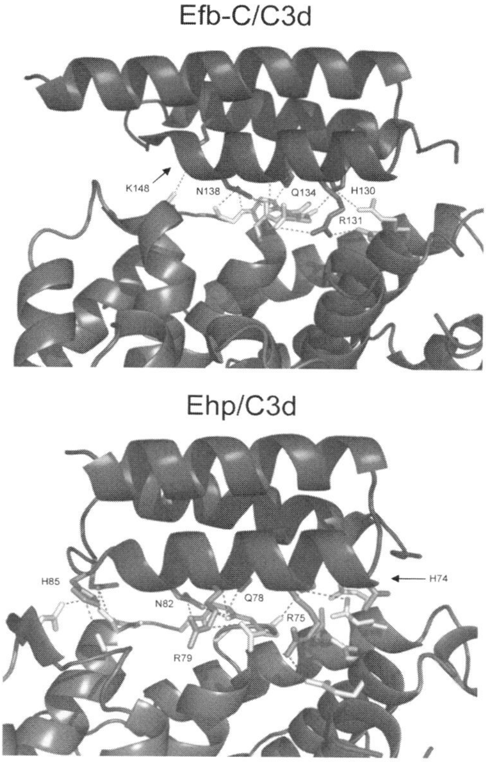

Fig. 2 Interfaces of the Efb-C/C3d and Ehp/C3d complexes as determined by X-ray crystallography. Refined crystal structures for the Efb-C/C3d and Ehp(N63E)/C3d complexes are shown in identical orientations where the amino terminus of the respective *S. aureus* component appears in the upper left-hand side of either panel. Residues positioned to form favorable contacts (*dashed lines*) are drawn as stick representations; the identities of the respective Efb-C and Ehp positions are indicated. Structural representations were drawn from the PDB entries 2GOX (Efb-C/C3d, (Hammel et al. 2007a)) and 2NOJ (Ehp(N63E)/C3d, (Hammel et al. 2007b))

interface of the Efb-C/C3d complex is comprised of a preponderance of basic residues donated by the bacterial component. The sidechains of these positions participate in a variety of charge-charge and/or polar interactions with both the sidechain and backbone atoms of its complement target, C3d.

The contributions of several of these Efb-C residues on both the thermodynamics and kinetics of the single-site, nanomolar-affinity Efb-C/C3(d) interaction have been probed by site directed mutagenesis using both isothermal titration calorimetry and surface plasmon resonance, respectively (Hammel et al.

2007b). Loss of R131 disrupted C3(d) binding and resulted in loss of alternative pathway inhibition, presumably by ablation of the series of interactions this sidechain forms with H1025, D1029, N1091, and L1092 of C3 and an ordered solvent molecule found nearby (Fig. 2, Top Panel). Similarly, loss of N138 also resulted in dramatically lowered affinity for C3(d) and decreased inhibition of the alternative pathway. In this case, these effects appear to be due to disruption of an intricate network of likely hydrogen bonds between the sidechain amide of N138 and the backbone atoms of residues V1090, I1093, and I1095 that comprise the H4-H5 loop region of C3d (Nagar et al. 1998).

Chen et al. (2008) have also reported analysis of Efb/C3(d) interactions using a complementary solution-based structural approach. Here, lysine-specific differentialcovalent modification of free vs. C3d-bound Efb was monitored by mass spectrometry to identify solvent-accessible sites protected upon Efb/C3(d) binding. The results of this study not only confirmed the interface observed in the Efb-C/C3d crystal structure, but also identified other potential sites of contact as well. In the Efb-C/C3d co-crystal, approximately 10% of the Efb-C polypeptide chain could not be modeled due to a lack of significant electron density in these regions following refinement. This amino-terminal region of Efb-C contains two lysine residues, K106 and K107, which are protected from modification in the presence of C3d. Similarly, positions K1105 and K1155 of C3d were protected from modification upon Efb-C binding. While these interactions were not detected by crystallographic analysis, the proximity of this amino-terminal portion of Efb-C to the corresponding region of C3d is nevertheless consistent with their existence (Fig. 2, Top Panel).

While all available data suggest that Efb-C binding to C3 requires either R131 or N138, the importance of the other residues and the interactions they form with C3(d) have yet to be assessed by direct experimental methods. On one hand, the lack of C3(d) binding by an Efb-C double mutant where both R131 and N138 were mutated to alanine suggests that any additional interactions are insufficient to drive formation of a stable Efb-C/C3(d) complex. Yet Efb-C has also been shown to alter the conformational and dynamics of its C3 target; this raises the possibility that apparently subtle or transient interactions may still be functionally important. While much is already known, unraveling the detailed nature and functional consequences of Efb-C binding to C3 and its activation products will require further analyses using both crystallographic and solution approaches.

3.2 Recognition of C3 by *S. aureus* Ehp

The recent discovery and characterization of the Efb homolog from *S. aureus* (denoted Ehp) has provided an exciting comparative basis for understanding the structure-function relationships of the Efb family (Hammel et al. 2007a). Ehp shares approximately 44% identity to Efb-C, adopts a similar three-helix bundle fold (Fig. 1), displays a binding preference for native forms of C3, and exerts its potent inhibitory effects primarily on the alternative pathway. Nevertheless, there are substantive differences between these two proteins.

Characterization of the interactions between Ehp and C3-derived fragments has demonstrated that Ehp has a unique ability to bind two molecules of C3. Thermodynamic analyses of Ehp/C3d binding using titration calorimetry are consistent with two, equimolar sites that are non-equivalent. The first binding site is entropically-favored, and displays an apparent dissociation constant on the order of 100 pM. In contrast, the second site evolves significantly greater enthalpy ($\Delta H_{site2} \approx -15.67 \pm 0.10$ vs. $\Delta H_{site1} \approx -7.94 \pm 0.06$ kcal/mol), but is entropically opposed and displays an approximately three-orders of magnitude lower affinity ($K_d \approx 100$ nM). It is worth noting that the concentration of C3 in the human circulatory system has been reported at 6.5 μM (Sahu and Lambris 2001). This is far greater than the K_d values for either of the C3-binding sites described above, and almost assuredly exceeds even the highest levels of Ehp produced and secreted by actively growing *S. aureus* cultures. In practical terms, this implies that Ehp exists predominantly as a ternary Ehp/C3 complex (i.e. C3·Ehp·C3) under typical physiological conditions.

The high level of sequence identity between Ehp and Efb-C makes the dramatic nature of their differences in C3 recognition even more interesting. Close inspection of sequence alignments, along with prior knowledge about the Efb-C/C3(d) interaction has provided an important basis for understanding the details of Ehp/C3 binding (Hammel et al. 2007a). First, Ehp residues R75 and N82 are in positions equivalent to R131 and N138 of Efb. Mutational analysis of these sidechains has confirmed that this "Efb-like" binding site contributes the higher-affinity C3-binding site (i.e. "site 1") in Ehp. Second, a peptide repeat that is closely related to that which defines the high-affinity binding site was identified in the amino-terminal, or α1 helix, of Ehp. This sequence, which spans residues 58–64, contains an equivalent asparagine residue (N63), although it notably lacks an arginine as the corresponding position is occupied by valine (V56). Nevertheless, site-directed mutagenesis of N63 revealed that this sequence does in fact represent the second, lower-affinity C3(d) binding site of Ehp.

While residues R75 and N82 in Ehp participate in an "Efb-like" C3(d)-binding site, the previously mentioned biochemical data suggest that there are still important differences in the nature of the interactions formed at these respective protein interfaces with C3(d) that lead to an approximately tenfold higher affinity for $\text{Ehp}_{\text{site1}}$ when compared to Efb-C. Many of these details were revealed by the 2.7 Å crystal structure of the N63E mutant of Ehp bound to C3d (Fig. 2, Bottom Panel). The overall structure of this complex shares many features in common with that of Efb-C/C3d (Fig. 2, Top Panel), including the extent and nature of the contacts formed by residues R75 and N82. However, there are two noteworthy differences. First, in terms of the overall number of contacting residues, there appear to be additional favorable interactions at the interface of Ehp/C3d when compared to the Efb-C/C3d complex. In particular, contacts are observed for two residues, H85 and Q87, which are found in the loop that joins the second and third helices of Ehp. Second, and perhaps more importantly, is that R79 of Ehp is perfectly positioned to serve a role in guiding, or orienting the interactions of the key residue R75. R79 effects the orientation of R75 indirectly by forming a hydrogen bond at its Nε position with the sidechain of residue N1091 of C3(d). This same sidechain from C3(d) can then interact with an Nω of R75, which by virtue of its conformation is poised to hydrogen bond with both D1029 and E1030 of C3(d). The overall result is a further intercalation of sidechains between Ehp and C3(d) that is impossible in the Efb-C/C3d complex. This is because the corresponding position in Efb-C is occupied by a lysine (K135), and lacks the central Nε atom of the arginine sidechain that makes this intricate series of interactions possible.

Less is currently known about the structural details of the second, lower-affinity C3(d)-binding site in Ehp. However, it is reasonable to expect some similarities to the site described in detail above. In particular, the role of N63 is likely equivalent to that of N82 in Ehp and N138 in Efb, since mutation of N63 can render this site non-functional (Hammel et al. 2007a). In contrast, the function of the corresponding residue to either R75 in Ehp or R131 in Efb is unclear, as this position is occupied by a valine. Though valine does not appear to sterically interfere with the binding site on C3(d), it obviously lacks the potential to form the favorable hydrogen bonds shown in Fig. 2. Clearly, fully understanding the nature of this lower affinity-binding site will require experimental structure determination in a C3(d)-bound state.

3.3 Inhibitory Mechanisms of the Efb Family

As a microorganism, the presence of *S. aureus* stimulates activation of the alternative pathway (Bredius et al. 1992; Kawasaki et al. 1987; Neth et al. 2002; Verbrugh et al. 1979; Wilkinson et al. 1978). It is appropriate, then, that both Efb and Ehp inhibit the function of the alternative pathway (Hammel et al. 2007a,b; Lee et al. 2004a). Subsequent analysis has demonstrated that the ability of Efb-C and Ehp to inhibit the alternative pathway is directly dependent on their affinity for C3(d), although the manner through which this inhibition is achieved is quite unusual.

In conjunction with direct biochemical analyses, studies using a series of conformation-specific monoclonal antibodies have shown that Efb-C and Ehp bind to native C3 and alter its confirmation to one that resembles an "active-like" state similar to that found in C3(H_2O) and C3b (Hammel et al. 2007a,b). Interestingly, the same serum-derived C3 bound to either Efb or Ehp also maintains its antigenicity to anti-C3a antibodies, thereby providing evidence that native C3 in this altered conformation cannot participate in downstream activation processes. Finally, conformational changes in actived C3b were also observed following Efb-C binding (Hammel et al. 2007b); while these have yet to be investigated for Ehp, it is very likely that such changes do occur.

The longstanding observation that C3 activation is accompanied by significant conformational changes suggests which molecules that alter C3 conformation and dynamics (such as Efb and Ehp) may also induce important changes in many of the complement activation and amplification processes where C3 or its fragments serve an essential role. Indeed, Jongerius et al. have recently provided evidence that both Efb and Ehp (herein denoted Ecb) inhibit the function of C3b-containing convertases (Jongerius et al. 2007). These observations explain why Efb and Ehp specifically inhibit the alternative pathway, since the classical and mannose-binding lectin pathway-derived C3 convertases contain C4b instead of C3b. Moreover, since all C5 convertases contain C3b, this work also suggests that Efb and Ehp serve as efficient inhibitors of C5a-dependent inflammation.

Can these observations regarding the Efb family of complement inhibitors be successfully integrated into a cohesive understanding of their structure/function relationships? The fact Ehp is a more potent inhibitor of the alternative pathway than Efb may hold important clues (Hammel et al. 2007a). On one hand, Ehp can bind twice as many C3 molecules as Efb, and thus it may simply be that Ehp binds, alters the conformation, and blocks the function of a larger amount of this essential complement component. On the other hand, this effect may be a manifestation of the altered conformations of the substrates or components of the C3b-containing convertases. Here, the recruitment of an additional C3b molecule by Ehp as compared to Efb may preclude efficient cleavage of additional native C3 by the convertases. Finally, the effect of Efb and Ehp on C5 convertases cannot be overlooked. In this respect, it is worth noting that the alternative pathway C5 convertase ($(C3b)_2Bb$) contains two molar-equivalents of C3b, which would seem to be an ideal target for inhibition through bivalent Ehp. As this would predict, Ehp is indeed a more potent inhibitor of C5 cleavage than is Efb (Jongerius et al. 2007). Though the available information is a promising start, a great deal of work is needed to thoroughly examine these possibilities and our appreciation of these proteins is bound to increase over the years to come.

4 Potential Therapeutic Applications of the Efb Family

The complement system plays a key role in the pathology of a continually-expanding list of inflammatory, autoimmune, and ischemic conditions. And while numerous attractive pharmacological targets in the complement cascades have been identified, there are only a limited number of anti-complement therapeutics approved for clinical use (Ricklin and Lambris 2007). The unique modes of C3 recognition by both Efb and Ehp may therefore constitute an important starting point for the design and optimization of a new class of therapeutic complement inhibitors.

Based upon the structure/function studies described above, a simplified scheme for the design of a new class of potential complement inhibitors is shown in Fig. 3. To begin, Ehp appears to be the logical choice of a "lead compound" since it is a more potent inhibitor of the complement activation than is Efb (Hammel et al. 2007a). Next, comparison of the biochemical and structural properties of Efb and Ehp suggests that the additional inhibitory potency of the latter protein is derived from

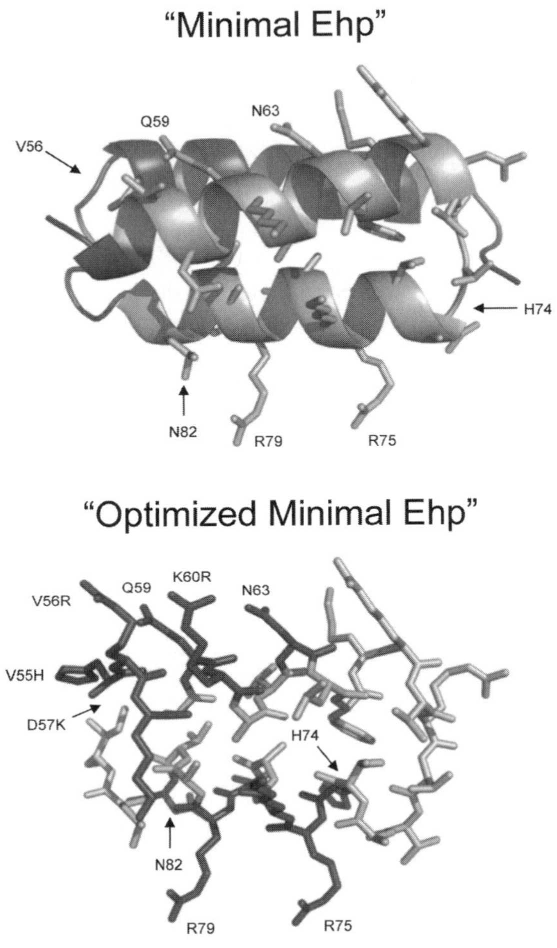

Fig. 3 Schematic approach for the optimization of complement inhibitory molecules derived from the Efb family. In the top panel, the Ehp structure is depicted with all contiguous residues that define both C3-binding sites as stick representations. Residues that are found (for the "Efb-like" site), or presumed to be located (for the lower-affinity site) at the interface of the C3 complex are labeled. In the bottom panel, a stick model is shown for a hypothetical peptide derived from the Ehp protein. This molecule consists of the 29 contiguous residues described in the top panel, but also contains a series of mutations (labeled) predicted to convert the low-affinity site (top of the drawing) into one with properties more in-line with the higher-affinity sites (bottom of the drawing). C3-contacting residues are drawn in black

its second C3 recognition site (Hammel et al. 2007a). Examination of the Ehp sequence reveals that both C3-binding sites are contained within a minimal 29-residue stretch that comprises helices α1 and α2 of the Ehp protein (Fig. 3, Top Panel). The fact that the second C3-binding site in Ehp is substantially lower in affinity than the canonical Efb-like site suggests that beneficial increases in both affinity and inhibitory potency may be gained by mutating select residues so that this lower-affinity site contains residues identical to the higher-affinity site (Fig. 3, Bottom Panel).

Although the family of compounds typified by the stick representation in Fig. 3 has yet to be validated for its complement inhibitory properties, there are important reasons to investigate such molecules further. First, while there are limitations to peptide-based therapeutics, the effective targeting of the alternative pathway by the peptidomimetic Compstatin suggests that this class of molecules can be utilized to successfully modulate C3 function (Morikis and Lambris 2005). Second, although the number of positions in these molecules that bind directly to C3 is limited, there are still a number of ancillary positions which could be screened by combinatoric optimization to potentially improve the complement inhibitory properties of the resulting molecules. Finally, the relatively small size of the key functional regions in these proteins suggests that effective mimics might possibly be found in the realm of pharmacologically-available small molecules. Together, these features make molecular mimics of the Efb family attractive candidates for therapeutic inhibition of the complement system.

5 Conclusions

Throughout their millennia of coexistence, higher organisms have evolved a number of very effective mechanisms for the detection and elimination of microbial intruders. Yet recent work has also demonstrated that a number of highly successful pathogens are equally adept at bypassing these mechanisms. While the interactions between *Staphylococcus aureus* and the complement system described here represent only a fraction of the fascinating process of immune evasion, study of these manifold and apparently unique mechanisms has greatly expanded our understanding of complement recognition, activation, and regulation. More importantly, they may ultimately hold important keys toward the development of long-awaited therapeutics for a number of complement-mediated human diseases.

References

Al-Shangiti, A.M., Naylor, C.E., Nair, S.P., Briggs, D.C., Henderson, B. and Chain, B.M. (2004) Structural relationships and cellular tropism of Staphylococcal supernantigen-like proteins. *Infect. Immun.* 72, 4261–4270

Bredius, R.G., Driedijk, P.C., Schouten, M.F., Weening, R.S. and Out, T.A. (1992) Compelment activation by polyclonal immunoglobulin G1 and G2 antibodies against

Staphylococcus aureus, *Haemophilus influenzae* type B, and tetanus toxoid. *Infect. Immun.* 60, 4838–4847

Cedergren, L., Andersson, R., Jansson, B., Uhlen, M. and Nilsson, B. (1993) Mutational analysis of the interaction between Staphylococcal protein A and human IgG1. *Protein Eng.* 6, 441–448

Chavakis, T., Wiechmann, K., Preissner, K.T. and Herrmann, M. (2005) *Staphylococcus aureus* interactions with the endothelium: the role of bacterial "Secreteable Expanded Repertoire Adhesive Molecules"(SERAM) in disturbing host defense systems. *Thromb. Haemost.* 94, 278–285

Chavakis, T., Preissner, K.T. and Herrmann, M. (2007) The anti-inflammatory activities of *Staphylococcus aureus*. *Trends Immunol.* 28, 408–418

Chen, H., Schuster, M.C., Sfyroera, G., Geisbrecht, B.V. and Lambris, J.D. (2008) Solution insights into the structure of the Efb/C3 complement inhibitory complex as revealed by lysine acetylation nand mass spectrometry. *J. Am. Soc. Mass. Spectrom.* 19, 55–65

Cunnion, K.M., Lee, J.C. and Frank, M.M. (2001) Capsule production and growth phase influence binding of complement to *Staphylococcus aureus*. *Infect. Immun.* 69, 6796–6803

Cunnion, K.M., Hair, P.S. and Buescher, E. (2004) Cleavage of complement C3b to iC3b on the surface of *Staphylococcus aureus* is mediated by serum complement factor I. *Infect. Immun.* 72, 2858–2863

Cunnion, K.M., Buescher, E.S. and Hair, P.S. (2005) Serum complement factor I decreases *Staphylococcus aureus* phagocytosis. *J. Lab. Clin. Med.* 146, 279–286

de Haas, C.J., Veldkamp, K.E., Peschel, A., Weerkamp, F., van Wamel, W.J., Heezius, E.C., Poppelier, M.J., van Kessel, K.P. and van Strijp, J.A. (2004) Chemotaxis inhibitory protein of *Staphylococcus aureus*, a bacterial antiinflammatory agent. *J. Exp. Med.* 199, 687–695

Foster, T.J. (2005) Immune evasion by Staphylococci. *Nat. Rev. Microbiol.* 3, 948–958

Frank, M.M. (2001) Annihilating host defense. *Nat. Med.* 7, 1285–1286

Geisbrecht, B.V., Hamaoka, B.Y., Perman, B., Zemla, A. and Leahy, D.J. (2005) The crystal structures of EAP domains from *Staphylococcus aureus* reveal an unexpected homology to bacterial superantigens. *J. Biol. Chem.* 280, 17243–17250

Gouda, H., Torigoe, H., Saito, A., Sato, M., Arata, Y. and Shimada, I. (1992) Three-dimensional solution structure of the B domain of Staphylococcal protein A: comparisons of the solution and crystal structures. *Biochemistry* 31, 9665–9672

Gros, P., Milder, F.J. and Janssen, B.J.C. (2008) Complement driven by conformational changes. *Nat. Rev. Immunol.* 8, 48–58

Haas, P.J., de Haas, C.J., Poppelier, M.J., van Kessel, K.P., van Strijp, J.A., Dijkstra, K., Scheek, R.M., Fan, H., Kruijtzer, J.A., Liskamp, R.M. and Kemmink, J. (2005) The structure of the C5a receptor-blocking domain of chemotaxis inhibitory protein of *Staphylococcus aureus* is related to a group of immune evasive molecules. *J. Mol. Biol.* 353, 859–872

Hammel, M., Sfyroera, G., Pyrpassopoulos, S., Ricklin, D., Ramyar, K.X., Pop, M., Jin, Z., Lambris, J.D. and Geisbrecht, B.V. (2007a) Characterization of Ehp: a secreted complement inhibitory protein from *Staphylococcus aureus*. *J. Biol. Chem.* 202, 30051–30061

Hammel, M., Sfyroera, G., Ricklin, D., Magotti, P., Lambris, J.D. and Geisbrecht, B.V. (2007b) A structural basis for complement inhibition by *Staphylococcus aureus*. *Nat. Immunol.* 8, 430–437

Janssen, B.J.C., Huizinga, E.G., Raaijmakers, H.C.A., Roos, A., Daha, M.R., Ekdahl-Nilsson, K., Nilsson, B. and Gros, P. (2005) Structures of complement component C3 provide insights into the function and evolution of immunity. *Nature* 437, 505–511

Janssen, B.J.C., Christodoulidou, A., McCarthy, A., Lambris, J.D. and Gros, P. (2006) Structure of C3b reveals conformational changes underlying complement activity. *Nature* 444, 213–216

Jongerius, I., Köhl, J., Pandey, M.K., Ruyken, M., van Kessel, K.P., van Strijp, J.A. and Rooijakkers, S.H. (2007) Staphylococcal complement evasion by various convertase-blocking molecules *J. Exp. Med.* 204, 2461–2471

Kawasaki, A., Takada, H., Kotani, S., Inai, S., Nagaki, K., Matsumoto, M., Yokogawa, K., Kawata, S., Kusumoto, S. and Shiba, T. (1987) Activation of the human complement cascade by bacterial cell walls, peptidoglycans, water-soluble peptidoglycan components, and synthetic muramylpeptides – studies on active components and structural requirements. *Microbiol. Immunol.* 31, 551–569

Lambris, J.D., Ricklin, D. and Geisbrecht, B.V. (2008) Complement evasion by human pathogens. *Nat. Rev. Microbiol.* 6, 132–142

Langley, R., Wines, B., WIlloughby, N., Basu, I., Proft, T. and Fraser, J.D. (2005) The Staphylococcal superantigen-like protein 7 binds IgA and complement C5 and inhibits IgA-FcaRI binding and serum killing of bacteria. *J. Immunol.* 174, 2926–2933

Lee, L.Y.L., Hook, M., Haviland, D., Wetsel, R.A., Yonter, E.O., Syribeys, P., Vernachio, J. and Brown, E.L. (2004a) Inhibition of complement activation by a secreted *Staphylococcus aureus* protein. *J. Infect. Dis.* 190, 571–579

Lee, L.Y.L., Liang, X., Hook, M. and Brown, E.L. (2004b) Identification and characterization of the C3 binding domain of the *Staphylococcus aureus* extracellular fibrinogen-binding protein (Efb). *J. Biol. Chem.* 279, 50710–50716

Lowy, F.D. (1998) *Staphylococcus aureus* infections. *N. Engl. J. Med.* 339, 520–532

Morikis, D. and Lambris, J.D. (2005) *Structure, Dynamics, Activity, and Function of Compstatin and Design of More Potent Analogs*. Taylor and Francis, Boca Raton, FL

Nagar, B., Jones, R.G., Diefenbach, R.J., Isenman, D.E. and Rini, J.M. (1998) X-ray crystal structure of C3d: a C3 fragment and ligand for complement receptor 2. *Science* 280, 1277–1281

Neth, O., Jack, D.L., Johnson, M., Klein, N.J. and Turner, M.W. (2002) Enhancement of complement activation and opsonophagocytosis by complexes of mannose-binding lectin with mannose-binding lectin-associated serine protease after binding to *Staphylococcus aureus*. *J. Immunol.* 169, 4430–4436

Nguyen, T., Ghebrehiwet, B. and Peerschke, E.I.B. (2000) *Staphylococcus aureus* protein A recognizes platelet gC1qR/p33: a nove mechanism for Staphylococcal interactions with platelets. *Infect. Immun.* 68, 2061–2068

O'Riordan, K. and Lee, J.C. (2004) *Staphylococcus aureus* capsular polysaccharides. *Clin. Microbiol. Rev.* 17, 218–234

Peterson, P.K., Kim, Y., Wilkinson, B.J., Schmeling, D., Michael, A.F. and Quie, P.G. (1978) Dichotomy between opsonization and serum complement activation by encapsulated Staphylococci. *Infect. Immun.* 20, 770–775

Postma, B., Poppelier, M.J., van Galen, J.C., Prossnitz, E.R., van Strijp, J.A., de Haas, C.J. and van Kessel, K.P. (2004) Chemotaxis inhibitory protein of *Staphylococcus aureus* binds specifically to the C5a and formylated peptide receptor. *J. Immunol.* 172, 6994–7001

Postma, B., Kleibeuker, W., Poppelier, M.J., Boonstra, M., van Kessel, K.P., van Strijp, J.A. and de Haas, C.J. (2005) Residues 10-18 within the C5a receptor N terminus compose a binding domain for chemotaxis inhibitory protein of *Staphylococcus aureus*. *J. Biol. Chem.* 280, 2020–2027

Ricklin, D. and Lambris, J.D. (2007) Complement-targeted therapeutics. *Nat. Biotechnol.* 25, 1265–1275

Rooijakkers, S.H., Ruyken, M., Roos, A., Daha, M.R., Presanis, J.S., Sim, R.B., van Wamel, W.J., van Kessel, K.P. and van Strijp, J.A. (2005a) Immune evasion by a Staphylococcal complement inhibitor that acts on C3 convertases. *Nat. Immunol.* 6, 920–927

Rooijakkers, S.H., van Kessel, K.P. and van Strijp, J.A. (2005b) Staphylococcal innate immune evasion. *Trends Microbiol.* 13, 596–601

Rooijakkers, S.H.M., Milder, F.J., Bardoel, B.W., Ruyken, M., van Strijp, J.A.G. and Gros, P. (2007) Staphylococcal complement inhibitor: structure and active sites. *J. Immunol.* 179, 2989–2998

Sahu, A. and Lambris, J.D. (2001) Structure and biology of complement protein 3, a connecting link between innate and acquired immunity. *Immunol. Rev.* 180, 35–48

Verbrugh, H.A., van Dijk, W.C., Peters, R., van der Tol, M.E. and Verhoef, J. (1979) The role of *Staphylococcus aureus* cell-wall peptidoglycan, teichoic acid, and protein A in the processes of complement activation and opsonization. *Immunology* 37, 615–621

Verbrugh, H.A., Peterson, P.K., Nguyen, B.-Y.T., Sisson, S.P. and Kim, Y. (1982) Opsonization of encapsulated *Stapylococcus aureus*: the role of specific antibody and complement. *J. Immunol.* 129, 1681–1687

Wilkinson, B.J., Kim, Y., Peterson, P.K., Quie, P.G. and Michael, A.F. (1978) Activation of complement by cell surface components of *Staphylococcus aureus*. *Infect. Immun.* 20, 388–392

Zhang, L., Jacobsson, K., Vasi, J., Lindberg, M. and Frykberg, L. (1998) A second IgG-binding protein in *Staphylococcus aureus*. *Microbiology* 144, 985–991

17. Human Astrovirus Coat Protein: A Novel C1 Inhibitor

Neel K. Krishna[1] and Kenji M. Cunnion[2]

[1]Department of Microbiology and Molecular Cell Biology, Eastern Virginia Medical School, Norfolk, VA 23507, USA, krishnnk@evms.edu
[2]Department of Pediatrics, Eastern Virginia Medical School and Children's Specialty Group, Norfolk, VA, USA, 23507

Abstract. C1 is a multimolecular complex that initiates the classical pathway of complement. It is composed of the pattern recognition component C1q and the serine proteases C1r and C1s. Activation of C1 elicits a series of potent effector mechanisms directed at limiting infection by invading pathogens as well as participating in other biological functions such as immune tolerance. While many molecules in addition to antibody have been demonstrated to activate C1, only a handful of C1 inhibitors have been described. Disregulated control of complement activation is associated with numerous autoimmune and inflammatory disease processes, thus tight regulation of C1 activation is highly desirable. We have recently discovered a novel inhibitor of C1, the coat protein of the human astroviruses, a family of enteric pathogens that infect young children. The astrovirus coat protein binds to the A-chain of C1q and inhibits spontaneous as well as antibody-mediated activation of the C1 complex resulting in suppression of classical pathway activation and complement-mediated terminal effector functions. This is the first description of a non-enveloped icosahedral virus inhibiting complement activation and the first description of a viral inhibitor of C1. The known inhibitors of C1 are reviewed and then discussed in the context of this novel viral C1 inhibitor. Additionally, the properties of this compound are elucidated highlighting its potential as an anti-complement therapeutic for the many diseases associated with inappropriate complement activation.

1 Introduction

Activation of the classical pathway of complement is initiated via C1, a multimolecular complex composed of the recognition component C1q and associated serine proteases C1r and C1s (Cooper 1985). The major mechanism for initiating classical pathway activation is C1q binding to antibody resulting in activation of C1. The cascading activation sequence triggers a number of robust inflammatory effector functions directed at limiting infection. In addition to host defense against invading pathogens, C1 is critical in the recognition and clearance of cellular debris, immune complexes and apoptotic cells (Kishore et al. 2004) and has been demonstrated to identify abnormal structures including beta-amyloid fibrils (Rogers et al. 1992; Tacnet-Delorme et al. 2001) and the pathological form

J.D. Lambris (ed.), *Current Topics in Complement II.*
doi: 10:1007/978-0-387-78952-1_17,

of the prion protein (Mabbott and Bruce 2001; Klein and Kaeser 2001). The ability of this molecule to distinguish self from non-self is critical for immune tolerance (Botto and Walport 2002).

Inappropriately controlled complement activation results in the damage and destruction of healthy host tissue contributing to a wide range of human diseases including systemic lupus erythematosus, rheumatoid arthritis, ischemia-reperfusion injury, myasthenia gravis, Alzheimer's disease and hyperacute xenograft rejection. Thus, a number of soluble and membrane-bound host regulators are present to modulate the complement cascade through interactions with specific factors at multiple points in the activation cascade, including C1. In addition to host-encoded complement regulators, it has been demonstrated that both viral and bacterial pathogens can evade or modulate the host complement system by encoding proteins that either inhibit complement components or mimic regulators (Bernet and Mullick 2003; Favoreel et al. 2003; Rooijakkers and van Strijp 2007). This review describes the recent discovery of a novel inhibitor of C1 expressed as the coat protein of astroviruses, a family of human enteric pathogens (Bonaparte et al. 2008). The properties of this virally-encoded C1 inhibitor will be presented in the context of the known inhibitors of C1 and discussed with respect to its potential as an anti-complement therapeutic.

2 C1, the First Complement Component

C1 initiates activation of the classical pathway and links the adaptive and innate immune systems by recognizing antibody-bound pathogens. C1 consists of a multi-molecular complex comprised of subunits C1q, a charge pattern recognition protein that is associated with a calcium-dependent tetramer of the serine proteases C1r and C1s ($C1r_2C1s_2$) (Fig. 1a) (Cooper 1985). Each subunit of C1q consists of the three protein chains A, B, and C that combine to form a two domain molecule; an N-terminal, triple-helical, collagen-like region (CLR, ~100 amino acid residues) followed by a globular head region (GHR) of ~120 amino acid residues (Fig. 1b). Interruptions to the repeating Gly-X-Y collagen sequence, act as a semi-flexible "hinge" (Poon et al. 1983), allowing the chains to diverge and form six individual "stems", each terminating in a globular head. This gives C1q the overall shape of a "bouquet of tulips". The GHR moiety is responsible for binding the conventional activating molecules, clustered IgG or a single IgM. However, other non-immunoglobulin molecules such as C-reactive protein, serum amyloid protein, LPS, fibromodulin and DNA, among others, have also been demonstrated to bind C1q and activate the complex (Cooper 1983; Trinder et al. 1993; Sjoberg et al. 2005). Binding of the globular heads of C1q to the activating molecule is thought to induce a conformational change via the hinge region of C1q that activates C1r, which in turn cleaves C1s leading to downstream classical pathway activation (Gaboriaud et al. 2004).

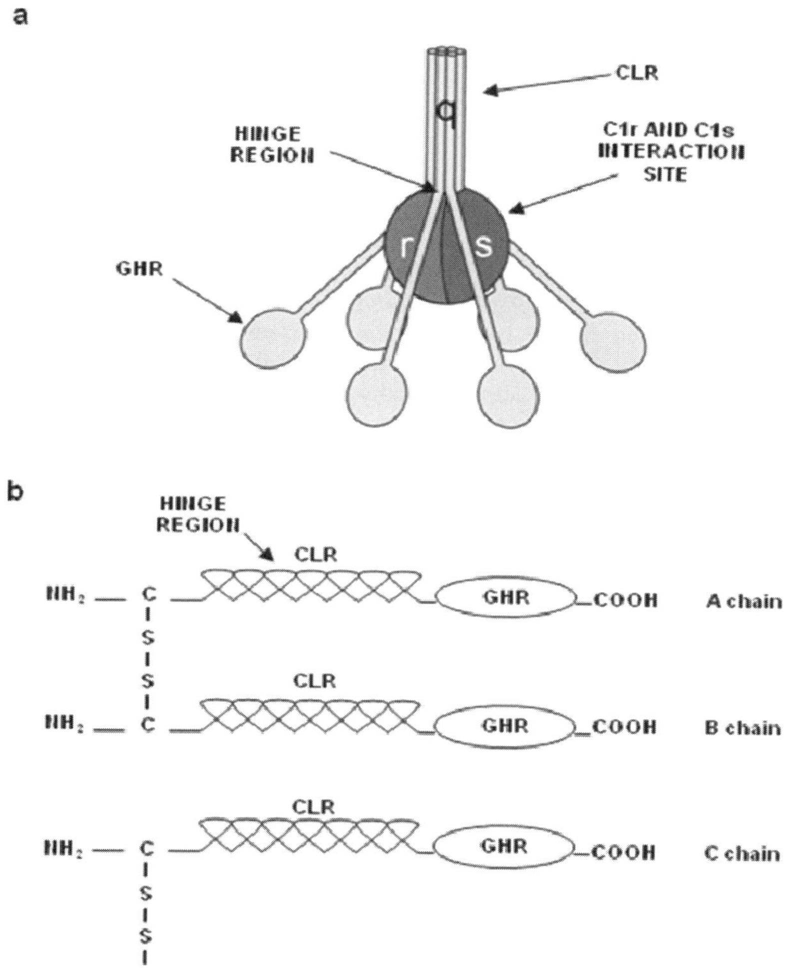

Fig. 1 Schematic representation of the structural organization of the human C1 complex. (**a**) Model showing the different components of C1 (C1q, C1r, C1s) and structural subunits of C1q (GLR, CLR, hinge region) and C1r/C1s binding site. (**b**) Illustration of the three chains of C1q (A, B, C) and their interactions. Six A–B and three C–C disulfide-bonded dimers form six identical heterotrimers through non-covalent interactions that compose the hexameric C1q structure

3 C1 Inhibitors

While many different substances in addition to antibody can bind and activate C1, only a handful of molecules that inhibit C1 activity have been characterized. The known inhibitors of C1 fall into four families which are described below.

3.1 C1-Inhibitor

C1-Inhibitor (C1-INH) is a heavily glycosylated serum protein and is the major inhibitor of activated C1 (Cooper 1983). In addition, it is also a known inhibitor of the coagulation system (Davis 1998), the kinin system (Gigli et al. 1970), the mannose-binding lectin pathway of complement (Matsushita et al. 2000) and has been reported to regulate the alternative pathway as well (Jiang et al. 2001). C1-INH protein is a member of the serine protease inhibitor (serpin) family (Levy and Lepow 1959). Upon activation of C1 by aggregated IgG, C1-INH rapidly binds stoichiometrically to the activated forms of C1r and C1s dissociating these subunits from C1q, thus blocking activation of the second cascade component, C4 (Sim et al. 1979; Ziccardi and Cooper 1979). Individuals that produce insufficient amounts of C1-INH or an inactive form of this protein suffer from hereditary angioedema, a disease characterized by recurrent episodes of severe skin and mucous membrane edema that can obstruct the trachea (Davis 2005). Purified C1-INH has been demonstrated to be an effective therapy for individuals with hereditary angioedema, however it must be prepared from human plasma and is exceptionally expensive (Ricklin and Lambris 2007). C1-INH is the only C1 inhibitor currently in use, but not FDA approved.

3.2 Decorin and Biglycan

Decorin and biglycan are extracellular matrix proteoglycans that have been demonstrated to bind C1q. Both bovine decorin (Krumdieck et al. 1992) and recombinant human decorin and biglycan (Groeneveld et al. 2005) bind to the GHR and to a lesser extent the CLR of C1q and inhibit classical pathway activation. In addition, these proteoglycans bind mannose-binding lectin with biglycan inhibiting this pathway (Groeneveld et al. 2005).

3.3 Neutrophil Defensins

The human neutrophil defensins are small (~30 amino acid residue), cationic peptides rich in cysteine and arginine involved in host antimicrobial defense. These peptides are members of the alpha-defensin family which can constitute up to 50% of the total protein content of azurophilic granules of neutrophils. Human neutrophil peptide-1 has been demonstrated to bind to C1q, most likely via the CLR and inhibit activation of the classical pathway (van den Berg et al. 1998). As with decorin and biglycan, human neutrophil peptide-1 is also able to bind mannose binding lectin and inhibit this pathway of complement activation (Groeneveld et al. 2007).

3.4 C1q Receptor Proteins

In addition to the soluble C1q inhibitors listed above, C1q receptor proteins have been identified on the surface of mammalian cells that can functionally inhibit C1q-initiated lytic activity (Ghebrehiwet et al. 1994; Kovacs et al. 1998). These include gC1qR/p33 which binds the GHR and cC1qR/calreticulin which binds the CLR (Ghebrehiwet and Peerschke 2004). A C1q binding protein found on the surface of *E. coli* cells has also been demonstrated to bind C1q via both the CLR and GHR, thus preventing the assembly of functional C1 from its constituent parts and inhibiting hemolytic activity (van den Berg et al. 1996).

A new type of inhibitor of C1, the first to be described from viral origin, is detailed below. Our laboratory has discovered that the coat protein of the human astroviruses, a causative agent of gastroenteritis in children, can potently suppress classical complement pathway activation by directly binding C1 and inhibiting its activation of the second cascade component, C4, and downstream inflammatory effector functions (Bonaparte et al. 2008). The astrovirus coat protein shares no homology to the proteins described above and thus represents a novel class of C1 inhibitors.

4 The Astroviruses

Human astroviruses (HAstVs) are a significant cause of acute gastroenteritis in young children. Second only to rotavirus in the incidence of virally-induced gastroenteritis in all children (Dennehy et al. 2001), HAstVs are recognized as the leading cause of viral diarrhea in infants (Shastri et al. 2004). Eight distinct HAstVs have been identified to date with serotype 1 being the most prevalent worldwide (Matsui and Greenberg 2001). In addition to humans, members of this virus family infect a variety of other young mammals and birds causing diseases ranging from diarrhea to nephritis (Matsui and Greenberg 2001). Astroviruses are small, non-enveloped, icosahedral particles with a single-stranded, messenger-sense RNA genome (~7 kb) that is organized into three open reading frames: open reading frames 1a and 1b encode non-structural proteins (Jiang et al. 1993; Lewis et al. 1994; Willcocks et al. 1994) whereas open reading frame 2 encodes the coat protein (CP) precursor (Lewis et al. 1994; Willcocks and Carter 1993). For HAstV serotype 1, the CP is 787 amino acid residues. CP precursors assemble into non-infectious particles, encapsidating the viral genome. In vitro cleavage of the viral capsid with trypsin renders the virions infectious (Bass and Qiu 2000; Méndez et al. 2002), however the proteolytic enzyme(s) required to cleave the particles in vivo are currently unknown. Detailed ultrastructural studies of trypsin-cleaved HAstV virions have revealed icosahedral particles with an array of spikes protruding from the surface of the virion (Matsui and Greenberg 2001; Matsui and Kiang 2001; Risco et al. 1995).

The pathogenesis of and immunity to the HAstVs is poorly understood. Virions have been identified in intestinal epithelial cells of children with diarrhea, correlating with fecal shedding of the virus (Moser and Schultz-Cherry 2005). However, in contrast to other enteric pathogens like rotavirus (Kapikian et al. 2001) HAstV-induced diarrhea appears not to result in significant cell death or inflammation in humans (Sebire et al. 2004) and an avian astrovirus animal model has suggested that the innate immune system may contribute to pathogenesis (Koci et al. 2003). These observations, in addition to the immature complement system of infants (Johnston 1986), led us to speculate that astroviruses may act upon the complement system during infection.

5 Inhibition of Complement Activity by Human Astrovirus Coat Protein (HAstV CP)

Our findings demonstrating the suppression of serum complement by HAstV CP were recently published in the Journal of Virology (Bonaparte et al. 2008). Here we present a portion of this data from the perspective of HAstV CP as a C1 inhibitor.

5.1 HAstV CP Suppresses Classical Pathway Activity

Using a standard hemolytic complement assay, HAstV-1 virions were found to strongly suppress complement-mediated sheep erythrocyte lysis by normal human serum (NHS). This was also demonstrated for HAstV serotypes 2, 3 and 4 suggesting that suppression of serum complement activity is a conserved property of the HAstVs. As a positive control for the suppression of hemolysis, cobra venom factor (CVF) was utilized. A time course assessing HAstV suppression of complement-mediated hemolysis compared with CVF, demonstrated similar kinetics for each (Fig. 2). This was a remarkable finding given that CVF is extremely potent, depleting serum complement components and suppressing effector functions (Vogel et al. 1984).

As noted above, astrovirus particles are composed of only one structural protein. To determine if this protein was responsible for mediating the complement suppressing activity of this virus family, HAstV-1 CP precursor was overexpressed in a recombinant baculovirus system and purified using biochemical methodology. Soluble CP was isolated in the form of trimers and was determined to be necessary and sufficient in mediating the suppression of complement-mediated hemolysis. To establish whether the HAstV CP preferentially suppresses the classical or the alternative pathway we tested factor B-depleted and C2-depleted serum, respectively. While HAstV-1 CP had a very modest effect on the suppression of alternative pathway complement activation (data not shown), the viral CP more

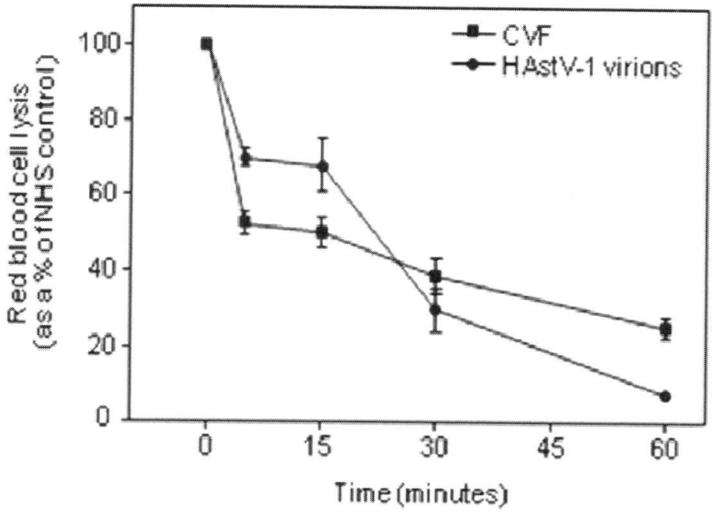

Fig. 2 HAstV-1 virions suppress complement activity in a hemolytic complement assay with similar kinetics to CVF. CaCo-2 cell lysates containing HAstV-1 virions (85 μl of cell lysate, corresponding to 2.92×10^8 genome copies) and 1 μg CVF were incubated from 0 to 60 min in the presence of 20 μl NHS. At 0, 5, 15, 30 and 60 min, aliquots were removed and incubated with sensitized sheep RBCs to test hemolytic complement activity. Data are the means from five independent experiments. Error bars denote SEM

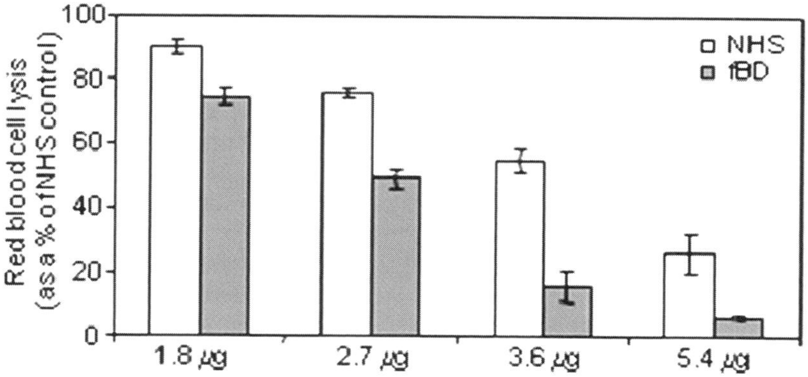

Fig. 3 HAstV-1 CP strongly suppresses classical pathway activity. Antibody-sensitized sheep RBCs were incubated with NHS (*white bars*) or factor B-depleted (fBD) sera (*shaded bars*) in the presence of the indicated amounts of CP. Hemolysis was standardized to 100% for each sera in the absence of CP. Data are the means of four independent experiments. Error bars denote SEM

efficiently suppressed classical pathway activation in factor B-depleted serum compared with NHS, where all pathways may activate, as assessed in a hemolytic assay (Fig. 3). The effect of the CP on the mannose-binding lectin pathway is unknown at this time.

5.2 HAstV CP Binds to the A-Chain of C1q

CP inhibition of classical pathway activation led us to speculate that CP interacts with one of the classical pathway factors (C1 complex, C4 or C2). To ascertain whether CP binds to specific complement factors, we utilized a modified virus overlay protein binding assay approach (Borrow and Oldstone 1992). Briefly, C1 complex and highly purified C2, C3 and C4 were separated by SDS–PAGE without reducing agents or boiling. Proteins were then transferred to nitrocellulose, blocked and probed with or without purified CP. After washing the membrane, CP binding was then detected with antisera to HAstV-1 followed by labeled secondary antibody. A specific band of approximately 59 kDa, consistent with a dimer of C1 chains, was present in the C1 lane while no binding was detected for C2–C4 (data not shown). The fact that C1 is a multimolecular complex of C1q, C1r and C1s under non-reducing conditions without boiling, made it difficult to determine exactly which component of C1 interacts with the viral CP. To address this, all three highly purified constituents of the C1 complex (C1q, C1r, C1s) were boiled, reduced and loaded onto SDS–PAGE gels, transferred to nitrocellulose and probed with or without CP, as above. The blot probed with CP detected a band of ~34 kDa in the C1 and C1q lanes whereas there was no signal for BSA, C1r or C1s

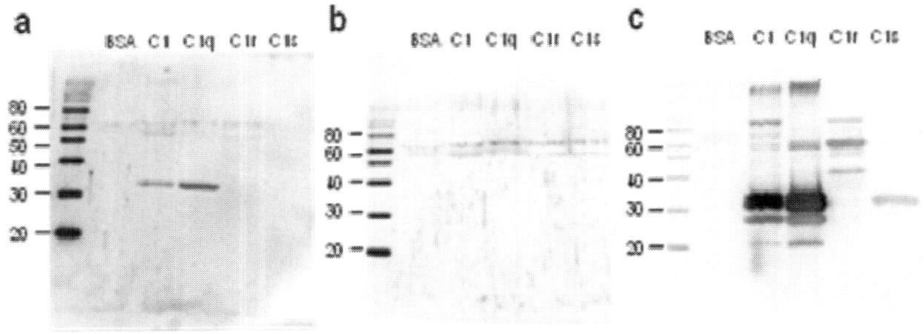

Fig. 4 HAstV-1 CP binds to the A-chain of C1q. (**a–c**) The indicated proteins were boiled, reduced, resolved on 12% SDS-PAGE gels and transferred to nitrocellulose. The blots were blocked with one blot subsequently receiving CP probe for 1 h (**a**) while the other did not (**b**). Blots were then washed and probed with antibody to HAstV-1 particles, followed by secondary antibody for detection. (**c**) Both overlay blots were stripped and reprobed with antibody to C1q, C1r and C1s. Only one reprobed blot is shown as both blots yielded identical results. The molecular weight markers (in kDa) are indicated to the left

(Fig. 4a). As expected, no signal was detected in the duplicate blot that did not receive the CP probe (Fig. 4b). The blots were stripped and reprobed with antisera that detects C1q, C1r and C1s to reveal the individual protein constituents of C1 (Fig. 4c). C1q is composed of six subunits, each of which contains three polypeptide chains A, B, and C (see Fig. 1b). As seen in the C1 and C1q lanes, all 3 C1q chains react with C1q antisera and under reducing conditions chain C runs at 27.5 kDa, chain B runs at 31.6 kDa and chain A runs at 34.8 kDa (Reid et al. 1972). CP was found to overlay precisely with the 34 kDa band corresponding to the A-chain of C1q in both the C1 and C1q lanes. A band detected by CP at approximately 59 kDa seen in the non-reduced C1 lanes (data not shown) corresponds with the predicted size of disulphide-bonded A-B C1q dimers. While the significance of CP binding the A-chain of C1q is unknown, it is of interest to note that the A chain has been demonstrated to preferentially bind a number of non-immunoglobulin substances such as C-reactive protein, serum amyloid P, LPS and DNA (Trinder et al. 1993).

5.3 HAstV CP Specifically Targets the C1 Complex

Based upon the overlay blot data, CP appears to interact with the A-chain of C1q. If C1 is interacting with CP, then additional exogenous C1 should reconstitute hemolytic complement activity. Thus, CP was added to NHS to suppress hemolysis by approximately half (Fig. 5), then exogenous purified C1 was added fully restoring complement-mediated lysis from 56 to 97% ($P = 0.0286$).

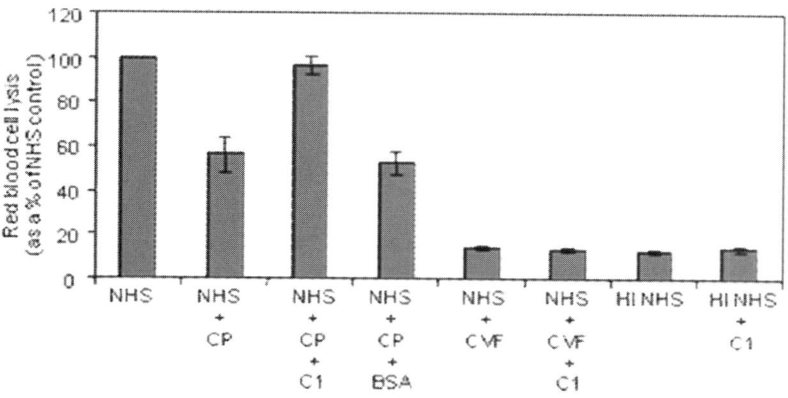

Fig. 5 Exogenous C1 reconstitutes hemolytic activity for CP-treated NHS. NHS was incubated alone, with 6.3 μg CP or 1 μg CVF for 1 h at 37°C. Heat inactivated NHS (HI-NHS) was used as an additional control. After the incubation, 2 μg of C1 or 10 μg BSA was added to the indicated samples. Sensitized RBCs were then added to all samples and hemolysis was determined. Data are the means of four independent experiments. Error bars denote SEM

Reconstitution of hemolytic activity did not occur when BSA was substituted for C1, nor when additional C1 was added to CVF-treated NHS or to heat-inactivated NHS (HI-NHS). These findings suggest that CP inhibits complement activation via C1 and the inhibition of complement activation can be overcome with exogenous C1. The reconstitution of complement activation in the presence of CP by the addition of C1 was confirmed by measuring the deposition of C3 on zymosan (data not shown).

5.4 HAstV CP Suppresses Complement Activation via an Inhibitory Mechanism

While HAstV CP was found to suppress complement in hemolytic assays and specifically target the C1 complex, it was unclear as to whether this effect was due to activation and depletion of serum complement components, as occurs with CVF, or the result of inhibition of activation. To test these competing hypotheses, we investigated whether CP suppressed the complement system at the level of C4, the second component of the classical pathway after C1. Upon activation of the C1 complex, C4 is cleaved then C2 is cleaved to form the classical pathway C3-convertase (Volanakis 1998). An ELISA that detects a specific by-product of C4 activation (C4d) demonstrated that in the presence of CP, serum generates very low levels of C4d in contrast to NHS alone at room temperature ($P = 0.0286$) (Fig. 6), suggesting inhibition of spontaneous classical pathway activation. Heat-aggregated

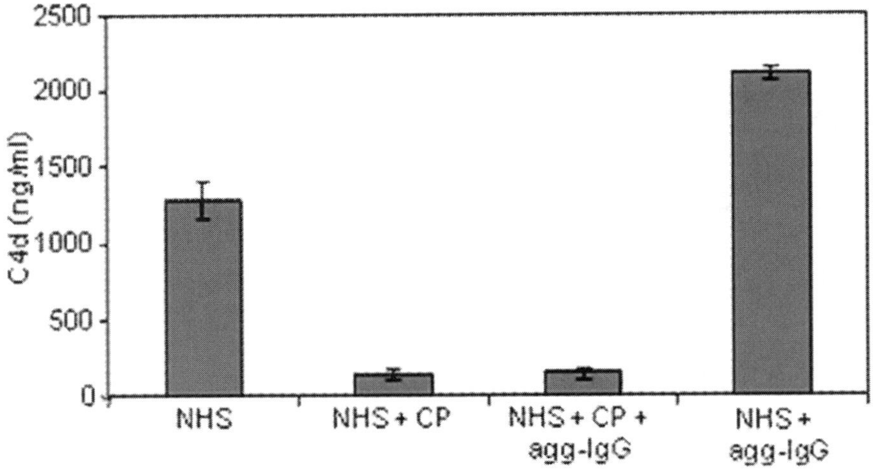

Fig. 6 HAstV-1 CP inhibits C4d formation. NHS was incubated for 1 h in the presence of heat-aggregated IgG (agg-IgG, a classical pathway activator) or CP, or both and measured for C4 cleavage by C4d ELISA. Standard curves were generated using purified C4d. Data are the means of 4 independent experiments for each ELISA. Error bars denote SEM

IgG, a potent activator of the classical pathway, greatly increased C4d generation. When CP was added simultaneously with heat-aggregated IgG to NHS, C4d formation was greatly inhibited ($P = 0.0286$) (Fig. 6). These results show that HAstV-1 CP can inhibit antibody-mediated classical pathway activation and spontaneous activation before C4 cleavage can occur. CP-treated NHS sera also inhibited the formation of iC3b and terminal complement cascade activation (SC5b-9) as assayed by ELISA (data not shown), suggesting inhibition of complement-mediated inflammatory effector functions. These results, demonstrate that CP is a powerful classical pathway inhibitor at C1, the inhibition suppresses complement-mediated eukaryotic membrane damage (hemolysis), and the suppression of complement activation by CP may be overcome with exogenous C1.

5.5 Hypothetical Mechanism of C1 Inhibition by HAstV CP

Although HAstV coat protein potently inhibits classical pathway activation at C1, the means by which this inhibition occurs is unknown. Unlike the C1 inhibitors described in the literature (see Sect. 3), HAstV CP does not share any apparent sequence homology with these proteins and peptides. We hypothesize that CP may inhibit activation by one of the following two methods: (A) CP binds to C1q dissociating some or all of the $C1r_2C1s_2$ complex from C1q or (B) CP binds to C1q functionally altering the conformation of C1 and inhibiting the activation of C1r and/or C1s. C1-INH acts via the first mechanism binding to the C1 complex and forming an extremely stable association with $C1r_2C1s_2$ which subsequently dissociates from C1q. We speculate that HAstV CP may not employ this mechanism given that it appears to bind C1q and not the $C1r_2C1s_2$ complex. Like HAstV CP, decorin and human neutrophil peptide-1 both bind C1q, but the mechanism of inhibition has not been published for either of these molecules. The mechanism of CP interaction with C1 is currently under investigation.

6 Human Astrovirus Coat Protein: Potential as a Therapeutic for Complement-Mediated Diseases

The complement system plays a critical role in immunoprotective and immunoregulatory functions. However, disregulated control of complement activation can result in severe autoimmune and inflammatory host tissue damage. Activation of the classical pathway of complement through autoantibodies and immune complexes has been implicated in autoimmune and inflammation-mediated diseases such as systemic lupus erythematosus, rheumatoid arthritis, ischemia-reperfusion injury, myasthenia gravis, Alzheimer's disease and hyperacute xenograft rejection. Currently there are a number of anti-complement therapeutics in pre-clinical development or in clinical trials, but only one is on the market, the humanized monoclonal antibody against C5a (Eculizumab/Soliris) (please see Ricklin and Lambris 2007 for an excellent overview of complement therapeutics).

In addition to its use as a therapeutic for the treatment of hereditary angioedema, C1-INH is in consideration for clinical trials targeting the prevention of ischemia-reperfusion injury after myocardial infarction (Ricklin and Lambris 2007). Specific inhibition of the classical pathway by C1q binding peptides has also been explored (Lauvrak et al. 1997; Roos et al. 2001), but as of yet none have progressed to clinical trials. Given the specificity of HAstV CP for C1 and its ability to potently inhibit antibody-mediated and spontaneous activation of downstream effector function (e.g., eukaryotic membrane perforation), we propose that HAstV CP may have potential as an anti-complement therapeutic for autoimmune and inflammatory conditions in which the classical pathway plays a major role. The specificity of HAstV CP for C1, its limited effects on alternative pathway function and reversibility of suppression are desirable traits suggesting the potential to inhibit complement-mediated host tissue damage without causing profound irreversible complement suppression, a state associated with frequent severe bacterial infections.

7 Conclusions

While viruses and bacteria that modulate or inhibit the human complement system have been described previously, to our knowledge this is the first example of a non-enveloped icosahedral virus inhibiting complement activation and the first description of a viral C1 inhibitor. Further defining the interaction of the astrovirus coat protein with the complement system will provide significant insights not only into the pathogenesis of this virus family but may well offer additional details into the regulation of C1. In addition, the potent suppression of the classical pathway and terminal effector functions displayed by astrovirus coat protein could find application as a novel anti-complement therapeutic for any of the myriad diseases mediated by disregulated complement activation.

Acknowledgments

This work was supported in part by National Institute of Health grant R21 AI060874 to NKK and by a grant from The Children's Hospital of The King's Daughters Research Endowment to KMC.

References

Bass DM, Qiu S (2000) Proteolytic processing of the astrovirus capsid. *J Virol* 74:1810–1814

Bernet J, Mullick J (2003) Viral mimicry of the complement system. *J Biosci* 28:249–264

Bonaparte RS, Hair PS et al. (2008) Human astrovirus coat protein inhibits serum complement activation via C1, the first component of the classical pathway. *J Virol* 82:817–827

Borrow P, Oldstone MBA (1992) Characterization of lymphocytic choriomeningitis virus-binding protein(s): a candidate cellular receptor for the virus. *J Virol* 66:7270–7281
Botto M, Walport MJ (2002) C1q, autoimmunity and apoptosis. *Immunobiology* 205:395–406
Cooper NR (1983) Activation and regulation of the first complement component. *Federation Proc* 42:134–138
Cooper NR (1985) The classical complement pathway: activation and regulation of the first complement component. *Adv Immunol* 37:151–216
Davis AE III (1998) C1-INH and hereditary angioedema. In: *The Human Complement System in Health and Disease*, 1st edn, Volanakis JE, Frank MM (eds.). Marcel Dekker Inc, New York
Davis AE III (2005) The pathophysiology of hereditary angioedema. *Clin Immunol* 114:3–9
Dennehy PH, Nelson SM et al. (2001) A prospective case-control study of the role of astrovirus in acute diarrhea among hospitalized young children. *J Infect Dis* 184:10–15
Favoreel HW, wan de Walle GR et al. (2003) Virus complement evasion strategies. *J Gen Virol* 84:1–15
Gaboriaud C, Thielens NM et al. (2004) Structure and activation of the C1 complex of complement: unraveling the puzzle. *Trends Immunol* 25:368–373
Ghebrehiwet B, Lim BL et al. (1994) Isolation, cDNA cloning, and overexpression of a 33-kD cell surface glycoprotein that binds to the globular "heads" of C1q. *J Exp Med* 179:1809–1821
Ghebrehiwet B, Peerschke EIB (2004) cC1q-R (calreticulin) and gC1q-R/p33: ubiquitously expressed multi-ligand binding cellular proteins involved in inflammation and infection. *Mol Immunol* 41:173–183
Gigli I, Mason JW et al. (1970) Interaction of plasma kallikrein with the C1 inhibitor. *J Immunol* 104:574–581
Groeneveld TWL, Oroszlan M et al. (2005) Interaction of the extracellular matrix proteoglycans decorin and biglycan with C1q and collectins. *J Immunol* 175:4715–4723
Groeneveld TWL, Ramwadhdoebé TH et al. (2007) Human neturophil peptide-1 inhibits both the classical and the lectin pathway of complement activation. *Mol Immunol* 44:3608–3614
Jiang B, Monroe SS et al. (1993) RNA sequence of astrovirus: Distinctive genomic organization and a putative retrovirus-like ribosomal frameshifting signal that directs the viral replicase synthesis. *Proc Natl Acad Sci U S A* 90:10539–10543
Jiang J, Wagner E et al. (2001) Complement 1 inhibitor is a regulator of the alternative pathway. *J Exp Med* 194:1609–1616
Johnston RB (1986) The complement system: physiology, disorders and activity in the newborn infant. *Mead Johnson Symp Perinat Dev Med* 24:7–12
Kapikian AZ, Hoshino Y et al. (2001) Rotaviruses. In: *Fields Virology*, 4th edn, Knipe DM, Howley PM et al. (eds). Lippincott, Williams & Wilkins, Philadelphia
Kishore U, Ghai R et al. (2004) Structural and functional anatomy of the globular domain of complement protein C1q. *Immunol Lett* 85:113–128
Klein MA, Kaeser PS (2001) Complement facilitates early prion pathogenesis. *Nat Med* 7:410–411
Koci MD, Moser LA et al. (2003) Astrovirus induces diarrhea in the absence of inflammation and cell death. *J Virol* 77:11798–11808
Kovacs H, Campbell ID et al. (1998) Evidence that C1q binds specifically to C_H2-like immunoglobulin gamma motifs present in the autoantigen calreticulin and interferes with complement activation. *Biochemistry* 37:17865–17874

Krumdieck R, Hook M et al. (1992) The proteoglycan decorin binds C1q and inhibits the activity of the C1 complex. *J Immunol* 149:3695–3701
Lauvrak V, Brekke OH et al. (1997) Identification and characterisation of C1q-binding phage displayed peptides. *Biol Chem* 378:1509–1519
Levy L, Lepow I (1959) Assay and properties of serum inhibitor of C'1 esterase. *Proc Soc Exp Biol Med* 101:608–611
Lewis TL, Greenberg HB et al. (1994) Analysis of astrovirus serotype 1 RNA, identification of the viral RNA-dependent RNA polymerase motif, and expression of a viral structural protein. *J Virol* 68:77–83
Mabbott NA, Bruce ME (2001) Temporary depletion of complement component C3 or genetic deficiency of C1q significantly delays onset of scrapie. *Nat Med* 7:485–487
Matsui SM, Greenberg HB (2001) Astroviruses. In: *Fields Virology*, 4th edn, Knipe DM, Howley PM et al. (eds.). Lippincott, Williams & Wilkins, Philadelphia
Matsui SM, Kiang D et al. (2001) Molecular biology of astroviruses: selected highlights. *Novartis Found Symp* 238:219–236
Matsushita MS, Thiel S et al. (2000) Proteolytic activities of two types of mannose-binding lectin-associated serine protease. *J Immunol* 165:2637–2642
Méndez E, Fernández-Luna T et al. (2002) Proteolytic processing of a serotype 8 human astrovirus ORF2 polyprotein. *J Virol* 76:7996–8002
Moser LA, Schultz-Cherry S (2005) Pathogenesis of astrovirus infection. *Viral Immunol* 18:4–10
Poon PH, Schumaker VN et al. (1983) Conformation and restricted segmental flexibility of C1, the first component of human complement. *J Mol Biol* 168:563–577
Reid KBM, Lowe DM et al. (1972) Isolation and characterization of C1q, a subcomponent of the first component of complement. *Biochem J* 130:749–763
Ricklin D, Lambris JD (2007) Complement-targeted therapeutics. *Nat Biotech* 25:1265–1275
Risco C, Carrascosa JL et al. (1995) Ultrastructure of human astrovirus serotype 2. *J Gen Virol* 76:2075–2080
Rogers J, Cooper NR et al. (1992) Complement activation by beta-amyloid in Alzheimer disease. *Proc Natl Acad Sci U S A* 89:10016–10020
Rooijakkers SH, van Strijp JA (2007) Bacterial complement evasion. *Mol Immunol* 44:23–32
Roos A, Nauta AJ et al. (2001) Specific inhibition of the classical complement pathway by C1q-binding peptides. *J Immunol* 167:7052–7059
Sebire NJ, Malone M et al. (2004) Pathology of astrovirus associated diarrhoea in a paediatric bone marrow transplant recipient. *J Clin Pathol* 57:1001–1003
Shastri S, Doane AM et al. (1998) Prevalence of astroviruses in a children's hospital. *J Clin Microbiol* 36:2571–2574
Sim RB, Arlaud G et al. (1979) C1-inhibitor-dependent dissociation of human complement component C1 bound to immune complexes. *Biochem J* 179:449–457
Sjoberg A, Onnerfjord P et al. (2005) The extracellular matrix and inflammation: fibromodulin activates the classical pathway of complement by directly binding C1q. *J Biol Chem* 280:32301–32308
Tacnet-Delorme, Chevallier S et al. (2001) Beta-amyloid fibrils activate the C1 complex of complement under physiological conditions. Evidence for a binding site for Ab on the C1q globular regions. *J Immunol* 167:6374–6381
Trinder PKE, Maeurer MJ et al. (1993) Functional domains of the human C1q A-chain. *Behring Inst Mitt* 93:180–188

van den Berg RH, Faber-Krol MC et al. (1996) Inhibition of hemolytic activity of the first component of complement C1 by an *Escherichia coli* C1q binding protein. *J Immunol* 156:4466–4473

van den Berg RH, Faber-Krol MC et al. (1998) Inhibition of activation of the classical pathway of complement by human neutrophil defensins. *Blood* 92:3898–3903

Vogel CW, Smith CA et al. (1984) Cobra venom factor: structural homology with the third component of complement. *J Immunol* 133:3235–3241

Volanakis JE (1998) Overview of the complement system. In: *The Human Complement System in Health and Disease*, 1st ed, Volanakis JE, Frank MM (eds.). Marcel Dekker Inc, New York

Willcocks MM, Brown TDK et al. (1994) The complete sequence of a human astrovirus. *J Gen Virol* 75:1785–1788

Willcocks MM, Carter MJ (1993) Identification and sequence determination of the capsid protein gene of human astrovirus serotype 1. *FEMS Microbiol Lett* 114:1–8

Ziccardi RJ, Cooper N (1979) Active disassembly of the first complement component C1 by C1 inactivator. *J Immunol* 123:788–792

18. Hypothesis: Combined Inhibition of Complement and CD14 as Treatment Regimen to Attenuate the Inflammatory Response

Tom Eirik Mollnes[1,2], Dorte Christiansen[3], Ole-Lars Brekke[2], and Terje Espevik[4]

[1]Institute of Immunology, University of Oslo, and Rikshospitalet University Hospital, Oslo, Norway
[2] Department of Laboratory Medicine, Nordland Hospital Bodo, and Institute of Medical Biology, University of Tromsø, Tromsø, Norway
[3]Department of Laboratory Medicine, Nordland Hospital, Bodø, Norway
[4]Department of Cancer Research and Molecular Medicine, Norwegian University of Science and Technology, Trondheim, Norway

Abstract. Pattern recognition is an essential event in innate immunity. Complement and Toll-like receptors (TLR), including the CD14 molecule, are two important upstream components of the innate immune system, recognizing exogenous structures as well as endogenous ligands. They act partly independent in the inflammatory network, but also have several cross-talk mechanisms which are under current investigation. Complement is an essential part of innate immunity protecting the host against infection. However, it is a double-edged sword since inappropriate activation may damage the host. Uncontrolled systemic activation of complement, as seen in severe sepsis, may contribute to the breakdown of homeostatic mechanisms leading to the irreversible state of septic shock. Complement inhibition is promising for protection of lethal experimental sepsis, but clinical studies are missing. Lipopolysaccharide (LPS) has been implicated in the pathogenesis of gram-negative sepsis by inducing synthesis of pro-inflammatory cytokines through binding to CD14 and the TLR4/MD-2 complex. Neutralization of LPS or blocking of CD14 has been effective in preventing LPS-induced lethal shock in animal studies, but results from clinical studies have been disappointing, as for most other therapeutic strategies. Based on some recently published data and further pilot data obtained in our laboratory, we hypothesize that inhibition of complement combined with neutralization of CD14 may attenuate the uncontrolled inflammatory reaction which leads to breakdown of homeostasis during sepsis. We further postulate this regimen as an approach for efficient inhibition of the initial innate recognition, exogenous as well as endogenous, to prevent downstream activation of the inflammatory reaction in general.

1 Introduction

The complement system and Toll-like receptors (TLRs) are two important upstream components of the innate immune system, recognizing exogenous and endogenous ligands initiating a complex downstream activation of the inflammatory

J.D. Lambris (ed.), *Current Topics in Complement II.*

doi: 10:1007/978-0-387-78952-1_18, © Springer Science+Business Media, LLC 2008

network. The two systems act partly independently, but several recent studies indicate a cross-talk between these major branches of pattern recognition. LPS-binding to CD14 induced up-regulation of the complement receptor (CR) 3 (Weingarten et al. 1993) and initiated a complex formation between CD14 and CR3 (Zarewych et al. 1996). A cross-talk between CD14 and CR3 was documented to enhance phagocytosis of mycobacteria (Sendide et al. 2005) and the sequential recognition by CR3 and TLR4 was essential for uptake and killing of Salmonella (van Bruggen et al. 2007). Complement activation may induce up-regulation of CD14 (Marchant et al. 1996) and the C1q component was found to modulate LPS/TLR-induced cytokine production (Yamada et al. 2004). Recently, a comprehensive in vivo study in mice documented a substantial role for complement activation, i.e. activation of the C3a and C5a receptors, in the enhancement of TLR-induced proinflammatory cytokine production (Zhang et al. 2007). Finally, both complement and TLRs are regulators of adaptive immunity and the interaction between these systems were recently reviewed (Hawlisch and Kohl 2006; Kohl 2006).

The cross-talk between complement and TLR may imply that intervention of one of the systems may influence the other, either by enhancing or attenuating their effects, depending on the experimental or pathophysiological circumstances. Thus, synergistic effects may be observed both during activation and inhibition. On the other hand, these two systems are definitely acting as upstream and partly independent branches of pattern recognition. They are part of the redundancy in host defence; i.e. if one of the systems do not function sufficiently the other part will compensate for this. Thus, surprisingly many knock-out mice are doing quite well. The inflammatory reaction induced by a foreign structure or an altered self-structure is complex and includes innumerable secondary mediator systems, like cytokines, chemokines, growth factors, reactive oxygen metabolites, arachidonic acid metabolites, proteolytic enzymes and matrix metalloproteinases to mention some of them. Notably, these mediators need to get a signal to be released or synthesized. These signals are dependent on upstream activation of a system which can recognize a pattern or danger signal, of which complement and TLRs are probably the two most important, not to ignore the scavenger receptor systems. This is important to bring in mind, since in particular complement is frequently and misleadingly listed among secondary inflammatory systems.

When talking about "the inflammatory network", both the upstream complement and TLRs, and the downstream systems are included. The primary and secondary response mechanisms further cross-talk and positive and negative feed-back loops are innumerable. For example, complement activation with release of C5a induces IL-6, while IL-6 then increases C5aR expression (Riedemann et al. 2003a,b). Therefore, in order to study the interaction of the inflammatory actors, it is essential to have models where all of the candidates are present and are mutually able to cross-talk. Animal models are crucial to reveal such cross-talk. However, to approach a human pathophysiological condition we have to use human material. We realized that whole blood models were in use, but with anticoagulants that interfere with a number of the inflammatory systems, e.g. the calcium chelators EDTA and citrate, which interfere with most biological reaction, or heparin, which has numerous adverse effects on leukocytes and platelets, and therefore excluded physiological relevant cross-talk. Thus, we developed a human whole blood model using a highly specific thrombin inhibitor, lepirudin, which is the recombinant analogue of hirudin. In this model we showed that, except for thrombin inhibition, which is a prerequisite to work with whole blood in vitro, the inflammatory systems including complement were free to cross-talk (Mollnes et al. 2002a). Using this model we have been able to study the role of complement and TLRs – in particular the CD14 molecule – in the induction of secondary inflammatory mediators. This has made a platform for investigation of the effect of inhibiting the two main upstream systems in innate immunity.

2 Complement

Complement, as an essential component of the immune system, is of substantial relevance for the destruction of invading micro-organisms and for maintaining tissue homeostasis including the protection against autoimmune diseases (Walport 2001a,b). However, excessive or uncontrolled complement activation significantly contributes to undesired tissue damage. Following complement activation, biologically active peptides, such as C5a and C3a elicit a number of pro-inflammatory effects, including the recruitment of leukocytes, degranulation of phagocytic cells, mast cells and basophils, smooth muscle contraction, and increase of vascular permeability. Upon complement-dependent cell activation, the inflamematory response is further amplified by subsequent generation of toxic oxygen radicals and the induction of synthesis and release of arachidonic acid metabolites and cytokines. Consequently, complement activation presents a considerable risk of harming the host by directly and indirectly mediating inflammatory tissue destruction.

Under physiological conditions, activation of complement is effectively controlled by the coordinated action of soluble as well as membrane-associated regulatory proteins. Soluble complement regulators, such as C1 inhibitor, anaphylatoxin inhibitor (serum carboxypeptidase N), C4b binding protein (C4BP), factors H and I, clusterin and S-protein (vitronectin), restrict the action of complement in body fluids at multiple sites of the cascade reaction. In addition, each individual cell is protected against the attack of homologous complement by surface proteins, such as the complement receptor 1 (CR1, CD35), the membrane cofactor protein (MCP, CD46) as well as by the glycosylphosphatidylinositol (GPI)-anchored proteins, decay-accelerating factor (DAF, CD55) and CD59. When complement is improperly activated, these regulatory mechanisms may be overwhelmed, resulting in tissue destruction and disease.

Clinical and experimental evidence underlines the prominent role of complement in the pathogenesis of numerous inflammatory diseases. The list of such conditions is rapidly growing, including immune complex diseases such as rheumatoid arthritis and systemic lupus erythematosus, ischemia-reperfusion (I/R) injury locally manifested as infarctions or systemically as a post-ischemic inflammatory syndrome, systemic inflammatory response syndrome (SIRS) and acute respiratory distress syndrome (ARDS), septic shock, trauma, burns, acid aspiration to the lungs, renal diseases, inflammatory and degenerative diseases in the nervous system, arteriosclerosis, transplant rejection and inflammatory complications seen after cardiopulmonary bypass and haemodialysis. In principle, when inflammation is involved in the pathogenesis, complement has to be considered as a possible mediator in the disease process and the list of conditions where complement is not involved may be very limited (Mollnes et al. 2002b)

In recent years, great progress has been made in complement analysis to better define disease severity, evolution and response to therapy. Modern diagnostic technologies which focus on the quantification of complement activation products now provide a comprehensive insight into the activation state of the system (Mollnes et al. 2007). Thus, to focus on complement inhibition appears to be a logical approach in order to arrest the process of inflammatory disorders.

3 CD14 and Toll-Like Receptors

3.1 Toll-Like Receptors are Essential Membrane Receptors in the Inflammatory Response

Since 1985 it has been a substantial improvement in our understanding of a class of pattern recognition receptors called TLRs. Most of the research on TLRs has been linked to antimicrobial defence mechanisms. TLRs are type I membrane proteins that recognize molecules most frequently found in microbes. The horseshoe shaped ectodomain consists of leucin rich repeats responsible for ligand binding, either directly or indirectly by binding to co-receptors. The crystal structure for the TLR3 ectodomain has been determined and shows extensive glycosylation, except at one face that is thought to mediate ligand binding and dimerization (Choe et al. 2005). Recently, the crystal structure of TLR4-MD-2 (myeloid differentiation-2) has also been reported (Kim et al. 2007a). TLR1/2 and 2/6 heterodimers recognize lipoproteins, peptidoglycans and glycans from bacteria, fungi and some parasites.

TLR4 is the signalling receptor for LPS, however, three additional proteins are necessary for induction of the response. LPS-binding protein (LBP) binds the surface of Gram-negative bacteria and extracts LPS from the outer membrane and delivers monomers to CD14. The CD14 molecule is either a soluble protein in microgram/ml quantities in human serum (Lien et al. 1998) or a GPI anchored protein and both forms enhance cellular responses to LPS (Frey et al. 1992; Latz et al. 2002). The third protein that is necessary for TLR4 activation by LPS is MD-2. TLR4 interacts with MD-2 and the heterodimer is the minimal recognition site for LPS that results in a cellular response. MD-2 is a soluble protein that binds LPS with nM affinity and CD14 has an important role in transferring LPS to MD-2. The crystal structure of MD-2 shows that the protein has a deep hydrophobic cavity in which the acyl chains of LPS are embedded (Ohto et al. 2007). Upon LPS binding, it is likely that TLR4-MD-2 dimerizes and then starts the signalling cascade (Kim et al. 2007a). TLR4 signalling in human cells by LPS is potently inhibited by lipid IVa. Crystal structures of MD-2 and lipid IVa suggests that the four acyl chains of the antagonist occupy almost all available space in the MD-2 pocket and thereby blocks the binding of LPS to MD-2 (Kim et al. 2007a; Ohto et al. 2007).

3.2 TLR Signalling

TLR2 and TLR4 meet their ligands on the plasma membrane and signalling occurs along the endocytic pathway from the membrane to the early endosome (Husebye et al. 2006). The TLR3, 7, 8, and 9, are located in the endoplasmic reticulum and are transported to endosomes where they recognize nucleic acids (Latz et al. 2004; Johnsen et al. 2006). The TLRs contain a cytoplasmic domain homologues to the IL-1/IL-18 receptor that has been termed the Toll/Interleukin-1 receptor (TIR) domain. The TIR domain is also present on the cytosolic TIR adaptors (MyD88, MAL, TRAM, and TRIF), that are recruited to the TLR TIR domains and mediate signalling (for review see (Uematsu and Akira 2007)). All TLRs, except TLR3, signal through the MyD88 adaptor pathway, leading to activation of the transcription factor NFκB. The MyD88-dependent pathway proceeds through phosphorylation of interleukin-1 receptor-associated kinases (IRAK1, 2 and 4), with subsequent activation of TRAF6, phosphorylation of IKK, and release and nuclear translocation of NFκB. In parallel, the mitogen-activated protein kinases (MAPK) mediate activation of activator protein 1 (AP-1). NFκB then promotes production of proinflammatory cytokines and microbicidal proteins, modulation of the phagocytic-, endocytic- and exocytic machinery, motility and homing preferences. Activation of TLR7 or TLR9 in the endosomes may in addition result in interactions of interferon regulatory factor (IRF) 7 with MyD88, which results in production of antiviral type I IFNs. The TRIF adaptor is more restricted than MyD88 and signals from TLR4 and TLR3 leading to activation of IRFs, and production of type I interferons. In addition, non-transcriptional activation programs are initiated from both, such as phosphoinositide kinase activity, which affects endocytosis, phagocytosis, trafficking, motility and antigen presentation. The MyD88-independent pathway from TLR3 and 4 proceeds through activation of TRIF, c-Src, TBK1/IKKε and the transcription factors IRF-3 and STAT-1 (Johnsen et al. 2006; Uematsu and Akira 2007) to yield IFNβ. This activation is negatively regulated by a number of proteins.

4 Rational for Combined Complement and CD14 Inhibition

4.1 *Escherichia coli* Bacteria

Given that complement and TLRs, with CD14 as an essential molecule, are the most important upstream recognition systems and that they to some extent cross-talk, we used the whole blood model described above to test the hypothesis that a

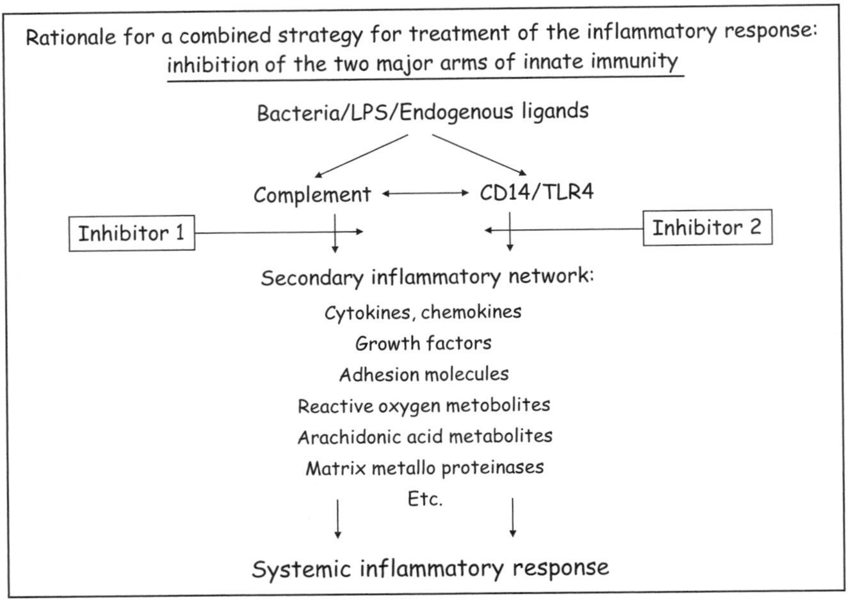

Fig. 1 Simplified illustration of complement- and CD14/TLR4 activation in sepsis. Whole Gram-negative bacteria activate both complement (various surface structures) and CD14/TLR4 (mainly LPS). Complement and CD14/TLR4 acts upstream, partly independently and partly dependently, to induce the secondary inflammatory responses. They comprise the two single and quantitatively most important inflammatory triggers of systemic inflammation in gram-negative sepsis and in many non-infectious conditions. Thus, in order to attenuate the systemic response, both systems have to be inhibited independently

combined inhibition of complement and CD14 could reduce the inflammatory response induced by *Escherichia coli.* Inhibition of CD14 with appropriate antibodies may stop the transfer of LPS to MD-2 and inhibition of TLR4 signalling. Surprisingly we found that the combination completely abrogated the up-regulation of CD11b, phagocytosis and oxidative burst, whereas each of the inhibitors only partially or marginally inhibited the response (Brekke et al. 2007). The individual effect was different on granulocytes and monocytes, but consistently the combined treatment abrogated the response in both cell types. This was the first observation of the very efficient effect of the combined treatment, but it was limited to CD11b-upregulation and the subsequent phagocytosis and burst. We therefore extended the studies to include a broad panel of cytokines, chemokines and growth factors and found that approximately 15 of these biomarkers which increased markedly by addition of *Escherichia coli* were virtually completely abolished by the combined therapy, despite the fact that each of them in many cases only modestly decreased the signal (Brekke et al. 2008).

4.2 Endotoxin (LPS)

LPS-models are frequently used to simulate sepsis. In our opinion, LPS is a very useful agent, in particular when ultrapure phenol extracted LPS is used to study the specific interaction with TLR4. When it comes to Gram-negative sepsis we argue that these modes are of very limited relevance. In this context we would like to emphasize that LPS as such is a poor complement activator compared with other bacterial membrane constituents of Gram-negative bacteria. Thus, we have shown that LPS is of limited importance for activation of complement by *Neisseria meninigitidis* (Sprong et al. 2004) and *Escherichia coli* (Brekke et al. 2007). Furthermore, we documented that purified LPS needed logs higher concentration for activation of complement in the fluid-phase, compared with whole *Escherichia coli* bacteria measured in LPS-equivalent amounts (Brekke et al. 2007). Thus, a combined inhibition of complement and CD14 in a pure LPS setting is not rational, but a selective inhibition of CD14/TLR4/MD2 would completely inhibit this response.

4.3 Gram Negative Sepsis

The prognosis of septic shock has hardly improved after the introduction of antibiotics 60–70 years ago, when the lethality decreased from close to 100% to approximately 50%. Innumerable of treatment strategies have been tested experimentally and a number of inflammatory inhibitors were found promising, including those neutralizing LPS, TNF-α, IL-1 and IL-6. However, when tested clinically, all failed. Experimental models of intestinal bacterial sepsis using complement inhibitors (Ward 2004) or TLR4/MD-2 inhibitors (Daubeuf et al. 2007) significantly improved survival. Neither of these therapeutic strategies has been tested in controlled clinical trials. Probably would each of these strategies not be sufficiently efficient. Based on our concept that complement and CD14/TLRs are the two main upstream inducers of the inflammatory reaction, we propose a strategy of combined inhibition of these systems in conditions where there is a risk of development of SIRS or sepsis. Notably, no treatment of sepsis will have success if the patient has come to “the point of no return” with complete breakdown of homeostasis. Specific treatment for SIRS and sepsis, as suggested by our regimen of inhibition of the main recognition systems, should be focused on the patients at risk and those who are about to develop the disease.

4.4 General Principles for the Inflammatory Reaction: Role of Endogenous Ligands

Activation of complement and TLRs by pattern recognition is not limited to microbial structures, but also occurs by a number of endogenous substances. Ischemia reperfusion injury activates complement through several mechanisms, including naturally occurring antibodies reacting with structures exposed through a damaged endothelium (Zhang and Carroll 2007) or through lectin pathway recognition by MBL (Collard et al. 2000). Inhibition of complement has in a number of experimental studies reduced the ischemia-reperfusion injury, e.g. in myocardial infarction (Mollnes and Kirschfink 2006). Similarly, a reduced mycocardial infarct size were observed in TLR4 deficient mice (Kim et al. 2007b; Chong et al. 2004) and in mice treated with the TLR4 inhibitor eritoran (Shimamoto et al. 2006). Furthermore, TLR4 was found to be essential for the ischemia-reperfusion injury during experimental heart transplantation (Kaczorowski et al. 2007). Thus, it might be hypothesised that a combined inhibition of complement and TLR4 will have a beneficial effect on ischemia-reperfusion injury compared with blocking each of them.

Meconium, the first stool dispatched from the infant bowel after delivery, is an interesting "endogenous" substance. It is composed of a number of various host agents and is normally sterile before delivery. It is, however, extracorporeal since it is located in intra-intestinally. If meconium is aspired to the lungs during delivery, a serious condition called meconium aspiration syndrome may develop. It is characterized by a local pulmonary inflammation which may lead to development of a systemic inflammatory reaction. Meconium is a potent activator of complement in vitro (Castellheim et al. 2004) and in vivo (Lindenskov et al. 2004), the latter associated with a systemic inflammatory response (Castellheim et al. 2005). Recently, we hypothesised that meconium-induced cytokine activation could be induced through a CD14-dependent mechanism and contribute together with complement to the inflammatory response. This was indeed showed in a recent study where we examined the effect of inhibition of complement and CD14 and a combination thereof, on meconium-induced cytokine response (Salvesen et al. 2008). Human adult and cord blood was incubated with meconium and the combined inhibition of complement and CD14 virtually abolished all of the cytokines, chemokines and growth factors which were produced. Each of the treatments differentially reduced the production to various degrees, but consistently an enhanced and very efficient reduction was obtained by the combination of them. The LPS content in the meconium could not explain the CD14-induced cytokines and it is tempting to speculate that endogenous CD14 ligands contributed to this effect. Recent data indicate that the effect is mediated via TLR4 (Espevik et al. unpublished data).

The main exogenous ligand for CD14/MD2/TLR4 complex is LPS. Interestingly, a number of endogenous ligands have recently been described (Gay and Gangloff 2007), including heparan sulphate (Tang et al. 2007), which is released from activated endothelial cells in conditions like SIRS and sepsis. Evolving evidence suggest that "pattern recognition" might be equally important to maintain endogenous tissue homeostasis as to protect the host against external foreigners.

5 Conclusion

Complement and TLRs are the two main upstream innate immune systems recognizing "foreign" patterns. These are either pathogen associated or altered self structures. For the host these danger signals are potentially lethal since they may induce a systemic inflammatory response incompatible with life, in particular in the case of sepsis. If we assume that the bacteria can be killed by appropriate antibiotics, inhibition of the overwhelming inflammatory response during sepsis is the main remaining problem. We hypothesize that a main goal for the host is to inhibit this response upstream and that inhibition of complement and TLRs, in particularly the CD14 molecule, might currently be the best approach to reach the goal.

References

Brekke, O.L., Christiansen, D., Fure, H., Fung, M., and Mollnes, T.E. (2007) The role of complement C3 opsonization, C5a receptor, and CD14 in *E. coli*-induced up-regulation of granulocyte and monocyte CD11b/CD18 (CR3), phagocytosis, and oxidative burst in human whole blood. *J Leukoc Biol* 81, 1404–1413

Brekke, O.L., Christiansen, D., Fure, H., Pharo, A., Fung, M., Riesenfeld, J., and Mollnes, T.E. (2008) Combined inhibition of complement and CD14 abolish *E. coli*-induced cytokine-, chemokine- and growth factor-synthesis in human whole blood. *Mol Immunol* 45, 3804–3813

Castellheim, A., Lindenskov, P.H., Pharo, A., Fung, M., Saugstad, O.D., and Mollnes, T.E. (2004) Meconium is a potent activator of complement in human serum and in piglets. *Pediatr Res* 55, 310–318

Castellheim, A., Lindenskov, P.H., Pharo, A., Aamodt, G., Saugstad, O.D., and Mollnes, T.E. (2005) Meconium aspiration syndrome induces complement-associated systemic inflammatory response in newborn piglets. *Scand J Immunol* 61, 217–225

Choe, J., Kelker, M.S., and Wilson, I.A. (2005) Crystal structure of human toll-like receptor 3 (TLR3) ectodomain. *Science* 309, 581–585

Chong, A.J., Shimamoto, A., Hampton, C.R., Takayama, H., Spring, D.J., Rothnie, C.L., Yada, M., Pohlman, T.H., and Verrier, E.D. (2004) Toll-like receptor 4 mediates ischemia/reperfusion injury of the heart. *J Thorac Cardiovasc Surg* 128, 170–179

Collard, C.D., Vakeva, A., Morrissey, M.A., Agah, A., Rollins, S.A., Reenstra, W.R., Buras, J.A., Meri, S., and Stahl, G.L. (2000) Complement activation after oxidative stress – role of the lectin complement pathway. *Am J Pathol* 156, 1549–1556

Daubeuf, B., Mathison, J., Spiller, S., Hugues, S., Herren, S., Ferlin, W., Kosco-Vilbois, M., Wagner, H., Kirschning, C.J., Ulevitch, R., and Elson, G. (2007) TLR4/MD-2 Monoclonal antibody therapy affords protection in experimental models of septic shock. *J Immunol* 179, 6107–6114

Frey, E.A., Miller, D.S., Jahr, T.G., Sundan, A., Bazil, V., Espevik, T., Finlay, B.B., and Wright, S.D. (1992) Soluble CD14 participates in the response of cells to lipopolysaccharide. *J Exp Med* 176, 1665–1671

Gay, N.J. and Gangloff, M. (2007) Structure and function of toll receptors and their ligands. *Annu Rev Biochem* 76, 141–165

Hawlisch, H. and Kohl, J. (2006) Complement and Toll-like receptors: key regulators of adaptive immune responses. *Mol Immunol* 43, 13–21

Husebye, H., Halaas, O., Stenmark, H., Tunheim, G., Sandanger, O., Bogen, B., Brech, A., Latz, E., and Espevik, T. (2006) Endocytic pathways regulate toll-like receptor 4 signaling and link innate and adaptive immunity. *EMBO J* 25, 683–692

Johnsen, I.B., Nguyen, T.T., Ringdal, M., Tryggestad, A.M., Bakke, O., Lien, E., Espevik, T., and Anthonsen, M.W. (2006) Toll-like receptor 3 associates with c-Src tyrosine kinase on endosomes to initiate antiviral signaling. *EMBO J* 25, 3335–3346

Kaczorowski, D.J., Nakao, A., Mollen, K.P., Vallabhaneni, R., Sugimoto, R., Kohmoto, J., Tobita, K., Zuckerbraun, B.S., Mccurry, K.R., Murase, N., and Billiar, T.R. (2007) Toll-like receptor 4 mediates the early inflammatory response after cold ischemia/reperfusion. *Transplantation* 84, 1279–1287

Kim, H.M., Park, B.S., Kim, J.I., Kim, S.E., Lee, J., Oh, S.C., Enkhbayar, P., Matsushima, N., Lee, H., Yoo, O.J., and Lee, J.O. (2007a) Crystal structure of the TLR4-MD-2 complex with bound endotoxin antagonist eritoran. *Cell* 130, 906–917

Kim, S.C., Ghanem, A., Stapel, H., Tiemann, K., Knuefermann, P., Hoeft, A., Meyer, R., Grohe, C., Knowlton, A.A., and Baumgarten, G. (2007b) Toll-like receptor 4 deficiency: smaller infarcts, but no gain in function. *BMC Physiol* 7, 5

Kohl, J. (2006) The role of complement in danger sensing and transmission. *Immunol Res* 34, 157–176

Latz, E., Visintin, A., Lien, E., Fitzgerald, K.A., Monks, B.G., Kurt-Jones, E.A., Golenbock, D.T., and Espevik, T. (2002) Lipopolysaccharide rapidly traffics to and from the Golgi apparatus with the toll-like receptor 4-MD-2-CD14 complex in a process that is distinct from the initiation of signal transduction. *J Biol Chem* 277, 47834–47843

Latz, E., Schoenemeyer, A., Visintin, A., Fitzgerald, K.A., Monks, B.G., Knetter, C.F., Lien, E., Nilsen, N.J., Espevik, T., and Golenbock, D.T. (2004) TLR9 signals after translocating from the ER to CpG DNA in the lysosome. *Nat Immunol* 5, 190–198

Lien, E., Aukrust, P., Sundan, A., Muller, F., Froland, S.S., and Espevik, T. (1998) Elevated levels of serum-soluble CD14 in human immunodeficiency virus type 1 (HIV-1) infection: correlation to disease progression and clinical events. *Blood* 92, 2084–2092

Lindenskov, P.H., Castellheim, A., Aamodt, G., Saugstad, O.D., and Mollnes, T.E. (2004) Complement activation reflects severity of meconium aspiration syndrome in newborn pigs. *Pediatr Res* 56, 810–817

Marchant, A., Tielemans, C., Husson, C., Gastaldello, K., Schurmans, T., De, G.D., Duchow, J., Vanherweghem, L., and Goldman, M. (1996) Cuprophane haemodialysis induces upregulation of LPS receptor (CD14) on monocytes: role of complement activation. *Nephrol Dial Transplant* 11, 657–662

Mollnes, T.E. and Kirschfink, M. (2006) Strategies of therapeutic complement inhibition. *Mol Immunol* 43, 107–121

Mollnes, T.E., Brekke, O.L., Fung, M., Fure, H., Christiansen, D., Bergseth, G., Videm, V., Lappegard, K.T., Kohl, J., and Lambris, J.D. (2002a) Essential role of the C5a receptor in *E. coli*-induced oxidative burst and phagocytosis revealed by a novel lepirudin-based human whole blood model of inflammation. *Blood* 100, 1869–1877

Mollnes, T.E., Song, W.C., and Lambris, J.D. (2002b) Complement in inflammatory tissue damage and disease. *Trends Immunol* 23, 61–64

Mollnes, T.E., Jokiranta, T.S., Truedsson, L., Nilsson, B., Rodriguez de, C.S., and Kirschfink, M. (2007) Complement analysis in the 21st century. *Mol Immunol* 44, 3838–3849

Ohto, U., Fukase, K., Miyake, K., and Satow, Y. (2007) Crystal structures of human MD-2 and its complex with antiendotoxic lipid IVa. *Science* 316, 1632–1634

Riedemann, N.C., Guo, R.F., Hollmann, T.J., Gao, H., Neff, T.A., Reuben, J.S., Speyer, C.L., Sarma, J.V., Wetsel, R.A., Zetoune, F.S., and Ward, P.A. (2003a) Regulatory role of C5a in LPS-induced IL-6 production by neutrophils during sepsis. *FASEB J* 370–372

Riedemann, N.C., Neff, T.A., Guo, R.F., Bernacki, K.D., Laudes, I.J., Sarma, J.V., Lambris, J.D., and Ward, P.A. (2003b) Protective effects of IL-6 blockade in sepsis are linked to reduced C5a receptor expression. *J Immunol* 170, 503–507

Salvesen, B., Fung, M., Saugstad, O.D., and Mollnes, T.E. (2008) The role of complement and CD14 in meconium-induced cytokine formation. *Pediatrics* 121, e496–e505

Sendide, K., Reiner, N.E., Lee, J.S., Bourgoin, S., Talal, A., and Hmama, Z. (2005) Cross-talk between CD14 and complement receptor 3 promotes phagocytosis of Mycobacteria: regulation by phosphatidylinositol 3-kinase and cytohesin-1. *J Immunol* 174, 4210–4219

Shimamoto, A., Chong, A.J., Yada, M., Shomura, S., Takayama, H., Fleisig, A.J., Agnew, M.L., Hampton, C.R., Rothnie, C.L., Spring, D.J., Pohlman, T.H., Shimpo, H., and Verrier, E.D. (2006) Inhibition of toll-like receptor 4 with eritoran attenuates myocardial ischemia-reperfusion injury. *Circulation* 114, I270–I274

Sprong, T., Moller, A.S., Bjerre, A., Wedege, E., Kierulf, P., van der Meer, J.W., Brandtzaeg, P., van Deuren, M., and Mollnes, T.E. (2004) Complement activation and complement-dependent inflammation by Neisseria meningitidis are independent of lipopolysaccharide. *Infect Immun* 72, 3344–3349

Tang, A.H., Brunn, G.J., Cascalho, M., and Platt, J.L. (2007) Pivotal advance: endogenous pathway to SIRS, sepsis, and related conditions. *J Leukoc Biol* 82, 282–285

Uematsu, S. and Akira, S. (2007) Toll-like receptors and type I interferons. *J Biol Chem* 282, 15319–15323

van Bruggen, R., Zweers, D., van, D.A., van Dissel, J.T., Roos, D., Verhoeven, A.J., and Kuijpers, T.W. (2007) Complement receptor 3 and Toll-like receptor 4 act sequentially in uptake and intracellular killing of unopsonized *Salmonella enterica serovar typhimurium* by human neutrophils. *Infect Immun* 75, 2655–2660

Walport, M.J. (2001a) Advances in immunology: complement (First of two parts). *N Engl J Med* 344, 1058–1066

Walport, M.J. (2001b) Advances in immunology: complement (Second of two parts). *N Engl J Med* 344, 1140–1144

Ward, P.A. (2004) The dark side of C5A in sepsis. *Nat Rev Immunol* 4, 133–142

Weingarten, R., Sklar, L.A., Mathison, J.C., Omidi, S., Ainsworth, T., Simon, S., Ulevitch, R.J., and Tobias, P.S. (1993) Interactions of lipopolysaccharide with neutrophils in blood via CD14. *J Leukoc Biol* 53, 518–524

Yamada, M., Oritani, K., Kaisho, T., Ishikawa, J., Yoshida, H., Takahashi, I., Kawamoto, S., Ishida, N., Ujiie, H., Masaie, H., Botto, M., Tomiyama, Y., and Matsuzawa, Y. (2004) Complement C1q regulates LPS-induced cytokine production in bone marrow-derived dendritic cells. *Eur J Immunol* 34, 221–230

Zarewych, D.M., Kindzelskii, A.L., Todd, R.F., III, and Petty, H.R. (1996) LPS induces CD14 association with complement receptor type 3, which is reversed by neutrophil adhesion. *J Immunol* 156, 430–433

Zhang, M. and Carroll, M.C. (2007) Natural IgM-mediated innate autoimmunity: a new target for early intervention of ischemia-reperfusion injury. *Expert Opin Biol Ther* 7, 1575–1582

Zhang, X., Kimura, Y., Fang, C., Zhou, L., Sfyroera, G., Lambris, J.D., Wetsel, R.A., Miwa, T., and Song, W.C. (2007) Regulation of toll-like receptor-mediated inflammatory response by complement in vivo. *Blood* 110, 228–236

19. Targeting Classical Complement Pathway to Treat Complement Mediated Autoimmune Diseases

Erdem Tüzün[1], Jing Li[2], Shamsher S. Saini[3], Huan Yang[4], and Premkumar Christadoss[5]

[1]Department of Neurology, University of Istanbul, Istanbul, Turkey, drerdem@yahoo.com
[2]Department of Neurology, University of Central South China University, P.R. China, jing_neurology@hotmail.com
[3]Department of Microbiology and Immunology, University of Texas Medical Branch, Galveston, TX, USA, sssaini@utmb.edu
[4]Department of Neurology, University of Central South China Universiy, P.R. China, yangh69@yahoo.com
[5]Department of Microbiology and Immunology, University of Texas Medical Branch, Galveston, TX, USA, pchrista@utmb.edu

Abstract. Mice deficient for classical complement pathway (CCP) factor C4 are resistant to antibody and complement mediated experimental autoimmune myasthenia gravis (EAMG). Anti-C1q antibody administration before or following acetylcholine receptor immunization suppresses EAMG development by reducing lymph node cell IL-6 production and neuromuscular junction IgG, C3 and C5b-C9 deposition. This effect is achieved by treating mice with 10 µg of anti-C1q antibody, twice weekly for 4 weeks. Treatment with a higher amount of anti-C1q antibody gives rise to increased serum anti-acetylcholine receptor antibody, immune complex and C3 levels, facilitates kidney C3 and IgG deposits and thus reduces the treatment efficacy. C4 KO and anti-C1q antibody treated mice display normal immune system functions and intact antibody production capacity. Furthermore, CCP inhibition preserves alternative complement pathway activation, which is required for host defense against microorganisms. Therefore, CCP inhibition might constitute a specific treatment approach for not only myasthenia gravis but also other complement mediated autoimmune diseases.

1 Complement System in Myasthenia Gravis

Myasthenia gravis (MG) is a T cell dependent and antibody mediated autoimmune disease. MG patients develop fluctuating muscle weakness and fatiguability as a result of impaired neuromuscular junction (NMJ) transmission due to malfunctioning or reduction of acetylcholine receptors (AChR) (Vincent 2006). Data obtained from clinical studies and experiments performed in animal model of MG, experimental autoimmune myasthenia gravis (EAMG), have provided insight on the complex immunological mechanisms causing NMJ destruction in MG.

J.D. Lambris (ed.), *Current Topics in Complement II.*
doi: 10:1007/978-0-387-78952-1_19,

Autoimmunity against AChR evolves following integrated interactions of antigen presenting cells (presenting AChR epitopes to T cells), T cells (activating B cells) and B cells (producing anti-AChR antibodies). All these cell types also contribute to MG pathogenesis by secreting cytokines (e.g. IL-6, IL-10, TNF-α, IL-12) required for fine functioning and maturation of immune cells. Additionally, MHC class II molecules and costimulatory molecules (e.g. ICOS and B7) are required for antigen presentation and AChR specific T and B cell activation (Tuzun and Christadoss 2006; Christadoss et al. 1995). Anti-AChR antibody is the major pathogenic factor in MG and interferes with AChR functions by increasing its degradation and initiating the complement-dependent lysis of the postsynaptic membrane (Howard et al. 1987; Engel and Fumagalli 1982). In both MG and EAMG, minimal or no cellular infiltration is observed in the muscle tissue and therefore anti-AChR antibody and the complement system are the major pathogenic factors.

Baseline evidence for the involvement of the complement system in MG pathogenesis comes from clinical studies reporting altered levels of complement factors in MG sera and immunohistochemistry experiments demonstrating IgG and colocalization of C3/C9 deposit at the NMJ postsynaptic membrane (Engel et al. 1977, 1981; Sahashi et al. 1980). Moreover, MG patients with higher serum anti-AChR antibody levels (but not anti-titin or anti-RyR antibodies) display lower serum C3 and C4 concentrations (Romi et al. 2005) and plasma levels of C3c, a C3 breakdown product, correlate with MG severity (Kamolvarin et al. 1991), suggesting an in vivo complement consumption in MG (Romi et al. 2005). These data imply that activation of the complement system by anti-AChR antibody deposits results in formation of membrane attack complex (MAC) and subsequent destruction of the NMJ and reduction in muscle AChR content. In line with this assumption, muscle weakness of mice with EAMG induced by passive transfer of anti-AChR antibodies improves following treatment by complement inhibitors, such as cobra venom factor (CVF), soluble complement receptor 1 (sCR1) and anti-C6 antibody (Lennon et al. 1978; Engel et al. 1979; Biesecker and Gomez 1989) CVF and sCR1 both inhibit classical and alternative complement pathways,while anti-C6 antibody interferes with MAC formation.

EAMG induced by immunization of mice with AChR closely mimics the clinical and pathogenic features of MG. Following AChR immunization, 70–80% of mice develop generalized muscle weakness with varied severity. Induction of muscle weakness is coupled with the elevation of anti-AChR antibody, C1q, C3 and circulating immune complex levels in sera in a time dependent manner (Tuzun et al. 2006a). In postmortem frozen muscle specimens of mice with passively or actively induced EAMG, IgG deposits overlap with C1q, C3 and MAC deposits (Tuzun et al. 2006a; Graus et al. 1993; Engel et al. 1979; Tuzun et al. 2003). Thus data obtained from the animal model of MG also support the role of anti-AChR antibody mediated complement activation in NMJ destruction and consequent development of muscle weakness. Moreover, the pathogenic significance of various immunological factors in MG pathogenesis have been first demonstrated in the

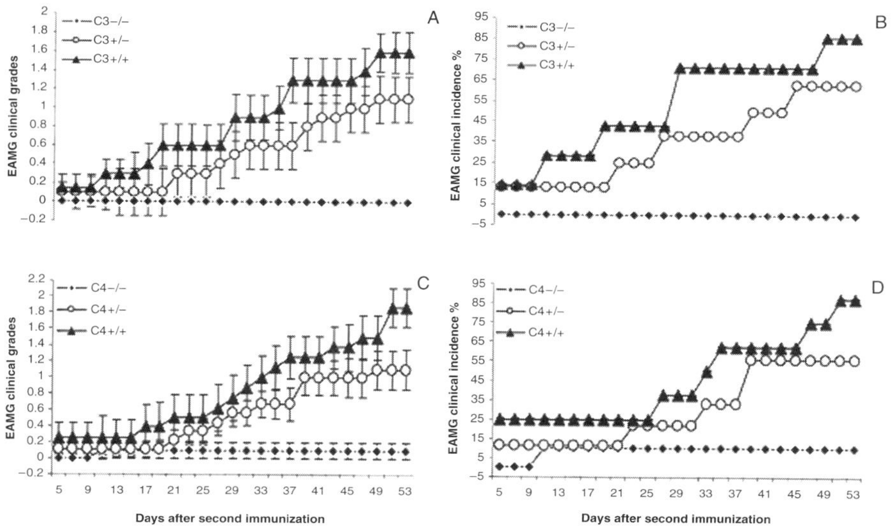

Fig. 1 AChR-immunized C3−/− and C4−/− mice are resistant to clinical EAMG. The clinical scores (**a, b**) and incidences (**b, d**) are significantly lower in C3−/− and C4−/− mice, as compared with that of control littermates. Bars indicate standard errors (Reprinted from Tuzun et al. (2003). Copyright 2003 The American Association of Immunologists, Inc.)

EAMG model. Many of these factors appear to exert their functions at least partially by activating the production of complement factors. EAMG resistance of IL-6, IL-12, IL-5, ICOS, FcγRIII KO and IL-1 receptor antagonist treated mice was associated with reduced serum and/or NMJ C3 content (Yang et al. 2005; Deng et al. 2002; Karachunski et al. 2000; Poussin et al. 2002; Tuzun et al. 2006b; Scott et al. 2004).

In an approach to find more direct evidence for the involvement of the complement system, particularly MAC formation, in EAMG induction, B10.D2/nSn (C5 sufficient) and B10.D2/oSn (C5 deficient) mice were immunized with AChR. C5 deficient mice had significantly lower EAMG incidence (5.6%) as compared to C5 sufficient mice (78.3%). Notably, despite EAMG resistance, serum anti-AChR antibody levels of C5 deficient mice were equivalent to those of C5 sufficient mice (Christadoss 1988). This study did not only show that complement activation and the consequent MAC formation are crucial for EAMG induction but also implied that anti-AChR antibody requires complement factors to induce muscle weakness at least in the animal model of the disease.

2 Classical Complement Pathway in EAMG

As can be deduced from the observation that C1q deposits located at the NMJ (Tuzun et al. 2006a) and well-known complement activation ability of antibodies, classical complement pathway (CCP) was long assumed to be the key component of the complement system in the initiation of MAC formation at NMJ. To obtain direct genetic evidence for the role of CCP in EAMG, we immunized $C4^{-/-}$, $C4^{+/-}$ and $C4^{+/+}$ littermates in the C57BL/6 (B6) background with AChR in CFA along with $C3^{-/-}$, $C3^{+/-}$ and $C3^{+/+}$ mice. $C3^{-/-}$ and $C4^{-/-}$ mice were highly resistant to EAMG induction with 0% and 5.8% disease incidence, respectively. $C3^{+/-}$ (63%) and $C4^{+/-}$ (56%) mice displayed an intermediary disease incidence as compared to $C3^{+/+}$ (86%) and $C4^{+/+}$ (88%) mice (Fig. 1).

In conclusion, deficiency of C4 (a CCP factor) protected mice from EAMG induction, and therefore CCP activation by anti-AChR antibody was crucial for EAMG induction (Tuzun et al. 2003).

Notably, C4 gene deficiency did not reduce serum anti-AChR antibody levels except for a marginal decrease in anti-AChR IgG2b level. Furthermore, double staining of frozen muscle sections with rhodamine-conjugated α-bungarotoxin (BTx) and FITC-conjugated antibodies against IgG, C3 or MAC revealed that $C4^{-/-}$ mice had only antibody deposits but no C3 or MAC deposits at the NMJ, further supporting the notion that complement mediated membrane lysis is required for EAMG induction in addition to anti-AChR antibody deposits. $C3^{-/-}$ mice had neither complement nor antibody deposits and wild-type control littermates had all types of deposits. Additionally, $C4^{-/-}$ mice appeared to have an intact and fully functioning immune system with normal lymph node cell proliferation and cytokine production to AChR challenge and lymph node cell ratios comparable to those of wild-type mice (Tuzun et al. 2003).

As is the case with all KO mice, the influence of the inherent deficiency of C4 gene on the immune system functioning could have influenced EAMG resistance. C4 is involved in tolerance induction to self antigens and its deficiency increases the elimination of B cells with negative selection via apoptosis associated mechanisms (Carroll 1999). In line with previous findings, C4 KO mice exhibited increased apoptosis inducer molecules (Fas ligand and CD69) and apoptotic B cells in lymph nodes following AChR immunization (Tuzun et al. 2003). This phenomenon could have influenced EAMG resistance of $C4^{-/-}$ mice, but this is unlikely since C4 deficient mice had significant anti-AChR antibody response and IgG deposits in NMJ.

3 Anti-C1q Antibody Prevents and Treats EAMG

To show that acquired deficiency of CCP may also render mice resistant to EAMG, we attempted to ameliorate myasthenic muscle weakness by inhibiting CCP using anti-C1q antibodies. In the preliminary experiments, we showed that anti-C1q

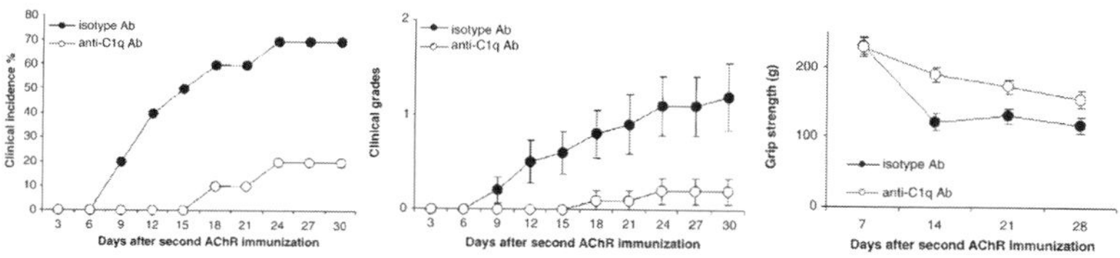

Fig. 2 Kinetics of severity and accumulated clinical incidence of EAMG for anti-C1q antibody and isotype antibody-treated mice. Bars indicate standard errors (Reprinted from Tuzun et al. 2006b. Copyright 2006 with permission from Elsevier.)

antibody was capable of susbtantially reducing serum C1q levels (Tuzun et al. 2006). In prevention experiments, B6 mice were treated i.p. with anti-C1q antibody or isotype antibody (negative control) with two loading doses of 200 μg given on days -7 and -4 before first AChR immunization (day 0), followed by 100 μg/injection, twice weekly for 5 weeks. As compared to isotype antibody treated mice, anti-C1q antibody treated mice had lower EAMG incidence (70% vs. 20%), lower EAMG severity (average clinical grades 1.2 vs. 0.3) and higher average grip strength (170 g vs. 120 g as measured by a dynamometer) (55, Fig. 2). Anti-C1q antibody prevented EAMG by reducing NMJ C3, IgG and MAC deposits and lymph node cell IL-6 production in response to AChR and immunodominant peptide α146-162 challenge (Tuzun et al. 2006a). By this experiment, we demonstrated for the first time that an autoimmune disease could be prevented by CCP inhibition.

In treatment experiments, following EAMG induction by AChR immunization, B6 and RIIIS/J mice were treated i.p. with either 10 or 100 μg of anti-C1q antibody or 100 μg of isotype antibody twice weekly for 4 weeks. B6 mice treated with 10 μg anti-C1q antibody had significantly reduced clinical grades and improved grip strength as compared to isotype antibody treated mice. Interestingly, a higher concentration (100 μg) of anti-C1q antibody was not effective in B6 mice. In 10 μg treatment group, 75% of mice improved and none of the mice clinically deteriorated. Both 10 μg and 100 μg anti-C1q antibody-treated RIIIS/J mice showed significant clinical improvement. In both anti-C1q antibody treatment groups, muscle weakness of 60% of mice clinically improved. Alternatively, improvement rate of isotype antibody treated mice ranged between 20 and 33%. Three mice (20%) in 100 μg treatment group and one mouse (9.1%) in isotype antibody treatment group developed severe EAMG (Tuzun et al. 2007).

This treatment effect of the anti-C1q antibody was associated with reduced lymph node cell IL-6 production and lymph node $CD4^+$ and $CD8^+$ T cell populations. Administration of 100 μg anti-C1q antibody increased serum anti-AChR antibody, immune complex and C3 levels and induced a mild nephropathy characterized with kidney C3 and IgG deposits. These hazardous side effects might have been associated with reduced treatment efficacy (Tuzun et al. 2007). Nevertheless, our data showed that anti-C1q antibody treatment was capable of

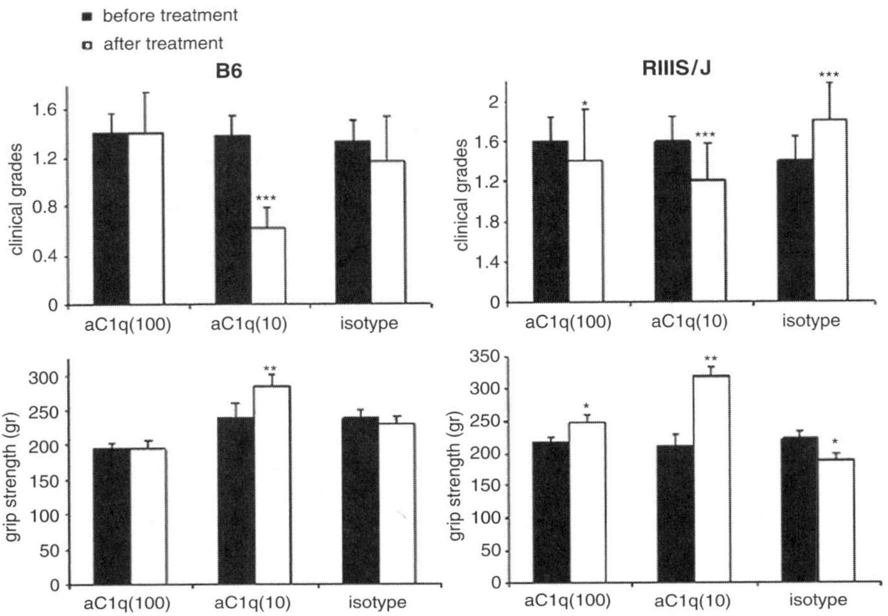

Fig. 3 Clinical grades and grip strengths of B6 and RIIIS/J mice with EAMG before and after anti-C1q antibody treatment. *, indicates $p < 0.05$, **, $p < 0.01$ and ***, $p < 0.001$, bars indicate standard errors (Reprinted from Tuzun et al. (2007). Copyright 2007 with permission from Elsevier.)

ameliorating myasthenic muscle weakness (Fig. 3) and an autoimmune disease could be treated by CCP inhibition.

4 Conclusion

MG patients are currently treated with steroids and/or other non-specific immunosuppressive agents. Other treatment methods like plasmapheresis and intravenous immunoglobulin administration are primarily used for treatment of acute symptoms. MG and other autoimmune diseases therefore require new treatment methods specifically targeting key pathogenic factors and without causing global immune suppression.

In this context, prevention and treatment of EAMG by CCP manipulation is a promising finding for future treatment of not only MG but also other complement associated autoimmune diseases (e.g. dermatomyositis, autoimmune glomerulonephritis and autoimmune haemolytic anemia). Our results show that CCP inhibition or genetic deficiency of a CCP factor C4 affect other immune system functions marginally and do not cause a significant immune suppression. Also, specific CCP inhibition would plausibly leave the alternative pathway intact to cope with invading microorganisms and prevent infections.

C1 inhibitors have been used effectively to treat both patients and experimental animals with non-autoimmune diseases such as hereditary angioedema, burn, shock, sepsis and stroke (De Serres et al. 2003; Henze et al. 1997; Dickneite 1993; Zeerleder et al. 2003; De Simoni et al. 2003). We believe that C1 manupulation might also be considered in treatment refractory MG patients or as an adjunct immunotherapy method to steroids and azathioprine.

References

Biesecker, G. and Gomez, C.M. (1989) Inhibition of acute passive transfer experimental autoimmune myasthenia gravis with Fab antibody to complement C6. *J. Immunol.* 142, 2654–2659

Carroll, M. (1999) Negative selection of self-reactive B lymphocytes involves complement. *Curr. Top. Microbiol. Immunol.* 246, 21–27

Christadoss, P. (1988) C5 gene influences the development of murine myasthenia gravis. *J. Immunol.* 140, 2589–2592

Christadoss, P., Kaul, R., Shenoy, M. and Goluszko, E. (1995) Establishment of a mouse model of myasthenia gravis which mimics human myasthenia gravis pathogenesis for immune intervention. *Adv. Exp. Med. Biol.* 383, 195–199

De Serres, J., Groner, A. and Lindner, J. (2003) Safety and efficacy of pasteurized C1 inhibitor concentrate (Berinert P) in hereditary angioedema: a review. *Transfus. Apher. Sci.* 29, 247–254

De Simoni, M.G., Storini, C., Barba, M., Catapano, L., Arabia, A.M., Rossi, E. and Bergamaschini L. (2003) Neuroprotection by complement (C1) inhibitor in mouse transient brain ischemia. *J. Cereb. Blood Flow. Metab.* 23, 232–239

Deng, C., Goluszko, E., Tuzun, E., Yang, H. and Christadoss, P. (2002) Resistance to experimental autoimmune myasthenia gravis in IL-6-deficient mice is associated with reduced germinal center formation and C3 production. *J. Immunol.* 169, 1077–1083

Dickneite, G. (1993) Influence of C1-inhibitor on inflammation, edema and shock. *Behring Inst. Mitt.* 93, 299–305

Engel, A.G. and Fumagalli, G. (1982) Mechanisms of acetylcholine receptor loss from the neuromuscular junction. *Ciba Found. Symp.* 90, 197–224

Engel, A.G., Lambert, E.H. and Howard, F.M. (1977) Immune complexes (IgG and C3) at the motor end-plate in myasthenia gravis: ultrastructural and light microscopic localization and electrophysiologic correlations. *Mayo Clin. Proc.* 52, 267–280

Engel, A.G., Sakakibara, H., Sahashi, K., Lindstrom, J.M., Lambert, E.H. and Lennon, V.A. (1979) Passively transferred experimental autoimmune myasthenia gravis. Sequential and quantitative study of the motor end-plate fine structure and ultrastructural localization of immune complexes (IgG and C3), and of the acetylcholine receptor. *Neurology* 29, 179–188

Engel, A.G., Sahashi, K. and Fumagalli, G. (1981) The immunopathology of acquired myasthenia gravis. *Ann. N. Y. Acad. Sci.* 377, 158–174

Graus, Y.M., Verschuuren, J.J., Spaans, F., Jennekens, F., van Breda Vriesman, P.J. and De Baets, M.H. (1993) Age-related resistance to experimental autoimmune myasthenia gravis in rats. *J. Immunol.* 150, 4093–4103

Henze, U., Lennartz, A., Hafemann, B., Goldmann, C., Kirkpatrick, C.J. and Klosterhalfen, B. (1997) The influence of the C1-inhibitor BERINERT and the protein-free haemodialysate ACTIHAEMYL20% on the evolution of the depth of scald burns in a porcine model. *Burns* 23, 473–477

Howard, F.M. Jr., Lennon, V.A., Finley, J., Matsumoto, J. and Elveback, L.R. (1987) Clinical correlations of antibodies that bind, block, or modulate human acetylcholine receptors in myasthenia gravis. *Ann. N. Y. Acad. Sci.* 505, 526–538

Kamolvarin, N., Hemachudha, T., Ongpipattanakul, B., Phanthumchinda, K. and Sueblinvong, T. (1991) Plasma C3c in immune-mediated neurological diseases: a preliminary report. *Acta. Neurol. Scand.* 83, 382–387

Karachunski, P.I., Ostlie, N.S., Monfardini, C. and Conti-Fine, B.M. (2000) Absence of IFN-gamma or IL-12 has different effects on experimental myasthenia gravis in C57BL/6 mice. *J. Immunol.* 164, 5236–5244

Lennon, V.A., Seybold, M.E., Lindstrom, J.M., Cochrane, C. and Ulevitch, R. (1978) Role of complement in the pathogenesis of experimental autoimmune myasthenia gravis. *J. Exp. Med.* 147, 973–983

Poussin, M.A., Goluszko, E., Franco, J.U. and Christadoss P. (2002) Role of IL-5 during primary and secondary immune response to acetylcholine receptor. *J. Neuroimmunol.* 125, 51–58

Romi, F., Kristoffersen, E.K., Aarli, J.A. and Gilhus, N.E. (2005) The role of complement in myasthenia gravis: serological evidence of complement consumption in vivo. *J. Neuroimmunol.* 158, 191–194

Sahashi, K., Engel, A.G., Lambert, E.H. and Howard, F.M. Jr. (1980) Ultrastructural localization of the terminal and lytic ninth complement component (C9) at the motor end-plate in myasthenia gravis. *J. Neuropathol. Exp. Neurol.* 39, 160–172

Scott, B.G., Yang, H., Tuzun, E., Dong, C., Flavell, R.A. and Christadoss, P. (2004) ICOS is essential for the development of experimental autoimmune myasthenia gravis. *J. Neuroimmunol.* 153, 16–25

Tuzun, E. and Christadoss, P. (2006) Unraveling myasthenia gravis immunopathogenesis using animal models. *Drug Discov Today Dis Models* 3, 15–20

Tuzun, E., Scott, B.G., Goluszko, E., Higgs, S. and Christadoss, P. (2003) Genetic evidence for involvement of classical complement pathway in induction of experimental autoimmune myasthenia gravis. *J. Immunol.* 171, 3847–3854

Tuzun, E., Saini, S.S., Ghosh, S., Rowin, J., Meriggioli, M.N. and Christadoss, P. (2006a) Predictive value of serum anti-C1q antibody levels in experimental autoimmune myasthenia gravis. *Neuromuscul. Disord.* 16, 137–143

Tuzun, E., Saini, S.S., Yang, H., Alagappan, D., Higgs, S. and Christadoss, P. (2006b) Genetic evidence for the involvement of Fcgamma receptor III in experimental autoimmune myasthenia gravis pathogenesis. *J. Neuroimmunol.* 174, 157–167

Tuzun, E., Li, J., Saini, S.S., Yang, H. and Christadoss, P. (2007) Pros and cons of treating murine myasthenia gravis with anti-C1q antibody. *J. Neuroimmunol.* 182, 167–176

Vincent, A. (2006) Immunology of disorders of neuromuscular transmission. *Acta Neurol. Scand. Suppl.* 183, 1–7

Yang, H., Tuzun, E., Alagappan, D., Yu, X., Scott, B.G., Ischenko, A. and Christadoss, P. (2005) IL-1 receptor antagonist-mediated therapeutic effect in murine myasthenia gravis is associated with suppressed serum proinflammatory cytokines, C3, and anti-acetylcholine receptor IgG1. *J. Immunol.* 175, 2018–2025

Zeerleder, S., Caliezi, C., van Mierlo, G., Eerenberg-Belmer, A., Sulzer, I., Hack, C.E. and Wuillemin, W.A. (2003) Administration of C1 inhibitor reduces neutrophil activation in patients with sepsis. *Clin. Diagn. Lab. Immunol.* 10, 529–535

20. Compstatin: A Complement Inhibitor on its Way to Clinical Application

Daniel Ricklin[1] and John D. Lambris[2]

[1]Department of Pathology and Laboratory Medicine, University of Pennsylvania, Philadelphia, PA, USA, ricklin@mail.med.upenn.edu
[2]Department of Pathology and Laboratory Medicine, University of Pennsylvania, Philadelphia, PA, USA, lambris@mail.med.upenn.edu

Abstract. Therapeutic modulation of the human complement system is considered a promising approach for treating a number of pathological conditions. Owing to its central position in the cascade, component C3 is a particularly attractive target for complement-specific drugs. Compstatin, a cyclic tridecapeptide, which was originally discovered from phage-display libraries, is a highly potent and selective C3 inhibitor that demonstrated clinical potential in a series of experimental models. A combination of chemical, biophysical, and computational approaches allowed a remarkable optimization of its binding affinity towards C3 and its inhibitory potency. With the recent announcement of clinical trials with a compstatin analog for the treatment of age-related macular degeneration, another important milestone has been reached on its way to a drug. Furthermore, the release of a co-crystal structure of compstatin with C3c allows a detailed insight into the binding mode and paves the way to the rational design of peptides and mimetics with improved activity. Considering the new incentives and the promising pre-clinical results, compstatin seems to be well equipped for the challenges on its way to a clinical therapeutic.

1 Tackling Complement at its Core

Therapeutic intervention in the human complement system has long been recognized as a promising strategy for the treatment of a series of ischemic, inflammatory and autoimmune diseases (Lambris and Holers 2000; Ricklin and Lambris 2007a). In principle, the large network of soluble and cell-surface-bound proteins, which builds the base of the complement cascade, offers a variety of potential drug targets. However, the quest for complement-specific therapeutics proved to be much more challenging than initially anticipated. With the therapeutic antibody eculizumab (Soliris®, Alexion Pharmaceuticals, Inc.) against paroxysmal nocturnal hemoglobinuria, the first drug with proven complement connectivity has been marketed only recently (Ricklin and Lambris 2007a; Rother et al. 2007). A second complement-associated compound, purified C1 esterase inhibitor (C1-INH), is available as a therapeutic option for the treatment of hereditary angioedema in several countries. However, its mechanism of action may be closer related to the bradykinin-kallikrein than the complement cascade (Davis 2006).

J.D. Lambris (ed.), *Current Topics in Complement II.*
doi: 10:1007/978-0-387-78952-1_20,

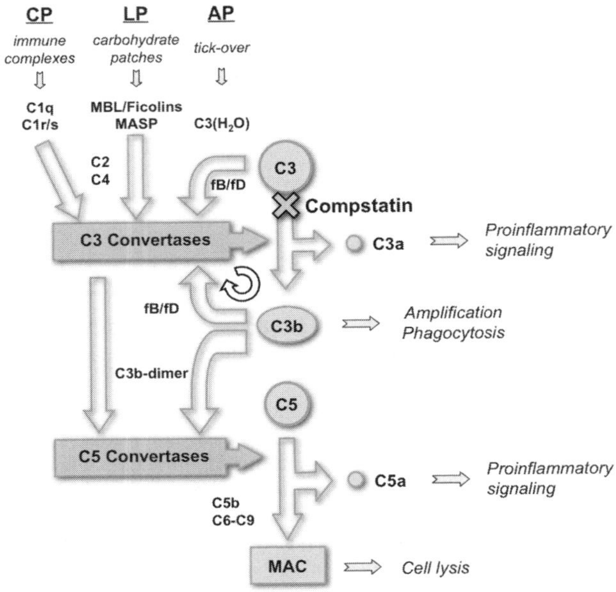

Fig. 1 Compstatin inhibits the cleavage of native C3 to its active fragments C3a and C3b. As a consequence, the deposition of C3b, the amplification of the alternative pathway and all downstream complement actions are prevented

Strikingly, both drugs cover relatively rare diseases and have been developed with the aid of orphan drug regulations. Yet, for many of the more common inflammatory or autoimmune conditions there are no complement drugs on the market. Any extension of the current complement-specific therapeutic arsenal is therefore highly desired.

Part of the problem in complement-directed drug discovery is the selection of the right target (Ricklin and Lambris 2007a). A controlled, localized modulation at the core of complement activation is considered to be the most promising approach in many cases. On a molecular level, inhibition at the level of C3, including the C3 convertases, is of particular interest since both the amplification of all initiation pathways and the generation of anaphylatoxins (C3a, C5a) and the membrane attack complex (MAC) are affected (Fig. 1). In this respect, C3 can be regarded as a central hub that mediates and controls the upstream activation and downstream effector functions of complement (Degn et al. 2007). Initial attempts in developing small molecule drugs for inhibiting the conversion of C3 focused on the various serine proteases that are involved in the convertase formation and activity. However, lack of potency and specificity as well as short half-lives so far limited clinical success of these compounds. Similarly, soluble forms of physiological complement regulators (e.g. complement receptor 1) have mostly been

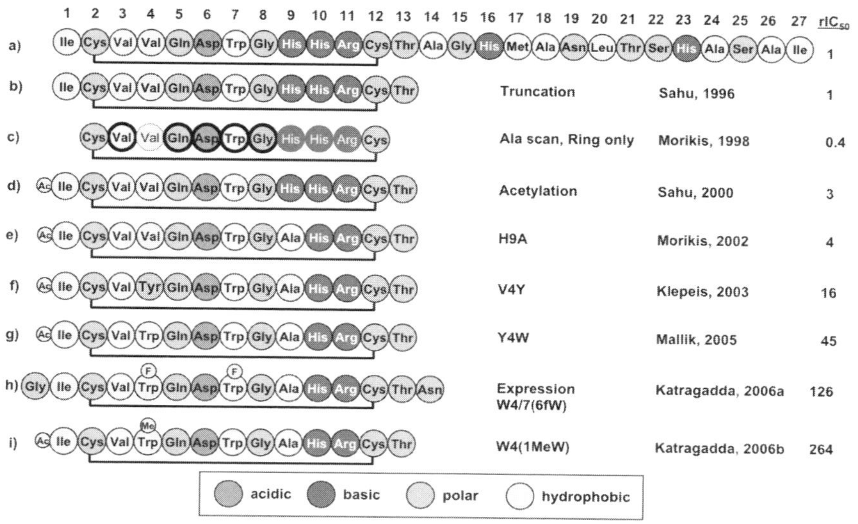

Fig. 2 Important achievements in the optimization of the compstatin peptide sequence as expressed by their relative activities (rIC_{50}) for inhibiting complement activation. The initial C3b-binding peptide from the phage display library (**a**) could be truncated to the *N*-terminal 13 residues without losing activity (**b**). While an alanine scan identified positions 3, 5, 6, 7, and 8 as essential for its functions (*bold rings*), the removal of the flanking amino acids caused a slight loss of activity (**c**). Acetylation of the N-terminus (**d**) and the exchange of His-9 by alanine (**e**) led to minor and the introduction of aromatic side chains at position 4 to large activity improvements (**f, g**). Beneficial effects due to modifications on the indole ring of tryptophan were observed for both compstatin expressed in *E. coli* (**h**) and the synthetic peptide (**i**), leading to a dramatic increase in the inhibitory activity

discontinued or put on hold (Ricklin and Lambris 2007a). Compared to these previous examples, which indirectly target activation of C3 via the convertases or C3b, the peptidic inhibitor compstatin exerts its function via direct binding to native C3 (Fig. 1). This unique mechanism, its comparatively small size, and promising pre-clinical results make compstatin an interesting candidate for further clinical development. In the following sections, we illustrate both the structural and functional, as well as the clinical properties of this versatile molecule.

2 Discovery and Initial Characterization

Compstatin has been discovered more than 10 years ago by screening phage-display libraries in the search for C3b-binding peptides (Sahu et al. 1996). In comparison with small drug molecules, peptides face some specific challenges (e.g. stability or short plasma half-lives) but also feature several benefits such as high selectivity and low toxicity (Sato et al. 2006). In addition, peptidic drugs are more successful in inhibiting large protein-protein interactions, which are common in the complement cascade. It is not surprising, therefore, that complement therapeutics in the pipelines almost exclusively encompass biopharmaceuticals like peptides, antibodies, and other soluble proteins (Ricklin and Lambris 2007a).

Phage-display technology represents an ideal technology for screening large peptide libraries of high structural diversity in an efficient manner (Kay et al. 2001; Ladner et al. 2004; Sarrias et al. 1999; Yu and Smith 1996). The method has proven especially important for drug discovery in the academic sector, since their access to small molecule libraries is usually largely restricted. In case of compstatin, a random 27-mer peptide library containing 2×10^8 unique clones had been screened for binding to C3b (Sahu et al. 1996). One selected clone showed strong affinity for C3, C3b, and C3c but not for the C3d fragment. Even more, the isolated peptide (Peptide I; Fig. 2a) was able to inhibit both the classical and alternative pathway of complement activation with IC_{50} values of 65 and 19 μM, respectively. No significant loss in activity was observed when the C-terminal half of the peptide was removed, leaving a cyclic 13-mer (ICVVQDWGHHRCT; Fig. 2b). Reduction and alkylation of the intramolecular disulfide bond between cysteins 2 and 12 resulted in a completely impaired functionality indicating the importance of the cyclic structure (Sahu et al. 1996). The cyclic peptide, which was later named compstatin, bound to C3 in a reversible manner and inhibited the convertase-mediated cleavage of C3 to C3a and C3b. A series of surface plasmon resonance (SPR) studies confirmed and quantified the selectivity and binding mode of compstatin (Ricklin and Lambris 2007b). While C3 and its hydrolyzed form C3(H_2O) featured similar affinities (K_D = 60–130 nM) for the immobilized peptide, the binding of C3b and C3c was found to be reduced by a factor of 20 and 70, respectively. Again, no binding to C3d could be detected (Sahu et al. 2000). This finding was rather surprising since the vast majority of physiological C3 ligands interact with its active form C3b, while only few ligands are known to bind native C3 (Sahu et al. 1998). In situ formation of the C3 convertase on a SPR sensor chip further revealed that compstatin indeed inhibited the formation and deposition of active C3b when it was co-injected with C3 (Nilsson et al. 1998). The activity of compstatin on the classical and alternative pathway has later been confirmed and further investigated by independent groups from both academic research and pharmaceutical industry (Furlong et al. 2000; Klegeris et al. 2002).

Since compstatin did only block enzymatic removal of the C3a/ANA domain by the C3 convertase but not by trypsin, a steric inhibition of the C3a cleavage site (Fig. 3b) seemed to be unlikely as functional explanation. Furthermore, the peptide did not destabilize the C3 convertase by interfering with the interaction between C3b and factor B, nor did it inhibit the binding to the convertase-stabilizing protein properdin (Sahu et al. 1996). As a consequence, the induction of a conformational change (Fig. 3c) or the steric hindrance of substrate binding (Fig. 3d) remained as

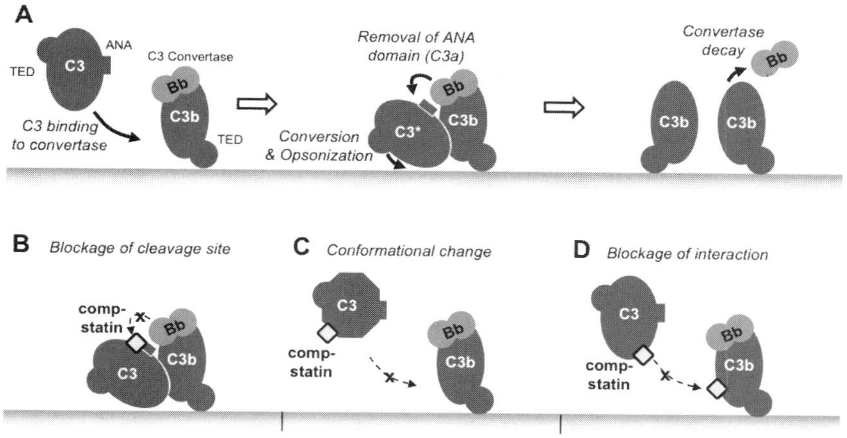

Fig. 3 In a current model of C3 activation, native C3 interacts with the surface-bound C3 convertase (C3bBb in case of the alternative pathway) and brings the anaphylatoxin domain (ANA) in close proximity to the enzymatic site on Bb. After enzymatic removal of ANA (i.e. C3a), the thioester domain (TED) undergoes a large relocation and exposes the active thioester, which leads the deposition of the newly generated C3b on the surface (**a**). In principle, compstatin may inhibit this process in several ways: (**b**) by sterically hinder the access of the convertase to the scissile bond; (**c**) by inducing a conformational change in C3 that prevents binding and/or cleavage by the convertase; or (**d**) by blocking the initial interaction between native C3 and the convertase-associated C3b

more plausible explanations. However, neither hypothesis could be undermined substantially without structural information (see Chap. 4).

Surprisingly, compstatin features a high selectivity for human and primate C3. While binding to C3 from baboons and other primates showed comparable interaction profiles as its human counterpart, no binding was detected for C3 from lower mammalian species such as mice, rats, guinea pigs, rabbits, or pigs (Sahu et al. 1996, 2003). While this selectivity has direct consequences for the clinical development of the peptide (see Chap. 6), the recent description of the molecular foundation for this selectivity may allow circumventing this issue (see Chap. 4).

Owing to its strong inhibition at the C3 level of complement, compstatin nowadays is not only a promising candidate for the development of complement-targeting drugs (see Chaps. 6 and 7) but also an indispensable tool for studying the role of complement in a variety of biological systems (examples in (Gronroos et al. 2005; Nielsen et al. 2007; Pedersen et al. 2007; Ritis et al. 2006)).

3 Tuning the Structure

While the original compstatin peptide represented a highly selective and potent lead structure itself, the combination of rational, combinatorial, and computational optimization methods was able to drastically improve its activity (Fig. 2). Early approaches in this direction have been reviewed previously (Holland et al. 2004; Morikis and Lambris 2002; Morikis et al. 2004). An initial alanine scan experiment identified Val-3 and the stretch between positions 5 and 8 as essential for complement inhibition (Fig. 2c). The study also showed that the flanking residues (Ile-1 and Thr-13) contribute to the overall activity since their removal led to a threefold decrease in activity (Morikis et al. 1998). The NMR-derived structure of compstatin (PDB code 1A1P; (Morikis et al. 1998)) offered first information about the conformation of the peptide in solution. The disulfide bond was found to form a hydrophobic cluster around the termini (Fig. 6a) and to restrain the conformational flexibility. As a consequence, residues at positions 5–8 (Gln-Asp-Trp-Gly) were arranged in a type I β-turn (Fig. 6a), which is highly prevalent in naturally occurring proteins and peptides and is often involved in molecular interactions (Rotondi and Gierasch 2006). Maintenance of this turn structure seemed to be essential for the functionality of compstatin, since both the linearization as well as the replacement of residues 5–8 by alanine led to a dramatic loss in activity (Morikis et al. 1998). Similarly, the introduction of D-amino acids into the compstatin ring in an attempt to reinforce the β-turn conformation led to a complete loss of activity in most cases (Furlong et al. 2000). Finally, a cycle size of 11 residues was shown to be mandatory since none of the shorter deletion analogs maintained activity (Sahu et al. 2000).

In contrast to the restriction concerning the ring structure, compstatin was found to be more tolerant to the replacement or modification of the terminal residues (Ile-1, Thr-13), which may be beneficial for improving the stability towards exopeptidases (Furlong et al. 2000; Sahu et al. 2000). Surprisingly, acetylation of the N-terminus did not only improve the degradation profile (see Chap. 7) but also increased complement inhibition potency by a factor of 3 (Fig. 2d) (Sahu et al. 2000). This effect has later been attributed to electrostatic effects, i.e. the reduction of charges at the N-terminus (Soulika et al. 2003). Extensive studies on the contributions of the turn-cluster arrangement revealed a more potent lead structure (Fig. 2e), in which His-9 had been replaced by alanine in order to reduce bulkiness and provide more conformational flexibility (Morikis et al. 2002). Using a two-stage computational protein design approach, the replacement of Val-4 by tyrosine was predicted to increase the fold stability and therefore improve peptide activity (Klepeis et al. 2004). Indeed, the synthesized peptide showed a sevenfold increase in potency (Fig. 2f). An impressive demonstration of rational design at residue 4 was presented, when this position was screened for aromatic amino acids and yielded a tryptophan analog (Fig. 2g) with 45-fold increased potency (Mallik et al. 2005).

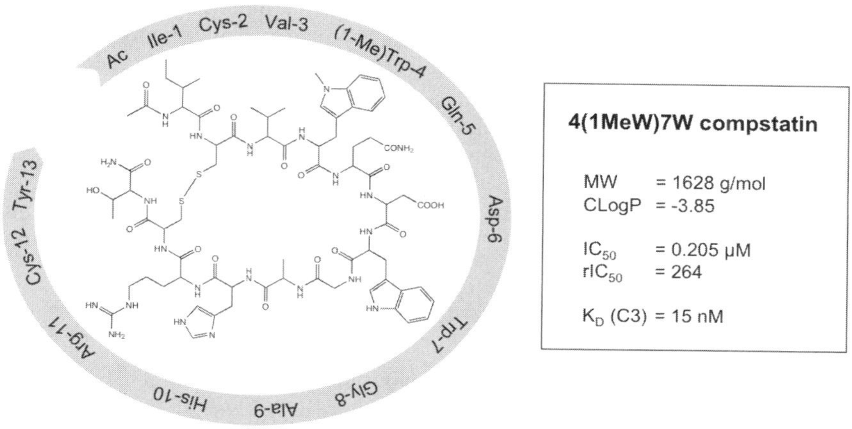

Fig. 4 Optimized lead structure of compstatin (Ac-ICV(1MeW)QDWGAHRCT-NH_2) with key biophysical properties. The tridecapeptide chain is synthesized using solid phase peptide synthesis and cyclized via a disulfide bond between the cysteine residues at positions 2 and 12. The IC_{50} values represent inhibition of the classical pathway of complement activation both as absolute value and relative to the original compstatin

By introducing hydrophobic, non-natural amino acids at the same position (e.g. 2-naphthylalanine), this effect could even be increased to a nearly 100-fold activity (Mallik et al. 2005). Further thermodynamic analyses indicated that the C3-compstatin interaction is largely driven by enthalpic contributions. In this context, the beneficial effects of a valine-to-tryptophan substitution at position 4 (and also histidine-to-alanine at position 9) could be linked to an increased enthalpy and binding affinity to C3 (Katragadda et al. 2004).

In an attempt to make the production of compstatin more economical, a bacterial expression system was successfully established in *Escherichia coli,* which generated high yields of compstatin analogs with an N-terminal glycine instead of the acetyl group (Katragadda and Lambris 2006). Incorporation of substituted tryptophan derivatives resulted in a highly active analog that carried two 6-fluorotryptophan residues at positions 4 and 7 (Fig. 2h). Encouraged by these improvements, the role of hydrophobic contributions and hydrogen bonding at position 4 and 7 was further elucidated using various tryptophan derivatives (and naphthylalanine) in combination with ELISA and calorimetric analysis (Katragadda et al. 2006). While increasing the hydrophobicity at position 4 was found to be generally beneficial, the introduction of 1-methyl tryptophan had the most prominent effect. With a C3 binding affinity of 15 nM and a 267-fold increased complement inhibition potency compared to the original peptide, 4(1MeW)7W compstatin is the most active analog published so far (Figs. 2i and 4).

The arsenal of active and inactive analogs in combination with the known solution structures led to the development of sophisticated in silico models based on QSAR, molecular dynamics, or conformational space annealing (Mallik et al. 2003; Mallik and Morikis 2005; Mulakala et al. 2007; Song et al. 2005; Tamamis et al. 2007). However, the most important breakthrough was reached with the release of the first co-crystal structure between a compstatin analog and C3c (see Chap. 4). These developments will augment our understanding about the binding mode and function of compstatin and pave the way to the development of analogs with even higher potency.

4 Exploring the Binding Site and Mode

The initial studies with compstatin already provided important hints for the localization of its binding site on C3, since both the active phage clone and the isolated peptide only bound to C3 fragments that contained the C3c portion of the protein (Sahu et al. 1996). Biophysical data confirmed a single binding site for compstatin on C3, while deviation from a simple kinetic 1:1 model in SPR indicated a conformational rearrangement of the peptide and/or protein upon binding (Katragadda et al. 2004; Sahu et al. 2000). Later, the binding region could be narrowed down to the C-terminal 40-kDa part of C3, which includes several macroglobulin domains (MG 3–6^{β}) as well as the linker (LNK) domain (Soulika et al. 2006). But even with this information, a clear description of the binding mode has proven difficult. The release of several key crystal structures such as C3, C3b, and C3c (Janssen et al. 2006; Janssen and Gros 2007; Wiesmann et al. 2006) offered an unprecedented insight into the complement activation process. After removal of the ANA domain (C3a) by the C3 convertase, the C3 structure undergoes a large structural rearrangement, in which the thioester-containing domain (TED) is completely relocated (Fig. 5). Interestingly, the core ring formed by the MG3-MG6 domains, which also included the proposed compstatin binding site, remained conformationally stable throughout these transformations. Hence, one of the most important breakthroughs in compstatin development was reached with the recent publication of a co-crystal structure of the compstatin analog 4W9A (Fig. 2g) and the C3c protein (PDB code 2QKI; Janssen et al. 2007). Alongside with indispensable information about this peptide-protein interaction, the structure also revealed a number of surprises.

First of all, the single binding site of compstatin could be localized to a shallow groove between the MG4 and MG5 domains in the stable β-ring of C3c. While this finding confirmed previous predictions of the binding area and a 1:1 interaction mode, it also had consequences for the peptide's potential mechanism of action. When assuming the same site in native C3, compstatin is too far distant from the ANA domain to allow a direct steric inhibition of C3a cleavage by the C3 convertase (Figs. 3b and 5). Similarly, an interference with the translocation of the TED domain is unlikely based on the location of the binding site. In addition, the

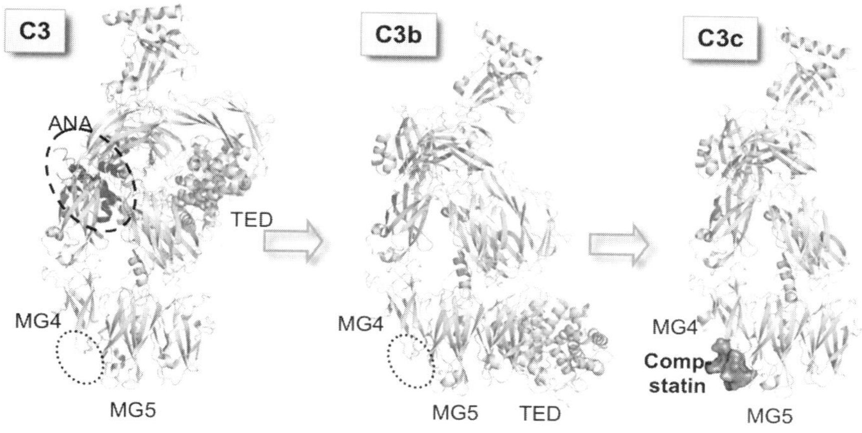

Fig. 5 The co-crystal structure of compstatin with C3c (right; PDB code 2QKI) shows that the peptidic inhibitor binds to a shallow site between macroglobulin domains 4 and 5 (MG4/5) in the core ring of C3c. Interestingly, the corresponding binding area (dotted circle) in native C3 (left; PDB code 2A73) is clearly distant from the C3a domain (ANA) that is cleaved during the activation of C3 to C3b. Furthermore, the site is nearly unaffected by the massive conformational changes during the conversion to C3b (middle; PBD code 2I07) that lead to a repositioning of the TED domain

discovery of a unique binding pocket away from other known C3 binding sites also confirms earlier observations that compstatin does not interfere with complement regulation proteins or formation of the convertase (Sahu et al. 1996).

Secondly, no large domain rearrangement could be detected between bound and unbound C3c (Janssen et al. 2007). While only a co-crystal of compstatin with native C3 may disqualify a conformational change of C3 as the functional mechanism (Fig. 3c), even the current C3c-based structure makes this option largely implausible. This leaves the steric hindrance of the C3 substrate binding to convertase-based C3b as the most likely explanation (Fig. 3d). Further support for this hypothesis came from the crystal of C3b, in which a C3b-C3b crystallographic symmetry-related contact area that includes the compstatin binding site was observed (Janssen et al. 2006, 2007). The authors therefore suggested, that this area may be involved in C3b-dimerization or C3-C3b binding, and that compstatin may interfere with these interactions (Janssen et al. 2007). A similar mechanism has been proposed for another complement inhibitor, the physiological receptor CRIg, that binds C3b at the same face of the β-chain (although via domains MG3/MG6) and selectively inhibits the alternative pathway convertases (Katschke et al. 2007; Wiesmann et al. 2006).

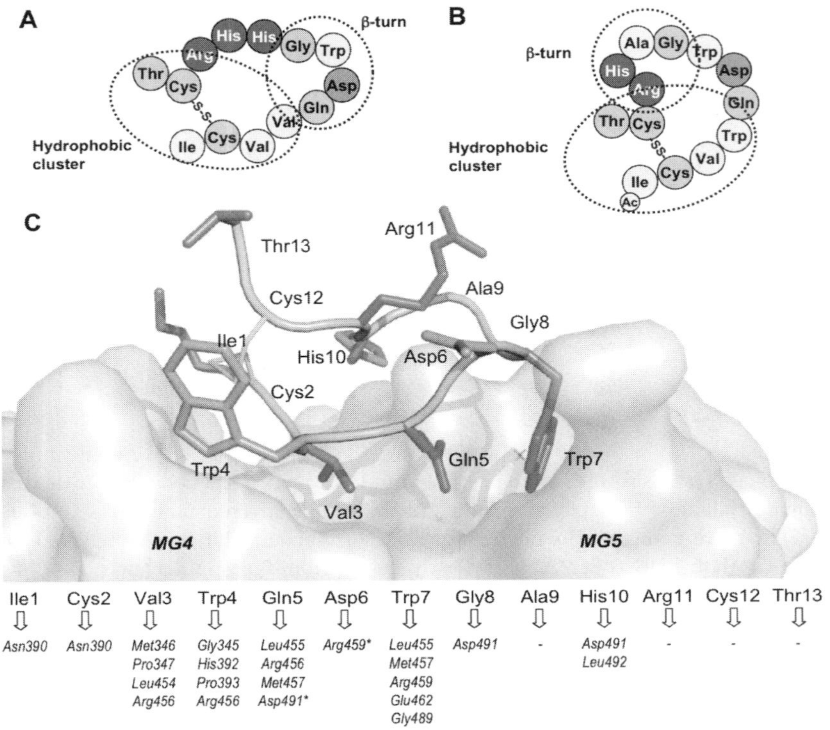

Fig. 6 The special arrangement and conformation of compstatin in solution (**a**) and in the bound form (**b**) are clearly distinct, which indicates an induced fit upon binding. During this process, the β-turn shifts from residues 5–8 to 8–11. In the co-crystal structure of 4W9A-compstatin and C3c (**c**; PDB code 2QKI), most of the residues in compstatin (main chain as ribbon, side chains as sticks) make contacts with C3c residues that form a shallow groove between domains MG4 and MG5 (surface representation), while the charged/polar amino acids Asp-6, Arg-11, and Thr-13 point out into the solvent. *Gln-5/Asp-491 and Asp-6/Arg-459 were mediated by water and bromine in the co-crystal, respectively

Although compstatin did not change the domain arrangement of C3c, there were still some structural adaptations within the binding site. Strikingly, the side chain arrangement did more closely resemble the local structure of native C3 than that of unbound C3c, which may also explain the higher relative affinity of compstatin to C3 (Janssen et al. 2007). Even more surprising, however, were the findings on the structure of bound compstatin, which was clearly distinct from the published solution structure (Janssen et al. 2007; Morikis et al. 1998). While bound compstatin still contains a β-turn element, it now encompasses residues 8–11 (instead of amino acids 5–8 as in the unbound peptide). As a consequence, the overall shape of the peptide changed significantly. The MG4/5 binding groove encloses 40% of the molecular surface of compstatin with Val-3, Trp-4, Gln-5, Trp-7, Gly-8, and His-10 pointing towards the C3c surface and making hydrogen bonds and hydrophobic contacts through both main and side chain atoms. On the other hand, the polar or charged amino acids Asp-6, Asp-11, and Thr-13 are pointing away from C3c towards the solvent (Fig. 6c). Additional interactions, i.e. Gln-5/Asp-491 and Asp-6/Arg-459, may be mediated through water or bromine molecules, respectively. However, these interactions are contradicting between the structures in the asymmetric unit or may be an artifact of the crystallization conditions (Janssen et al. 2007).

Analysis of the binding site also explains the species specificity of compstatin. C3c residues Gly-345, His-392, Pro-393, Leu-454, and Arg-459 all show direct interaction with compstatin and are highly conserved in primates but not in other species (Janssen et al. 2007). Unfortunately, these large variations in the binding sites of different animals makes the development of a 'universal', less species-specific compstatin analog that could be used in a wide panel of disease models rather unlikely. On the other hand, the creation of transgenic animals (e.g. mice) with 'humanized' compstatin binding sites may develop into a potential approach to circumvent this problem and could provide a compstatin-sensitive animal model.

Finally, our knowledge of the binding pocket and the contribution of individual compstatin residues to the interactions greatly facilitate further development and rational design approaches towards its human target. Despite the major differences between the bound and solution structure of compstatin, many predictions from previous structure-function-studies were in good agreement with the co-crystal (Janssen et al. 2007) and these models may further support development efforts.

5 First Steps Towards Therapeutic Applications

The potential to block the central step of complement activation makes compstatin an attractive candidate for a therapeutic use in a variety of clinical setups (Holland et al. 2004; Ricklin and Lambris 2007a). First steps in this direction were undertaken by investigating its effect on complement activation during extracorporeal blood circulation. Many artificial surfaces (e.g. plastics) and biomaterials are known to activate complement and induce inflammatory reactions (Nilsson et al. 2007). During cardiopulmonary bypass (CBP) surgery or haemodialysis, where whole blood is circulated extracorporeally, complement activation is a major complication and can influence the clinical outcome. When compstatin was added to blood circulating in PVC tubes, all readouts (C3a and MAC generation, C3b deposition, expression of CD11b on polymorphonuclear leukocytes) were clearly reduced when compared to the linear control peptide (Nilsson et al. 1998). By selective inhibition of the classical and alternative pathways using Mg-EGTA and compstatin, respectively, the major contribution of biomaterial-induced complement activation could be attributed to the alternative pathway (Andersson et al. 2005). Extended studies using the PVC tube model showed that expression of CD11b on both granulocytes and monocytes and the formation of platelet-granulocyte conjugates was attenuated and that induction of leukotriene B4 was reduced by acetyl-compstatin, while the formation of monocyte-platelet conjugates and the release of myeloperoxidase, lactoferrin, thrombospondin, thromboxane B4, and prostaglandin E2 was not affected (Lappegard et al. 2004, 2005). Very recently, the anti-inflammatory effect of compstatin in this model was impressively confirmed by monitoring a panel of 27 inflammation mediators during biomaterial-activation of complement. Compstatin efficiently reduced the formation of 12 of the 14 mediators (e.g. IL-8, FGF, VEGF) that were induced by PVC tubes (Lappegard et al. 2007). During CBP surgery, complement activation is not only triggered by blood contact with artificial surfaces and reperfusion of the arrested heart, but also by the postoperative administration of protamine in order to neutralize heparin and restore blood coagulation (Cavarocchi et al. 1985). In a primate model of protamine/heparin-induced complement activation, a combined injection and infusion of compstatin effectively inhibited the generation of C3 activation products without influencing heart rate, blood pressure or hematological parameters (Soulika et al. 2000). In this context, the study demonstrated for the first time both the efficacy and safety of compstatin as a complement inhibitory drug in an in vivo animal model. The use of this compound to prevent biomaterial-induced complement activation was further undermined by a study showing a largely reduced activation of neutrophils (as assessed by expression of the Mac-1 receptor) when the peptide was added to polymer-exposed blood (Schmidt et al. 2003).

Beneficial effects have also been described for an ex vivo model of xenotransplantation. Complement-mediated inflammation plays an essential role in the pathophysiology of hyperacute rejection. In the study, the survival time of pig kidneys that were perfused with human blood was remarkably increased in the presence of compstatin compared to the control (380 and 90 min, respectively). Furthermore, the study also confirmed that compstatin indeed acts at the level of C3 in vivo, since only the concentrations of C3 fragments and MAC but not C1 and C4 were decreased in the compstatin group (Fiane et al. 1999a,b). In another transplantation model, compstatin helped to elucidate the role of complement in delayed xenograft rejection. For this purpose, the expression of E-selectin on porcine aortic endothelial cells that were stimulated with human serum was monitored in the presence of compstatin, C1-inhibitor as well as C5, C7, and fD-specific antibodies. While attenuation of C1, C5, and C7 inhibited E-selectin expression completely, compstatin led to a markedly decreased expression level. No effect was observed for the anti-fD antibody. Together, these data suggested that complement activation via the classical pathway is predominantly involved in the pathophysiology and that MAC mediates endothelial cell activation (Solvik et al. 2001).

In a whole-blood model of *E. coli*-induced inflammation, compstatin not only reduced the formation of various complement-activation products, but also drastically inhibited oxidative burst in granulocytes and monocytes as well as formation of interleukin (IL)-8 (but not IL-6 and IL-10). The authors concluded that complement is a primary inducer of inflammation and that its inhibition could serve as a promising approach for treating sepsis or inflammatory diseases (Mollnes et al. 2002). Cytokine production (IL-12p40) was also influenced by compstatin-mediated complement inhibition in mononuclear cells that were stimulated by immune complexes (Tejde et al. 2004). In this respect, compstatin may be beneficial for treating immune complex-associated diseases such as rheumatoid arthritis or systemic lupus erythematosus.

6 From Bench to Bedside: Clinical Development

Regardless convincing concepts, potent lead compounds, and promising preclinical data, the road from the experimental entity to the marketed drug is usually long and bumpy. Several complement-targeting therapeutics have already taken this road and many of them have been discontinued (Ricklin and Lambris 2007a). Besides a high efficacy, several parameters such as stability, safety, toxicity, or pharmacokinetics have to be considered in the development process. Even though there are no clinical data available for compstatin so far, there are nevertheless many indicators from in vitro, ex vivo, and in vivo studies that constitute a high degree of drug-likeliness to the peptide.

Degradation, metabolism, and rapid (renal) excretion are some of the recurrent challenges in the development of peptidic drugs, which often lead to short plasma half-lives (Sato et al. 2006). In case of compstatin, the cyclic structure is largely responsible for a rather low degree of biotransformation in serum compared to previously investigated peptides for complement inhibition (Sahu et al. 2000). This high stability has been confirmed both in human and baboon plasma (Sahu et al. 2000; Soulika et al. 2000). Acetylation of the N-terminus even improved the stability in human plasma and resulted in a very low inactivation rate of 0.03%/min (Sahu et al. 2000). So far, no pharmacokinetic data regarding distribution and excretion are available for compstatin. However, since the peptide is likely to be used in acute phase situations as well as local or extracorporeal applications, a short plasma half-live may not be as limiting as in the case of chronic, systemic administration. On the other hand, unfavorable pharmacokinetic properties could potentially be modulated by chemical modifications such as PEG-ylation (Sato et al. 2006).

The high species specificity of compstatin so far limited the preclinical development in animal models. Complement inhibition has only been achieved in human and primate systems (e.g. baboon or rhesus monkey) but not for rodents (mouse, rat, guinea pig, rabbit) or pigs (Sahu et al. 2003). While residual activity has been observed in dog blood, compstatin activity was clearly reduced by two orders of magnitude (Furlong et al. 2000). The peptide was found to be highly specific for human C3 and did not affect blood clotting or the activity of serine proteases such as thrombin, trypsin, or elastase (Furlong et al. 2000). No adverse effects on heart rate or hematological parameters have been detected during in vivo experiments with baboons (Soulika et al. 2000). Furthermore, compstatin did not show any cytotoxicity towards polymorphonuclear cells (Furlong et al. 2000).

In its first attempt to enter clinical development, compstatin was licensed from the University of Pennsylvania by Potentia Pharmaceuticals Inc. in 1996 (Potentia 2006). Recently, the company announced the initiation of phase I clinical trials for the treatment of both dry and wet forms of age-related macular degeneration (AMD) (Potentia 2007). Since AMD represents the primary cause of blindness in industrialized nations, therapies that are able to prevent, retard, or even reverse this disease are highly desirable. While the formation of lipoprotein-rich deposits (drusen) and penetration of new blood vessels in the area between the choroids and the retinal pigmented epithelium has long been identified as the primary cause of the impaired vision, the contributions of different cellular pathways is still matter of investigation. Vascular endothelial growth factor (VEGF) has been found to be a key element in the neovascularization, and anti-VEGF antibody fragments (ranibizumab; Lucentis®, Genentech Inc.) or aptamers (pegaptanib; Macugen®, Eyetech Inc.) have already been introduced as treatment options for wet AMD (Blick et al. 2007). Although these drugs show high efficacy, they also require frequent injections into the eye and are very expensive. Despite an early intervention would be beneficial for reducing the risk of disease progression, there are currently no therapeutics available to treat the more prevalent dry form of AMD (Petrukhin 2007). The presence of several complement components (e.g. C3, C5, MAC) in the drusen, the correlation with mutations or polymorphisms in several complement genes (e.g. for factor H), and reported promoting effects of anaphylatoxins (C3a, C5a) to the neovascularization all indicate an involvement of the complement system in AMD pathogenesis (Jha et al. 2007; Markiewski and Lambris 2007; Petrukhin 2007). Local inhibition of complement activation is therefore considered a promising approach for treating both forms of AMD, and compstatin is the first complement-specific candidate that enters clinical trials for this indication. While other AMD drugs like Macugen or Lucentis have to be administered by monthly injections into the eye, Potentia is developing a intravitreal delivery system that should release a continuous therapeutic dose of a compstatin analog (POT-4) during a period of at least 12 months. Considering the novel form of administration and the first use of compstatin for human in vivo studies, the results from this clinical trial are keenly awaited.

7 Conclusions and Perspectives

Its target at the core of complement and its favorable properties as a rather small and stable peptide drug uniquely position compstatin in the repertoire of complement-specific therapeutics. The first clinical studies for an application in AMD is considered an important indicator for its potential as a drug and will certainly help collecting pharmacokinetic and toxicological parameters for this compound. Even though the outcome concerning efficacy is difficult to predict in view of the unknown impact of complement in the pathogenesis of AMD, a favorable safety profile may pave the way to the development of alternative clinical applications of compstatin. In this context, the prevention of complement-induced

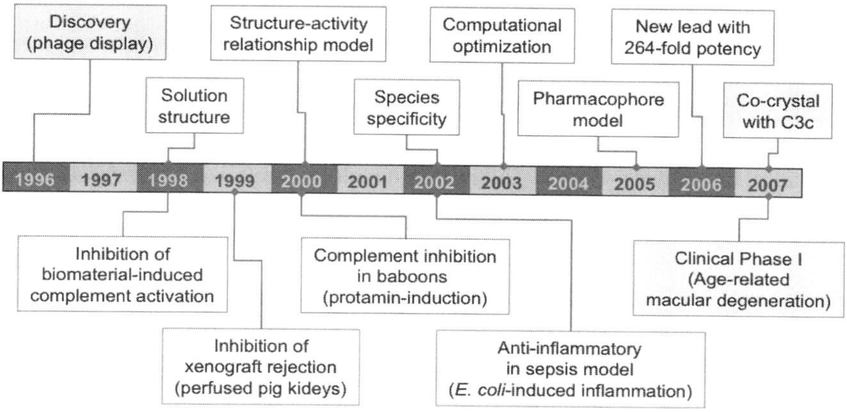

Fig. 7 Milestones in the first decade of compstatin development with an emphasis on structural/analytical (*upper part*) and clinical (*lower part*) improvements and findings

inflammation during haemodialysis or CBP surgery could be promising candidates, even more since compstatin already has proven efficacy in the baboon model and biomaterial-related studies (see above).

Meanwhile, the structural development has gained new impetus with the C3c/compstatin co-crystal. The detailed knowledge of the binding site and contact residues will allow a rational drug design. Alongside with an optimization of the peptide sequence, a transition to smaller, non-peptidic structures may allow to improve the drug-like properties of the compound. For this purpose, β-turn mimetics could provide a starting point: owing to their high prevalence in protein interaction sites, β-turn peptides and mimetics thereof are regarded as promising scaffolds in drug design (Kee and Jois 2003). Considering the proposed mechanism of action by blocking the initial interaction between C3 and the convertase, compstatin could be an interesting addition to the emerging class of low-molecular-weight protein-protein interaction inhibitor drugs (Wells and McClendon 2007). While the current molecular mass (1600 Da) is slightly higher than comparable compounds in this class (500–900 Da), compstatin already has a promising ligand efficiency (0.09 kcal per mole per non-hydrogen atom) and has certainly found a 'hotspot' in the C3 activation process.

Compared with other complement therapeutics in late preclinical development or in clinical trial, the way of acting directly on the native C3 protein is rather unique. However, a similar strategy has been described in case of complement evasion by human pathogens (Lambris et al. 2008). In particular, the extracellular fibrinogen-binding protein (Efb) and the related Efb-homologous protein (Ehp) from *Staphylococcus aureus* have been found to efficiently inhibit complement activation by binding to native C3 and all its fragments that contain the thioester domain (Hammel et al. 2007a,b). This binding induces a conformational change in both C3 and C3b and therefore influences the conversion of C3 and C5 by the corresponding convertases (Hammel et al. 2007b; Jongerius et al. 2007). While the bacterial proteins themselves may be too immunogenic for a direct use as therapeutics, they may be used as templates for developing complement-inhibiting drugs.

Finally, the success story of compstatin development illustrates the potential of academia-centered drug discovery efforts. Even without the financial and logistic resources of big pharmaceutical companies, the careful selection of screening methods, the combination of powerful techniques, and a diversified net of collaborations may produce valuable lead compounds for clinical development. Supporting incentives by governmental science foundations, such as the 'Rapid Access to Interventional Development' pilot program (RAID) by the National Institutes of Health (NIH), will clearly further these efforts in the future.

Acknowledgment

This work was supported by the National Institutes of Health grants GM062134, GM069736, EB003968, and AI068730.

References

Andersson, J., Ekdahl, K.N., Lambris, J.D. and Nilsson, B. (2005) Binding of C3 fragments on top of adsorbed plasma proteins during complement activation on a model biomaterial surface. *Biomaterials* 26, 1477–1485

Blick, S.K., Keating, G.M. and Wagstaff, A.J. (2007) Ranibizumab. *Drugs* 67, 1199–1206; discussion 1207–1199

Cavarocchi, N.C., Schaff, H.V., Orszulak, T.A., Homburger, H.A., Schnell, W.A., Jr. and Pluth, J.R. (1985) Evidence for complement activation by protamine-heparin interaction after cardiopulmonary bypass. *Surgery* 98, 525–531

Davis, A.E., III (2006) Mechanism of angioedema in first complement component inhibitor deficiency. *Immunol. Allergy Clin. North Am.* 26, 633–651

Degn, S.E., Thiel, S. and Jensenius, J.C. (2007) New perspectives on mannan-binding lectin-mediated complement activation. *Immunobiology* 212, 301–311

Fiane, A.E., Mollnes, T.E., Videm, V., Hovig, T., Hogasen, K., Mellbye, O.J., Spruce, L., Moore, W.T., Sahu, A. and Lambris, J.D. (1999a) Compstatin, a peptide inhibitor of C3, prolongs survival of ex vivo perfused pig xenografts. *Xenotransplantation* 6, 52–65

Fiane, A.E., Mollnes, T.E., Videm, V., Hovig, T., Hogasen, K., Mellbye, O.J., Spruce, L., Moore, W.T., Sahu, A. and Lambris, J.D. (1999b) Prolongation of ex vivo-perfused pig xenograft survival by the complement inhibitor compstatin. *Transplant. Proc.* 31, 934–935

Furlong, S.T., Dutta, A.S., Coath, M.M., Gormley, J.J., Hubbs, S.J., Lloyd, D., Mauger, R.C., Strimpler, A.M., Sylvester, M.A., Scott, C.W. and Edwards, P.D. (2000) C3 activation is inhibited by analogs of compstatin but not by serine protease inhibitors or peptidyl alpha-ketoheterocycles. *Immunopharmacology* 48, 199–212

Gronroos, J.O., Salonen, J.H., Viander, M., Nevalainen, T.J. and Laine, V.J. (2005) Roles of group IIA phospholipase A2 and complement in killing of bacteria by acute phase serum. *Scand. J. Immunol.* 62, 413–419

Hammel, M., Sfyroera, G., Pyrpassopoulos, S., Ricklin, D., Ramyar, K.X., Pop, M., Jin, Z., Lambris, J.D. and Geisbrecht, B.V. (2007a) Characterization of Ehp, a secreted complement inhibitory protein from *Staphylococcus aureus*. *J. Biol. Chem.* 282, 30051–30061

Hammel, M., Sfyroera, G., Ricklin, D., Magotti, P., Lambris, J.D. and Geisbrecht, B.V. (2007b) A structural basis for complement inhibition by *Staphylococcus aureus*. *Nat. Immunol.* 8, 430–437

Holland, M.C., Morikis, D. and Lambris, J.D. (2004) Synthetic small-molecule complement inhibitors. *Curr. Opin. Investig. Drugs* 5, 1164–1173

Janssen, B.J. and Gros, P. (2007) Structural insights into the central complement component C3. *Mol. Immunol.* 44, 3–10

Janssen, B.J., Christodoulidou, A., McCarthy, A., Lambris, J.D. and Gros, P. (2006) Structure of C3b reveals conformational changes that underlie complement activity. *Nature* 444, 213–216

Janssen, B.J., Halff, E.F., Lambris, J.D. and Gros, P. (2007) Structure of compstatin in complex with complement component C3c reveals a new mechanism of complement inhibition. *J. Biol. Chem.* 282, 29241–29247

Jha, P., Bora, P.S. and Bora, N.S. (2007) The role of complement system in ocular diseases including uveitis and macular degeneration. *Mol. Immunol.* 44, 3901–3908

Jongerius, I., Kohl, J., Pandey, M.K., Ruyken, M., van Kessel, K.P., van Strijp, J.A. and Rooijakkers, S.H. (2007) Staphylococcal complement evasion by various convertase-blocking molecules. *J. Exp. Med.* 204, 2461–2471

Katragadda, M. and Lambris, J.D. (2006) Expression of compstatin in *Escherichia coli*: incorporation of unnatural amino acids enhances its activity. *Protein Expr. Purif.* 47, 289–295

Katragadda, M., Morikis, D. and Lambris, J.D. (2004) Thermodynamic studies on the interaction of the third complement component and its inhibitor, compstatin. *J. Biol. Chem.* 279, 54987–54995

Katragadda, M., Magotti, P., Sfyroera, G. and Lambris, J.D. (2006) Hydrophobic effect and hydrogen bonds account for the improved activity of a complement inhibitor, compstatin. *J. Med. Chem.* 49, 4616–4622

Katschke, K.J., Jr., Helmy, K.Y., Steffek, M., Xi, H., Yin, J., Lee, W.P., Gribling, P., Barck, K.H., Carano, R.A., Taylor, R.E., Rangell, L., Diehl, L., Hass, P.E., Wiesmann, C. and van Lookeren Campagne, M. (2007) A novel inhibitor of the alternative pathway of complement reverses inflammation and bone destruction in experimental arthritis. *J. Exp. Med.* 204, 1319–1325

Kay, B.K., Kasanov, J. and Yamabhai, M. (2001) Screening phage-displayed combinatorial peptide libraries. *Methods* 24, 240–246

Kee, K.S. and Jois, S.D. (2003) Design of beta-turn based therapeutic agents. *Curr. Pharm. Des.* 9, 1209–1224

Klegeris, A., Singh, E.A. and McGeer, P.L. (2002) Effects of C-reactive protein and pentosan polysulphate on human complement activation. *Immunology* 106, 381–388

Klepeis, J.L., Floudas, C.A., Morikis, D., Tsokos, C.G., Argyropoulos, E., Spruce, L. and Lambris, J.D. (2003) Integrated computational and experimental approach for lead optimization and design of compstatin variants with improved activity. *J. Am. Chem. Soc.* 125, 8422–8423

Klepeis, J.L., Floudas, C.A., Morikis, D., Tsokos, C.G. and Lambris, J.D. (2004) Design of peptide analogues with improves activity using a novel de novo protein design approach. *Ind. Eng. Chem. Res.* 43, 3817–3826

Ladner, R.C., Sato, A.K., Gorzelany, J. and de Souza, M. (2004) Phage display-derived peptides as therapeutic alternatives to antibodies. *Drug Discov. Today* 9, 525–529

Lambris, J.D. and Holers, V.M. (Eds.) (2000) *Therapeutic Interventions in the Complement System*. Humana Press, Totowa, NJ, USA

Lambris, J.D., Ricklin, D. and Geisbrecht, B.V. (2008) Complement evasion by human pathogens. *Nat. Rev. Microbiol.* 6, 132–142

Lappegard, K.T., Fung, M., Bergseth, G., Riesenfeld, J., Lambris, J.D., Videm, V. and Mollnes, T.E. (2004) Effect of complement inhibition and heparin coating on artificial surface-induced leukocyte and platelet activation. *Ann. Thorac. Surg.* 77, 932–941

Lappegard, K.T., Riesenfeld, J., Brekke, O.L., Bergseth, G., Lambris, J.D. and Mollnes, T.E. (2005) Differential effect of heparin coating and complement inhibition on artificial surface-induced eicosanoid production. *Ann. Thorac. Surg.* 79, 917–923

Lappegard, K.T., Bergseth, G., Riesenfeld, J., Pharo, A., Magotti, P., Lambris, J.D. and Mollnes, T.E. (2007) The artificial surface-induced whole blood inflammatory reaction revealed by increases in a series of chemokines and growth factors is largely. complement dependent. *J. Biomed. Mater. Res. A* doi: 10.1002/jbm.a.31750

Mallik, B. and Morikis, D. (2005) Development of a quasi-dynamic pharmacophore model for anti-complement peptide analogues. *J. Am. Chem. Soc.* 127, 10967–10976

Mallik, B., Lambris, J.D. and Morikis, D. (2003) Conformational interconversion in compstatin probed with molecular dynamics simulations. *Proteins* 53, 130–141

Mallik, B., Katragadda, M., Spruce, L.A., Carafides, C., Tsokos, C.G., Morikis, D. and Lambris, J.D. (2005) Design and NMR characterization of active analogues of compstatin containing non-natural amino acids. *J. Med. Chem.* 48, 274–286

Markiewski, M.M. and Lambris, J.D. (2007) The role of complement in inflammatory diseases from behind the scenes into the spotlight. *Am. J. Pathol.* 171, 715–727

Mollnes, T.E., Brekke, O.L., Fung, M., Fure, H., Christiansen, D., Bergseth, G., Videm, V., Lappegard, K.T., Kohl, J. and Lambris, J.D. (2002) Essential role of the C5a receptor in *E. coli*-induced oxidative burst and phagocytosis revealed by a novel lepirudin-based human whole blood model of inflammation. *Blood* 100, 1869–1877

Morikis, D. and Lambris, J.D. (2002) Structural aspects and design of low-molecular-mass complement inhibitors. *Biochem. Soc. Trans.* 30, 1026–1036

Morikis, D., Assa-Munt, N., Sahu, A. and Lambris, J.D. (1998) Solution structure of compstatin, a potent complement inhibitor. *Protein Sci.* 7, 619–627

Morikis, D., Roy, M., Sahu, A., Troganis, A., Jennings, P.A., Tsokos, G.C. and Lambris, J.D. (2002) The structural basis of compstatin activity examined by structure-function-based design of peptide analogs and NMR. *J. Biol. Chem.* 277, 14942–14953

Morikis, D., Soulika, A.M., Mallik, B., Klepeis, J.L., Floudas, C.A. and Lambris, J.D. (2004) Improvement of the anti-C3 activity of compstatin using rational and combinatorial approaches. *Biochem. Soc. Trans.* 32, 28–32

Mulakala, C., Lambris, J.D. and Kaznessis, Y. (2007) A simple, yet highly accurate, QSAR model captures the complement inhibitory activity of compstatin. *Bioorg. Med. Chem.* 15, 1638–1644

Nielsen, E.W., Waage, C., Fure, H., Brekke, O.L., Sfyroera, G., Lambris, J.D. and Mollnes, T.E. (2007) Effect of supraphysiologic levels of C1-inhibitor on the classical, lectin and alternative pathways of complement. *Mol. Immunol.* 44, 1819–1826

Nilsson, B., Larsson, R., Hong, J., Elgue, G., Ekdahl, K.N., Sahu, A. and Lambris, J.D. (1998) Compstatin inhibits complement and cellular activation in whole blood in two models of extracorporeal circulation. *Blood* 92, 1661–1667

Nilsson, B., Ekdahl, K.N., Mollnes, T.E. and Lambris, J.D. (2007) The role of complement in biomaterial-induced inflammation. *Mol. Immunol.* 44, 82–94

Pedersen, E.D., Aass, H.C., Rootwelt, T., Fung, M., Lambris, J.D. and Mollnes, T.E. (2007) CD59 efficiently protects human NT2-N neurons against complement-mediated damage. *Scand. J. Immunol.* 66, 345–351

Petrukhin, K. (2007) New therapeutic targets in atrophic age-related macular degeneration. *Expert Opin. Ther. Targets* 11, 625–639

Potentia (2006) Press Release: Potentia Pharmaceuticals Licences Complement Inhibitor Compstatin from the University of Pennsylvania (http://www.potentiapharma.com/about/ news.htm#13)

Potentia (2007) Press Release: Potentia Pharmaceuticals Announces Initiation of Phase I Clinical Trials to Evaluate its Lead Compound for Age-Related Macular Degeneration (http://www.potentiapharma.com/about/news.htm#17)

Ricklin, D. and Lambris, J.D. (2007a) Complement-targeted therapeutics. *Nat. Biotechnol.* 25, 1265–1275

Ricklin, D. and Lambris, J.D. (2007b) Exploring the complement interaction network using surface plasmon resonance. *Adv. Exp. Med. Biol.* 598, 260–278

Ritis, K., Doumas, M., Mastellos, D., Micheli, A., Giaglis, S., Magotti, P., Rafail, S., Kartalis, G., Sideras, P. and Lambris, J.D. (2006) A novel C5a receptor-tissue factor cross-talk in neutrophils links innate immunity to coagulation pathways. *J. Immunol.* 177, 4794–4802

Rother, R.P., Rollins, S.A., Mojcik, C.F., Brodsky, R.A. and Bell, L. (2007) Discovery and development of the complement inhibitor eculizumab for the treatment of paroxysmal nocturnal hemoglobinuria. *Nat. Biotechnol.* 25, 1256–1264

Rotondi, K.S. and Gierasch, L.M. (2006) Natural polypeptide scaffolds: beta-sheets, beta-turns, and beta-hairpins. *Biopolymers* 84, 13–22

Sahu, A., Kay, B.K. and Lambris, J.D. (1996) Inhibition of human complement by a C3-binding peptide isolated from a phage-displayed random peptide library. *J. Immunol.* 157, 884–891

Sahu, A., Sunyer, J.O., Moore, W.T., Sarrias, M.R., Soulika, A.M. and Lambris, J.D. (1998) Structure, functions, and evolution of the third complement component and viral molecular mimicry. *Immunol. Res.* 17, 109–121

Sahu, A., Soulika, A.M., Morikis, D., Spruce, L., Moore, W.T. and Lambris, J.D. (2000) Binding kinetics, structure-activity relationship, and biotransformation of the complement inhibitor compstatin. *J. Immunol.* 165, 2491–2499

Sahu, A., Morikis, D. and Lambris, J.D. (2003) Compstatin, a peptide inhibitor of complement, exhibits species-specific binding to complement component C3. *Mol. Immunol.* 39, 557–566

Sarrias, M.R., Whitbeck, J.C., Rooney, I., Spruce, L., Kay, B.K., Montgomery, R.I., Spear, P.G., Ware, C.F., Eisenberg, R.J., Cohen, G.H. and Lambris, J.D. (1999) Inhibition of herpes simplex virus gD and lymphotoxin-alpha binding to HveA by peptide antagonists. *J. Virol.* 73, 5681–5687

Sato, A.K., Viswanathan, M., Kent, R.B. and Wood, C.R. (2006) Therapeutic peptides: technological advances driving peptides into development. *Curr. Opin. Biotechnol.* 17, 638–642

Schmidt, S., Haase, G., Csomor, E., Lutticken, R. and Peltroche-Llacsahuanga, H. (2003) Inhibitor of complement, compstatin, prevents polymer-mediated Mac-1 up-regulation of human neutrophils independent of biomaterial type tested. *J. Biomed. Mater. Res. A* 66, 491–499

Solvik, U.O., Haraldsen, G., Fiane, A.E., Boretti, E., Lambris, J.D., Fung, M., Thorsby, E. and Mollnes, T.E. (2001) Human serum-induced expression of E-selectin on porcine aortic endothelial cells in vitro is totally complement mediated. *Transplantation* 72, 1967–1973

Song, M.K., Kim, S.Y. and Lee, J. (2005) Understanding the structural characteristics of compstatin by conformational space annealing. *Biophys. Chem.* 115, 201–207

Soulika, A.M., Khan, M.M., Hattori, T., Bowen, F.W., Richardson, B.A., Hack, C.E., Sahu, A., Edmunds, L.H., Jr. and Lambris, J.D. (2000) Inhibition of heparin/protamine complex-induced complement activation by compstatin in baboons. *Clin. Immunol.* 96, 212–221

Soulika, A.M., Morikis, D., Sarrias, M.R., Roy, M., Spruce, L.A., Sahu, A. and Lambris, J.D. (2003) Studies of structure-activity relations of complement inhibitor compstatin. *J. Immunol.* 171, 1881–1890

Soulika, A.M., Holland, M.C., Sfyroera, G., Sahu, A. and Lambris, J.D. (2006) Compstatin inhibits complement activation by binding to the beta-chain of complement factor 3. *Mol. Immunol.* 43, 2023–2029

Tamamis, P., Skourtis, S.S., Morikis, D., Lambris, J.D. and Archontis, G. (2007) Conformational analysis of compstatin analogues with molecular dynamics simulations in explicit water. *J. Mol. Graph. Model.* 26, 571–580

Tejde, A., Mathsson, L., Ekdahl, K.N., Nilsson, B. and Ronnelid, J. (2004) Immune complex-stimulated production of interleukin-12 in peripheral blood mononuclear cells is regulated by the complement system. *Clin. Exp. Immunol.* 137, 521–528

Wells, J.A. and McClendon, C.L. (2007) Reaching for high-hanging fruit in drug discovery at protein-protein interfaces. *Nature* 450, 1001–1009

Wiesmann, C., Katschke, K.J., Yin, J., Helmy, K.Y., Steffek, M., Fairbrother, W.J., McCallum, S.A., Embuscado, L., DeForge, L., Hass, P.E. and van Lookeren Campagne, M. (2006) Structure of C3b in complex with CRIg gives insights into regulation of complement activation. *Nature* 444, 217–220

Yu, J. and Smith, G.P. (1996) Affinity maturation of phage-displayed peptide ligands. *Meth. Enzymol.* 267, 3–27

21. Derivatives of Human Complement Component C3 for Therapeutic Complement Depletion: A Novel Class of Therapeutic Agents

David C. Fritzinger[1], Brian E. Hew[1], June Q. Lee[1], James Newhouse[2], Maqsudul Alam[3], John R. Ciallella[4], Mallory Bowers[4], William B. Gorsuch[5], Benjamin J. Guikema[5], Gregory L. Stahl[5], and Carl-Wilhelm Vogel[1]

[1]Cancer Research Center of Hawaii, University of Hawaii at Manoa, Honolulu, HI 96813, USA, cvogel@crch.hawaii.edu
[2]Maui High Performance Computing Center, Kihei, HI 96753, USA
[3]Advanced Studies in Genomics, Proteomics, and Bioinformatics, University of Hawaii at Manoa, Honolulu, HI 96822, USA
[4]Melior Discovery Corporation, Exton, PA 19341, USA
[5]Brigham and Women's Hospital, Center for Experimental Therapeutics and Reperfusion Injury, Harvard University, Boston, MA 02115, USA

Abstract. To obtain proteins with the complement-depleting activity of Cobra Venom Factor (CVF), but with less immunogenicity, we have prepared human C3/CVF hybrid proteins, in which the C-terminus of the α-chain of human C3 is exchanged with homologous regions of the C-terminus of the β-chain of CVF. We show that these hybrid proteins are able to deplete complement, both in vitro and in vivo. One hybrid protein, HC3-1496, is shown to be effective in reducing complement-mediated damage in two disease models in mice, collagen-induced arthritis and myocardial ischemia/reperfusion injury. Human C3/CVF hybrid proteins represent a novel class of biologicals as potential therapeutic agents in many diseases where complement is involved in the pathogenesis.

1 Background and Concept

Cobra Venom Factor (CVF) has long been known to be a structural analog of a complement component C3 (Vogel 1991; Vogel et al. 1984, 1996). Both C3 and CVF are synthesized as single-chain pre-pro-proteins that are subsequently processed into the mature two-chain C3 protein, and the mature three-chain CVF protein, respectively (de Bruijn and Fey 1985; Fritzinger et al. 1994). The two proteins share extensive sequence similarity, at both the protein and DNA levels. CVF and human C3 are approximately 50% identical at the protein level and approximately 70% similar if one allows for conservative replacements (Fritzinger et al. 1994). The structural homology between cobra C3 and CVF is even greater;

J.D. Lambris (ed.), *Current Topics in Complement II.*
doi: 10:1007/978-0-387-78952-1_21, © Springer Science+Business Media, LLC 2008

Table 1 Diseases with complement pathogenesis

Rheumatoid arthritis
Lupus erythematosus
Macular degeneration
Hyperacute rejection (xenotransplantation)
Ischemia/reperfusion injury
Psoriasis
Myasthenia gravis
Bullous pemphigoid
Asthma

these two proteins exhibit an amino acids sequence identity of approximately 85% and a cDNA sequence identity of approximately 93% (Fritzinger et al. 1992, 1994).

CVF is a functional analog of C3b, the activated form of complement component C3. Both proteins can bind factor B, leading to the subsequent formation a bimolecular C3/C5 convertase of the alternative complement pathway (Vogel 1991; Vogel et al. 1996). The two C3/C5 convertases C3b,Bb and CVF,Bb share the molecular architecture, the active site-bearing Bb subunit, and the substrate specificity for the two complement proteins C3 and C5. However, despite the striking structural and functional similarities, the two convertases exhibit important functional differences. C3b,Bb is a surface-bound convertase, exhibiting rapid decay-dissociation into the respective subunits with a half-life of 1.5 min at 37 °C, and the C3b,Bb enzyme and C3b itself are subject to inactivation by the complement regulatory proteins factors H and I (Fritzinger et al. 2005, 2008; Hew et al. 2004; Medicus et al. 1976; Pangburn and Müller-Eberhard 1986; Pangburn et al. 1977; Whaley and Ruddy 1976). In contrast, the CVF-dependent convertase is a fluid-phase enzyme, exhibiting significantly greater physico-chemical stability with a half-life of approximately 7 h at 37 °C for its spontaneous decay-dissociation. Additionally, both the CVF,Bb enzyme and CVF itself are completely resistant to inactivation by factors H and I (Fritzinger et al. 2008; Lachmann and Halbwachs 1975; Medicus et al. 1976; Nagaki et al. 1978; Pangburn and Müller-Eberhard 1986; Pangburn et al. 1977; Vogel and Müller-Eberhard 1982). Due to its properties of physico-chemical stability and resistance to regulation, the CVF,Bb enzyme, once formed in serum, will continually hydrolyze C3 and C5, thereby significantly reducing the serum concentration of these two complement components and ultimately resulting in depletion of serum complement activity. CVF in purified form, despite of its origin from cobra venom, does not constitute a toxin and can be safely administered to laboratory animals (Cochrane et al. 1970; Maillard and Zarco 1968; Nelson 1966; Vogel 1991). This property of CVF has been exploited for approximately 40 years to decomplement laboratory animals, from small rodents to primates, by i.v. or i.p. injection ((Vogel 1991) and references therein). Complement depletion by CVF has become an important experimental tool, if not the gold standard, to study the biological role of

Table 2 Concepts of pharmacological complement intervention

COMPLEMENT INHIBITION (inhibition of activated or activatable complement components)

- e.g., antibodies, receptor antagonists, recombinant regulators

COMPLEMENT RESISTANCE (increased resistance to activated complement components)

- expression of membrane inhibitors (e.g., MCP, CD59)

COMPLEMENT DEPLETION (consumption of complement components)

- e.g., Cobra Venom Factor, human C3/Cobra Venom Factor hybrids

complement in host defense as well as its role in the pathogenesis of diseases (Hebell et al. 1991; Morgan and Harris 2003; Sahu and Lambris 2000).

The complement system is an important biological effector system, playing important roles in both innate and adaptive immunity as well as in the inflammatory response (Carroll 2004; Fujita et al. 2004; Mollnes and Kirschfink 2006). However, the complement system is also involved in the pathogenesis of many diseases (Table 1). Some important and frequent diseases with a well establish role for complement include rheumatoid arthritis, lupus erythematosus, myasthenia gravis, macular degeneration, and ischemia/reperfusion injury, to name a few. As a matter of fact, complement depletion with CVF in animal models of diseases has often served as an experimental tool to establish the role of complement as the pathogenetic mechanism.

Given the frequency, severity, and, in some cases, debilitating and life threatening nature of diseases with complement pathogenesis, the past decade has seen the development of multiple experimental therapies with the aim to counteract harmful complement activation (Hebell et al. 1991, Morgan and Harris 2003, Sahu and Lambris 2000). Conceptually, there are at least three different concepts for pharmacological intervention of complement (Table 2). The most frequently pursued concept is to develop agents for complement inhibition directed to inhibit the activation of an activitable complement component or to inhibit the action of an activated complement component. Examples include antibodies or antibody fragments to individual complement components (e.g., anti-C5), antagonists to complement receptors (e.g., anaphylatoxin receptor blockers), and recombinant complement regulatory proteins (e.g., soluble complement receptor 1 (CR1)) (Fodor et al. 1995, Mollnes and Kirschfink 2006, Weisman et al. 1990). This category includes the only antibody against a complement component approved for human use. It is an antibody against C5 that is used to inhibit its activation in patients with paroxysmal nocturnal hemoglobinuria (PNH) (Hillmen et al. 2004). A conceptually different pharmacological approach to mitigate or prevent the effects of unwanted complement activation is to increase the resistance of cells or organs to complement attack. This concept has been tried to overcome the hyperacute rejection of organs after xenotransplantation by creating transgenic pigs expressing

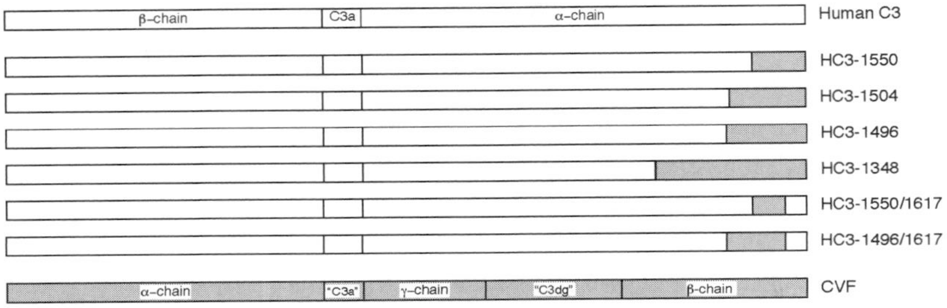

Fig. 1 Schematic representation of human C3/CVF hybrid proteins ("Humanized" CVF). The N-terminus is always to the left. Chain homologies between C3 and CVF are indicated. The nomenclature for hybrid proteins shows the amino acid sequence number of human C3 from which on it was replaced with CVF sequence until the C-terminus. For hybrid proteins for which C3 sequence was not replaced all the way to the C-terminus, the second number indicates the amino acid residue from which human C3 sequence continued to the C-terminus. Please note that even the stretches of CVF sequence contain many positions with identical amino acid residues between human C3 and CVF

human membrane-bound complement regulatory proteins such as MCP or CD59 that physiologically protect host cells against low-grade complement activation (Costa et al. 2002, Zhou et al. 2005). However, it appears that this approach has been abandoned.

The third conceptually different approach is complement depletion as exemplified by the activity of CVF. The depletion of complement in a clinical situation where complement activation contributes to the disease process will prevent harmful effects of complement activation. If complement is consumed before any harmful trigger of complement activation occurs, it will be without pathological consequences. The concept of pharmacological depletion of complement has been developed by nature itself (cobras of the *Naja* species); and CVF represents the lead example of a complement-depleting pharmacological agent. However, CVF itself appears unsuitable for therapeutic use in humans because of its high immunogenicity, both because of the immunogenic nature of its carbohydrate structures (Gowda et al. 1992, 1994, 2001; Grier et al. 1987) and the phylogenetic distance between cobra and humans. Furthermore, cobra venom as the natural source for CVF is in limited supply, even if the cobra would not be on the list of endangered species.

2 Development of C3 Derivatives for Therapeutic Complement Depletion

The development of a biological reagent for therapeutic complement depletion devoid of the inherent shortcomings of CVF will involve successful strategies for the recombinant production of CVF as well as for the reduction of its

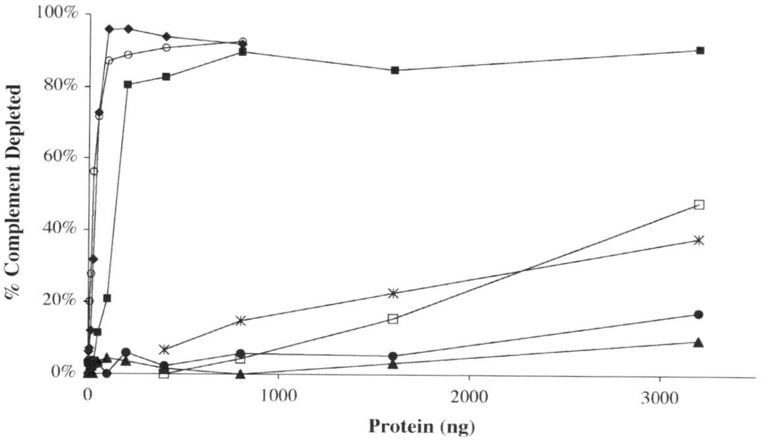

Fig. 2 Depletion of the complement activity in human serum by native CVF and different human C3/CVF hybrid proteins. Complement activity was determined in a hemolytic assay using sensitized sheep erythrocytes after incubation of human serum for 3 h with native CVF or hybrid proteins as described (Vogel and Müller-Eberhard 1984) (native CVF (◆), HC3-1496 (O), HC3-1348 (■), HC3-1550 (*), HC3-1504 (□), HC3-1496/1617 (●), HC3-1550/1617 (▲))

immunogenicity ("humanization" of CVF). The first goal of developing a system for recombinant production has been met with the successful generation of functionally active recombinant CVF using different systems for expression in eukaryotic cells (Kock et al. 2004; Vogel et al. 2004). Depending on the expression system and the experimental conditions used, the recombinant CVF can be expressed as single-chain pro-CVF, as a C3-like two-chain form, and as a C3b-like two-chain form, (Kock et al. 2004; Vogel et al. 2004). All three forms of recombinant CVF appear to be functionally identical and indistinguishable from native CVF. The recombinant production of CVF in eukaryotic expression systems should also reduce the immunogenicity from carbohydrate epitopes as unusual carbohydrate epitopes in CVF have been known to contribute to its immunogenicity (Gowda et al. 2001, Gowda et al. 1994, Gowda et al. 1992, Grier et al. 1987).

The strategy for humanization of CVF requires the identification of the important structural features in the CVF molecule that are responsible for its ability to form a physico-chemically stable convertase as well as its resistance to factors H and I. Previous work in our laboratory demonstrated that replacing the C-terminal 113 amino acid residues of the CVF β-chain with corresponding sequence from the

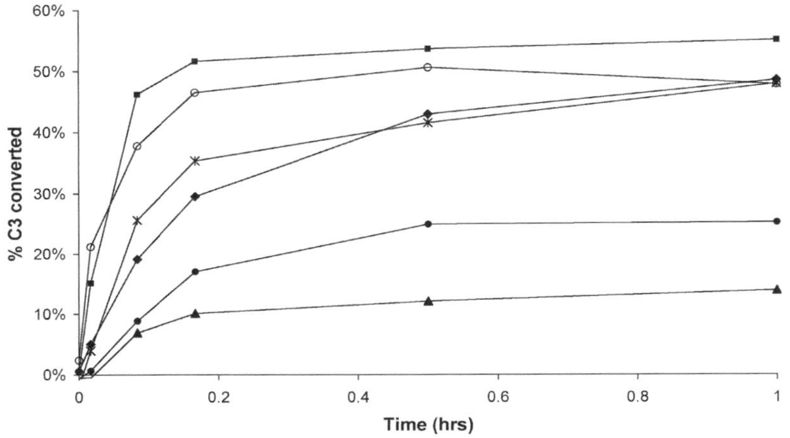

Fig. 3 C3-cleaving activity of convertases formed with native CVF and different human C3/CVF hybrid proteins. Shown is the percent of C3 converted to C3b after incubation of preformed convertases with purified human C3. At time intervals as indicated, the extent of C3 cleavage was determined by the amount of C3b α'-chain generated as measured by densitometry of SDS polyacrylamide gels under reducing conditions as described previously (Kock et al. 2004) (native CVF (◆), HC3-1496 (O), HC3-1348 (■), HC3-1550 (*), HC3-1496/1617 (●), HC3-1550/1617 (▲))

α-chain of cobra C3 resulted in a CVF/cobra C3 hybrid protein with substantially reduced complement-depleting activity (Wehrhahn et al. 2000), strongly suggesting that the C-terminal end of the CVF β-chain harbors crucial structures for the ability of CVF to form a stable convertase. Subsequent work in our laboratory (Fritzinger et al. 2005, 2006, 2007, 2008; Hew et al. 2004) and the laboratory of our previous co-workers (Kölln et al. 2004a,b, 2005) resulted in the generation of reverse hybrid proteins, in which small portions of the C-terminus of the α-chain of human C3 were replaced with homologous sequences from the β-chain of CVF. As shown in Fig. 1, the human C3/CVF hybrid proteins represent human C3 hybrids with a small stretch of CVF sequence at the C-terminus. Even without allowing for conservative replacements, the sequence homology between human C3 and the human C3/CVF hybrids is usually greater than 90%, and, in the case of hybrid HC3-1550, greater than 96% (Fritzinger et al. 2004, 2005, 2008; Hew et al. 2004; Vogel and Fritzinger 2008).

Figure 2 shows that the human C3/CVF hybrid proteins exhibit complement-depleting activity in human serum. The complement-depleting activity of several hybrid proteins resembles that of CVF whereas others exhibit a significantly

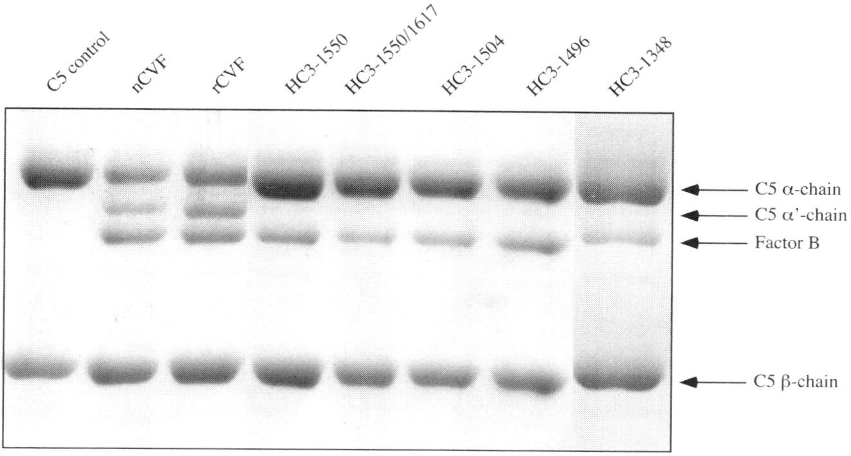

Fig. 4 Lack of C5-cleaving activity of convertases formed by human C3/CVF hybrid proteins. The C5-cleaving activity was determined by incubation of preformed convertases with purified human C5 for 24 h and subsequent densitometry of SDS polyacrylamide gels run under reducing conditions as described (Kock et al. 2004). Natural CVF (nCVF) and recombinant (rCVF) served as positive controls where C5-cleaving activity is evident from the appearance of the C5α'-chain. The left lane shows the substrate human C5

reduced ability to deplete human complement. The fact that these human C3 derivatives are able to deplete complement in serum was unexpected because all hybrid proteins contain all known binding sites for factor H and all three cleavage sites for factor I which would have led to the prediction that these hybrid proteins are readily inactivated by factors H and I. This result also indicates that the C-terminal region of the CVF β-chain not only contains the stability site but is also responsible for at least partial resistance to the regulatory actions of factors H and I.

Figure 3 shows the ability of convertases formed with several human C3/CVF hybrid proteins to activate C3. It is important to note that the C3-cleaving activity of some convertases exceeds the C3-cleaving activity of the convertase formed with CVF. Figure 4 shows that convertases formed with human C3/CVF hybrids do not exhibit cleaving activity for human C5, a particularly important property as the C5a anaphylatoxin generated by the cleavage of C5 is the most important pro-inflammatory anaphylatoxin generated during complement activation. Figure 5 demonstrates that human C3/CVF hybrid proteins are able to deplete complement in vivo in rats.

In order to address the therapeutic potential of human C3/CVF hybrid proteins they were tested in two established disease models in mice. Figure 6 shows the therapeutic effect of complement depletion using hybrid protein HC3-1496 in a model of collagen-induced arthritis (Brand et al. 1994). The therapeutic effect was

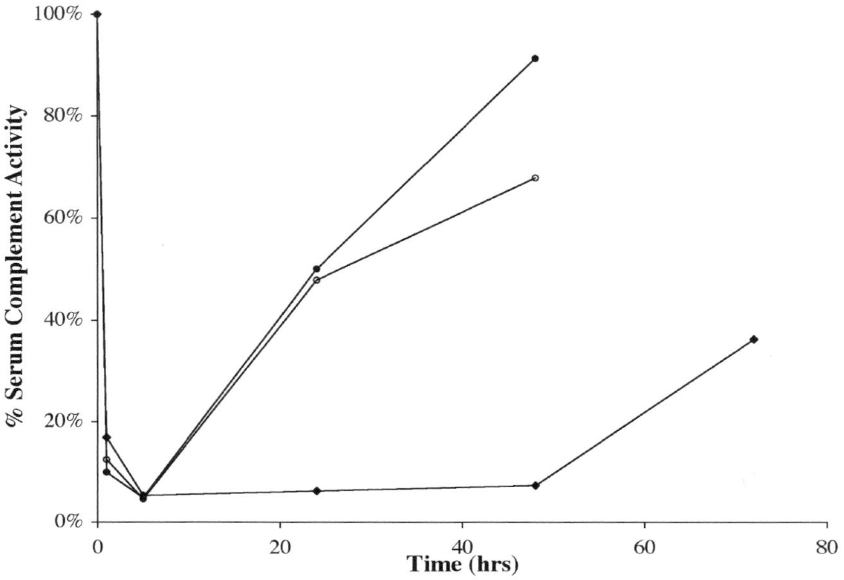

Fig. 5 In vivo complement depletion by human C3/CVF hybrid proteins. Adult Sprague-Dawley rats were injected i.p. with 280 μg/kg (O) or 760 μg/kg (•) HC3-1496 or 500 μg/kg native CVF (◆). Blood samples were taken at time intervals as indicated, and the hemolytic complement activity was determined as described (Vogel and Müller-Eberhard 1984)

pronounced in both a prophylactic regimen in which treatment with HC3-1496 started on the day after the booster immunization with collagen, and in a therapeutic regimen in which treatment began six days after the booster immunization. A significant reduction of the gross pathological changes at front and back paws and ankles was observed, with a significant reduction of the swelling. The therapeutic effectiveness of hybrid protein HC3-1496 was also tested in a murine model of myocardial ischemia/reperfusion injury (Walsh et al. 2005). As can be seen in Fig. 7, HC3-1496 or CVF virtually protected the myocardium from reperfusion injury after 30 min of ischemia as measured by echocardiographic determination of the ejection fraction.

3 Discussion and Outlook

Complement depletion represents a conceptually different approach for therapeutic intervention in clinical situations where complement activation is harmful. Complement depletion as a therapeutic concept appears promising. As much as many lead compounds of today's drugs were derived from natural compounds, the fact that CVF is a natural product may help to validate the concept. Human C3 derivatives with CVF-like function represent hit-and-run drugs which circumvents the necessity to maintain a high drug concentration for effective prevention of harmful complement activation over an extended period of time. This is particularly relevant for complement-mediated tissue damage, as harmful triggers

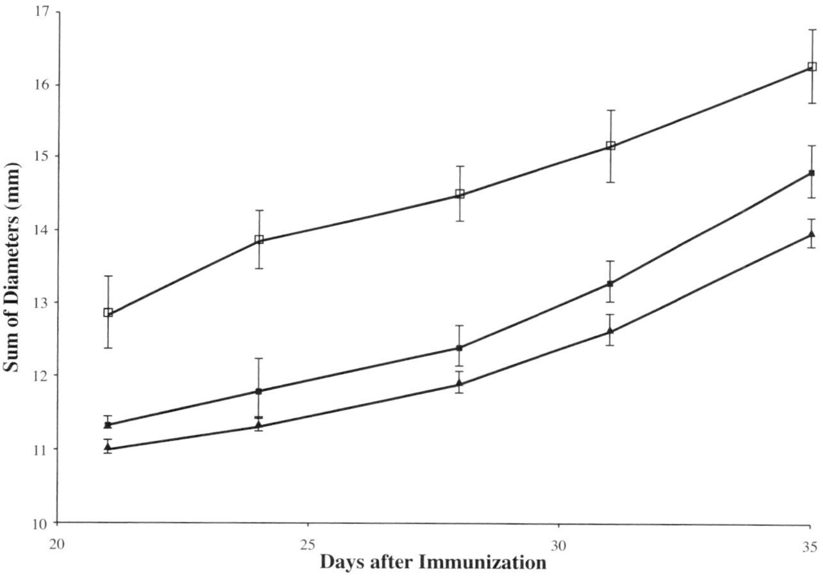

Fig. 6 Effect of complement depletion with HC3-1496 in a murine model of collagen-induced arthritis. DBA/1 Male mice 8 weeks of age were immunized with chicken CII collagen in incomplete Freund's adjuvant intradermally at the base of the tail. A booster immunization with complete Freund's adjuvant was given two weeks after the primary injection. A loading dose of 500 μg/kg HC3-1496 or PBS was administered i.p. beginning on the day after the booster immunization (prophylactic regimen) or 6 days after the booster immunization (therapeutic regimen). A maintenance dose of 250 μg/kg HC3-1496 was administered daily (5 days/week) after the initial loading dose in both regimens. Arthritis was monitored by measuring the diameters of hind paws, fore paws, and ankles using a micrometer. The data shown represents the sum of the three measurements (therapeutic regiment (▲), prophylactic regimen (■), PBS control (□))

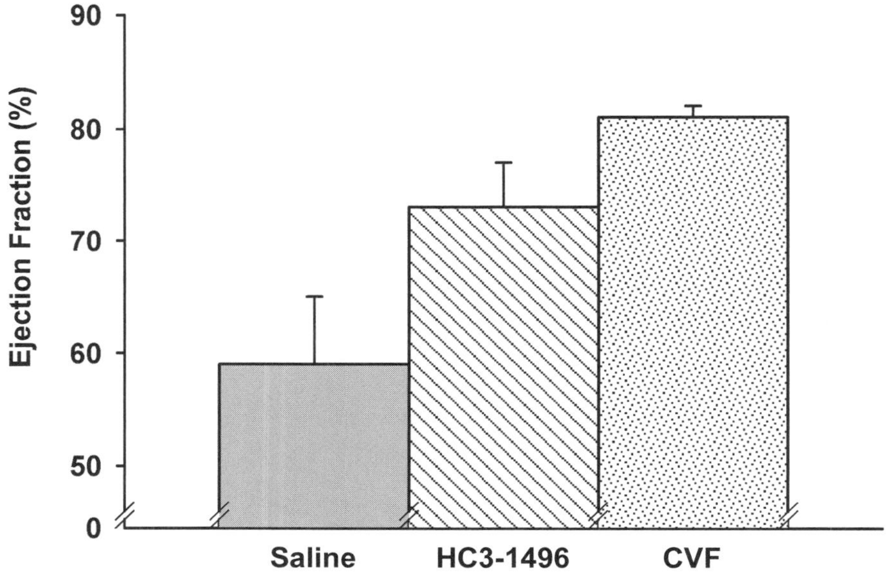

Fig. 7 Effect of complement depletion with HC3-1496 in a model of murine myocardial ischemia/reperfusion injury. CH57/BL/6 mice (8–10 weeks of age) were injected i.p. with 250 μg/kg HC3-1496, 250 μg/kg of native CVF, or sterile saline 2 h prior to induction of anesthesia. Subsequently the left anterior descending coronary artery was occluded with a nylon suture for 30 min to induce ischemia, and then reopened to begin reperfusion. After 3 h of reperfusion echocardiography was performed to determine the left ventricular ejection fraction (Walsh et al. 2005)

to activate complement are often localized and result in powerful bursts of complement activation which other complement inhibitory reagents may not be able to control (Shernan et al. 2004). However strong the complement activating signal may be, prior complement depletion will render it harmless. For the same reason, CVF has often served as the gold standard to evaluate the efficacy of other complement-inhibiting pharmaceuticals (Hebell et al. 1991, Morgan and Harris 2003, Sahu and Lambris 2000).

There are many reasons why complement depletion may represent a safe therapeutic approach. Ever since it has been shown that CVF could be to used safely deplete serum complement in laboratory animals (Cochrane et al. 1970, Maillard and Zarco 1968, Nelson 1966), numerous studies performed over a period of four decades have employed CVF to study the biological functions of complement and its role in the pathogenesis of disease without any indication of serious side effects. The sole known side effect appears to be a consequence of the initial boost of release of the C3a and C5a anaphylatoxins as massive complement activation has been shown to sequester neutrophils to the lungs with ensuing neutropenia and

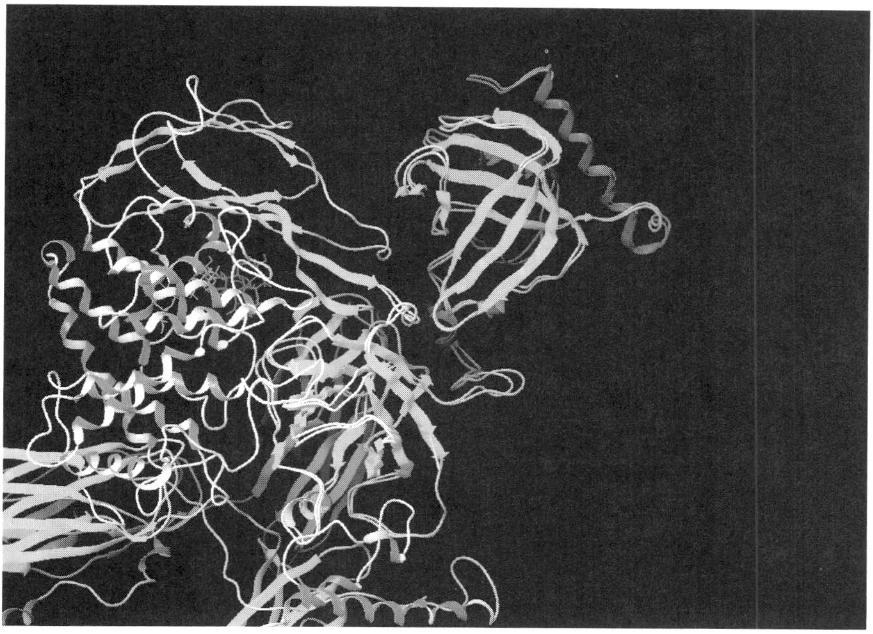

Fig. 8 Prediction of the three-dimensional structure of the C-terminal end (C345C region) of hybrid protein HC3-1348 compared to the corresponding region of human C3. The important C345C domain at the C-terminus of both molecules is pictured in the upper right quadrant of the figure. Please note that the two protein structures are virtually superimposable. The structural prediction of HC3-1348 was obtained using the homology modeling software Schrödinger Prime (Jacobson et al. 2004) via the Maestro interface by aligning the CVF sequence in the region to the human C3 sequence as a template (Janssen, et al. 2005). The new regions were homology-modeled, and complete proteins that included each substitution were created. The new homology model was allowed to relax (minimize) using Schrödinger MacroModel since the homology process ensures conformations in agreement with the X-ray crystal structure of the template, but the residues that differ may interact differently and the confirmation of the protein may change

temporary lung infiltration (Till et al. 1982, 1987), something that can be easily controlled by the dose, route, and pattern of a complement-depleting reagent. There is absolutely no evidence of any off-target toxicity of CVF; all effects of CVF administration appear to be the sole consequence of its ability to form a convertase with factor B with the ensuing complement activation and eventual depletion. Complement depletion by CVF, as well as acquired complement deficiencies found in a large number of human diseases, never completely eliminate C3, allowing for residual but important protection against infections. In addition, complement depletion does not affect the ability to synthesize C3 locally in tissues in responses to stimuli such as injury or penetrating microorganisms. The safety of long-term complement depletion has also been demonstrated by the generation of transgenic mice constitutively expressing CVF (Andrä et al. 2002). These transgenic mice exhibit low levels of serum complement activity and low levels of C3 in the serum, but show no abnormal phenotype (Andrä et al. 2002). The mice appear normal, and appear to have a normal lifespan. Lastly, human C3/CVF hybrid proteins are fortuitously devoid of C5-cleaving activity (Vogel and Fritzinger 2007), thereby preventing the generation of C5a, the most pro-inflammatory anaphylatoxin. C5a has also been shown to be the primary if not sole reagent responsible for neutropenia and neutrophil sequestration to the lungs after massive complement activation (Guo and Ward 2005, Till et al. 1982, 1987).

The question of the immunogenicity of human C3/CVF hybrid proteins is difficult to predict. However, based on the experience with other therpeutic recombinant biologicals, immunogenicity, even after repeated administration, is unlike to be a major concern. In addition, the CVF portion of the human C3/CVF hybrids is not a structurally unrelated protein. Rather, it is the structurally highly homologous portion of CVF which even contains identical amino acid residues to human C3 throughout its sequence. It is certain that the three dimensional structure of the very C-terminus of the β-chain of CVF will be highly homologous to the C345C region of complement component C3 (Fritzinger et al. 1994, Janssen et al. 2005, 2006). Indeed, computer analysis of the three-dimensional structure of the C-terminus of the CVF β-chain predicts its structure to be very similar to that of the C-terminus of the human C3 α-chain (Fig. 8). Collectively, these observations suggest that serious immunogenicity problems are unlikely.

In conclusion, the substitution of human C3 with homologous CVF sequences at the C-terminus of the C3 α-chain results in human C3 derivatives exhibiting CVF-like functions. These proteins represent a form of humanized CVF which can deplete complement in vivo and exert beneficial therapeutic effects in animal models of disease with complement pathogenesis. These human C3 derivatives represent a novel class of pharmacological agents potentially useful for therapeutic complement depletion in many human diseases.

Acknowledgments

Part of the research was supported by Incode Biopharmaceutics Corporation, Lahaina, Hawaii, USA.

References

Andrä, J., Halter, R., Kock, M. A., Niemann, H., Vogel, C.-W., and Paul, D. (2002). Generation and characterization of transgenic mice expressing cobra venom factor. *Mol Immunol* 39, 357–365

Brand, D. D., Myers, L. K., Terato, K., Whittington, K. B., Stuart, J. M., Kang, A. H., and Rosloniec, E. F. (1994). Characterization of the T cell determinants in the induction of autoimmune arthritis by bovine alpha 1(II)-CB11 in H-2q mice. *J Immunol* 152, 3088–3097

Carroll, M. C. (2004). The complement system in B cell regulation. *Mol Immunol* 41, 141–146

Cochrane, C. G., Müller-Eberhard, H. J., and Aikin, B. S. (1970). Depletion of plasma complement in vivo by a protein of cobra venom: its effect on various immunologic reactions. *J Immunol* 105, 55–69

Costa, C., Zhao, L., Burton, W. V., Rosas, C., Bondioli, K. R., Williams, B. L., Hoagland, T. A., Dalmasso, A. P., and Fodor, W. L. (2002). Transgenic pigs designed to express human CD59 and H-transferase to avoid humoral xenograft rejection. *Xenotransplantation* 9, 45–57

de Bruijn, M. H. and Fey, G. H. (1985). Human complement component C3: cDNA coding sequence and derived primary structure. *Proc Natl Acad Sci U S A* 82, 708–712

Fodor, W. L., Rollins, S. A., Guilmette, E. R., Setter, E., and Squinto, S. P. (1995). A novel bifunctional chimeric complement inhibitor that regulates C3 convertase and formation of the membrane attack complex. *J Immunol* 155, 4135–4138

Fritzinger, D. C., Petrella, E. C., Connelly, M. B., Bredehorst, R., and Vogel, C.-W. (1992). Primary structure of cobra complement component C3. *J Immunol* 149, 3554–3562

Fritzinger, D. C., Bredehorst, R., and Vogel, C.-W. (1994). Molecular cloning and derived primary structure of cobra venom factor. *Proc Natl Acad Sci U S A* 91, 12775–12779

Fritzinger, D. C., Hew, B. E., Thorne, M., and Vogel, C.-W. (2004). Functional characterization of cobra venom factor/cobra C3 hybrid proteins. *Mol Immunol* 41, 230

Fritzinger, D. C., Hew, B. E., Pangburn, M. K., and Vogel, C.-W. (2005). Generation of human C3 derivatives with CVF-like function for therapeutic complement depletion. *FASEB J* 19, A324

Fritzinger, D. C., Hew, B. E., Pangburn, M. K., Janssen, B. J. C., Gros, P., and Vogel, C.-W. (2006). Human C3/cobra venom factor hybrid proteins with potential therapeutic applications. *Mol Immunol* 43, 141–142

Fritzinger, D. C., Hew, B. E., Lee, J. Q., and Vogel, C.-W. (2007). Human C3/cobra venom factor hybrid proteins for therapeutic complement depletion: in vivo activity and fine mapping of important domains. *Mol Immunol* 44, 3945

Fritzinger, D. C., Hew, B. E., Thorne, M., Pangburn, M. K., Janssen, B. J. C., Gros, P., and Vogel, C.-W. (2008). Functional characterization of human C3/cobrar venom factor hybrid proteins for therapeutic complement depletion. *Dev Comp Immunol*, in press

Fujita, T., Matsushita, M., and Endo, Y. (2004). The lectin-complement pathway – its role in innate immunity and evolution. *Immunol Rev* 198, 185–202

Gowda, D. C., Schultz, M., Bredehorst, R., and Vogel, C.-W. (1992). Structure of the major oligosaccharide of cobra venom factor. *Mol Immunol* 29, 335–342

Gowda, D. C., Petrella, E. C., Raj, T. T., Bredehorst, R., and Vogel, C.-W. (1994). Immunoreactivity and function of oligosaccharides in cobra venom factor. *J Immunol* 152, 2977–2986

Gowda, D. C., Glushka, J., Halbeek, H., Thotakura, R. N., Bredehorst, R., and Vogel, C.-W. (2001). N-linked oligosaccharides of cobra venom factor contain novel alpha(1-3)galactosylated Le(x) structures. *Glycobiology* 11, 195–208

Grier, A. H., Schultz, M., and Vogel, C.-W. (1987). Cobra venom factor and human C3 share carbohydrate antigenic determinants. *J Immunol* 139, 1245–1252

Guo, R. F. and Ward, P. A. (2005). Role of C5a in inflammatory responses. *Annu Rev Immunol* 23, 821–852

Hebell, T., Ahearn, J. M., and Fearon, D. T. (1991). Suppression of the immune response by a soluble complement receptor of B lymphocytes. *Science* 254, 102–105

Hew, B. E., Thorne, M., Fritzinger, D. C., and Vogel, C.-W. (2004). Humanized cobra venom factor (CVF): Generation of Human C3 derivitives with CVF-like function. *Mol Immunol* 41, 244 –245

Hillmen, P., Hall, C., Marsh, J. C., Elebute, M., Bombara, M. P., Petro, B. E., Cullen, M. J., Richards, S. J., Rollins, S. A., Mojcik, C. F., and Rother, R. P. (2004). Effect of eculizumab on hemolysis and transfusion requirements in patients with paroxysmal nocturnal hemoglobinuria. *N Engl J Med* 350, 552–559

Jacobson, M. P., Pincus, D. L., Rapp, C. S., Day, T. J., Honig, B., Shaw, D. E., and Friesner, R. A. (2004). A hierarchical approach to all-atom protein loop prediction. *Proteins* 55, 351–367

Janssen, B. J., Huizinga, E. G., Raaijmakers, H. C., Roos, A., Daha, M. R., Nilsson-Ekdahl, K., Nilsson, B., and Gros, P. (2005). Structures of complement component C3 provide insights into the function and evolution of immunity. *Nature* 437, 505–511

Janssen, B. J., Christodoulidou, A., McCarthy, A., Lambris, J. D., and Gros, P. (2006). Structure of C3b reveals conformational changes that underlie complement activity. *Nature* 444, 213–216

Kock, M. A., Hew, B. E., Bammert, H., Fritzinger, D. C., and Vogel, C. W. (2004). Structure and function of recombinant cobra venom factor. *J Biol Chem* 279, 30836–30843

Kölln, J., Spillner, E., Andrä, J., Klensang, K., and Bredehorst, R. (2004a). Human C3 derivatives engineered for decomplementation by forming stable C3 convertases. *Mol Immunol* 41, 259

Kölln, J., Spillner, E., Andrä, J., Klensang, K., and Bredehorst, R. (2004b). Complement inactivation by recombinant human C3 derivatives. *J Immunol* 173, 5540–5545

Kölln, J., Bredehorst, R., and Spillner, E. (2005). Engineering of human complement component C3 for catalytic inhibition of complement. *Immunol Lett* 98, 49–56

Lachmann, P. J. and Halbwachs, L. (1975). The influence of C3b inactivator (KAF) concentration on the ability of serum to support complement activation. *Clin Exp Immunol* 21, 109–114

Maillard, J. L., and Zarco, R. M. (1968). [Decomplementization by a factor extracted from cobra venom. Effect on several immune reactions of the guinea pig and rat]. *Ann Inst Pasteur* (Paris) 114, 756–774

Medicus, R. G., Götze, O., and Müller-Eberhard, H. J. (1976). The serine protease nature of the C3 and C5 convertases of the classical and alternative complement pathways. *Scand J Immunol* 5, 1049–1055

Mollnes, T. E. and Kirschfink, M. (2006). Strategies of therapeutic complement inhibition. *Mol Immunol* 43, 107–121

Morgan, B. P. and Harris, C. L. (2003). Complement therapeutics; history and current progress. *Mol Immunol* 40, 159–170

Nagaki, K., Iida, K., Okubo, M., and Inai, S. (1978). Reaction mechanisms of beta1H globulin. *Int Arch Allergy Appl Immunol* 57, 221–232

Nelson, R. A., Jr. (1966). A new concept of immunosuppression in hypersensitivity reactions and in transplantation immunity. *Surv Ophthalmol* 11, 498–505

Pangburn, M. K., Schreiber, R. D., and Müller-Eberhard, H. J. (1977). Human complement C3b inactivator: isolation, characterization, and demonstration of an absolute requirement for the serum protein beta1H for cleavage of C3b and C4b in solution. *J Exp Med* 146, 257–270

Pangburn, M. K. and Müller-Eberhard, H. J. (1986). The C3 convertase of the alternative pathway of human complement. Enzymic properties of the bimolecular proteinase. *Biochem J* 235, 723–730

Sahu, A. and Lambris, J. D. (2000). Complement inhibitors: a resurgent concept in anti-inflammatory therapeutics. *Immunopharmacology* 49, 133–148

Shernan, S. K., Fitch, J. C., Nussmeier, N. A., Chen, J. C., Rollins, S. A., Mojcik, C. F., Malloy, K. J., Todaro, T. G., Filloon, T., Boyce, S. W., Gangahar, D. M., Goldberg, M., Saidman, L. J., and Mangano, D. T. (2004). Impact of pexelizumab, an anti-C5 complement antibody, on total mortality and adverse cardiovascular outcomes in cardiac surgical patients undergoing cardiopulmonary bypass. *Ann Thorac Surg* 77, 942–949; *discussion* 949–950

Till, G. O., Johnson, K. J., Kunkel, R., and Ward, P. A. (1982). Intravascular activation of complement and acute lung injury. Dependency on neutrophils and toxic oxygen metabolites. *J Clin Invest* 69, 1126–1135

Till, G. O., Morganroth, M. L., Kunkel, R., and Ward, P. A. (1987). Activation of C5 by cobra venom factor is required in neutrophil-mediated lung injury in the rat. *Am J Pathol* 129, 44–53

Vogel, C.-W. (1991). Cobra venom factor, the complement-activating protein of cobra venom. In: *Handbook of Natural Toxins: Reptile and Amphibian Venoms*, Vol. 5. Anthony Tu (ed.). Marcel Dekker, New York, pp. 147–188

Vogel, C.-W., and Müller-Eberhard, H. J. (1982). The cobra venom factor-dependent C3 convertase of human complement. A kinetic and thermodynamic analysis of a protease acting on its natural high molecular weight substrate. *J Biol Chem* 257, 8292–8299

Vogel, C.-W., and Müller-Eberhard, H. J. (1984). Cobra venom factor: improved method for purification and biochemical characterization. *Journal of Immunology Methods* 73, 203–220

Vogel, C.-W. and Fritzinger, D. C. (2007). Humanized cobra venom factor: experimental therapeutics for targeted complement activation and complement depletion. *Curr Pharm Des* 13, 2916–2926

Vogel, C.-W., Smith, C. A., and Müller-Eberhard, H. J. (1984). Cobra venom factor: structural homology with the third component of human complement. *J Immunol* 133, 3235–3241

Vogel, C.-W., Bredehorst, R., Fritzinger, D. C., Grunwald, T., Ziegelmüller, P., and Kock, M. A. (1996). Structure and function of cobra venom factor, the complement-activating protein in cobra venom. *Adv Exp Med Biol* 391, 97–114

Vogel, C. W., Fritzinger, D. C., Hew, B. E., Thorne, M., and Bammert, H. (2004). Recombinant cobra venom factor. *Mol Immunol* 41, 191–199

Walsh, M. C., Bourcier, T., Takahashi, K., Shi, L., Busche, M. N., Rother, R. P., Solomon, S. D., Ezekowitz, R. A., and Stahl, G. L. (2005). Mannose-binding lectin is a regulator of inflammation that accompanies myocardial ischemia and reperfusion injury. *J Immunol* 175, 541–546

Wehrhahn, D., Meiling, K., Fritzinger, D. C., Bredehorst, R., Andrä, J., and Vogel, C.-W. (2000). Analysis of the structure/function relationship of Cobra Venom Factor (CVF) and C3: generation of CVF/cobraC3 hybrid proteins. *Immunopharmacology* 49, 94

Weisman, H. F., Bartow, T., Leppo, M. K., Marsh, H. C., Jr., Carson, G. R., Concino, M. F., Boyle, M. P., Roux, K. H., Weisfeldt, M. L., and Fearon, D. T. (1990). Soluble human complement receptor type 1: in vivo inhibitor of complement suppressing post-ischemic myocardial inflammation and necrosis. *Science* 249, 146–151

Whaley, K., and Ruddy, S. (1976). Modulation of the alternative complement pathways by beta1H globulin. *J Exp Med* 144, 1147–1163

Zhou, C. Y., McInnes, E., Copeman, L., Langford, G., Parsons, N., Lancaster, R., Richards, A., Carrington, C., and Thompson, S. (2005). Transgenic pigs expressing human CD59, in combination with human membrane cofactor protein and human decay-accelerating factor. *Xenotransplantation* 12, 142–148

Index